I0882583

Contemporary Developments in High-Frequency Photonic Devices

Siddhartha Bhattacharyya
RCC Institute of Information Technology, India

Pampa Debnath
RCC Institute of Information Technology, India

Arpan Deyasi
RCC Institute of Information Technology, India

Nilanjan Dey
Techno India College of Technology, India

A volume in the Advances in Computer and Electrical Engineering (ACEE) Book Series

IGI Global
DISSEMINATOR OF KNOWLEDGE

Published in the United States of America by
IGI Global
Engineering Science Reference (an imprint of IGI Global)
701 E. Chocolate Avenue
Hershey PA, USA 17033
Tel: 717-533-8845
Fax: 717-533-8661
E-mail: cust@igi-global.com
Web site: http://www.igi-global.com

Copyright © 2019 by IGI Global. All rights reserved. No part of this publication may be reproduced, stored or distributed in any form or by any means, electronic or mechanical, including photocopying, without written permission from the publisher.
Product or company names used in this set are for identification purposes only. Inclusion of the names of the products or companies does not indicate a claim of ownership by IGI Global of the trademark or registered trademark.

Library of Congress Cataloging-in-Publication Data

Names: Bhattacharyya, Siddhartha, 1975- editor. | Debnath, Pampa, 1981- editor. | Deyasi, Arpan, 1978- editor. | Dey, Nilanjan, 1984- editor.
Title: Contemporary developments in high-frequency photonic devices / Siddhartha Bhattacharyya, Pampa Debnath, Arpan Deyasi, and Nilanjan Dey, editors.
Description: Hershey, PA : Engineering Science Reference, [2020] | Includes bibliographical references.
Identifiers: LCCN 2018055424| ISBN 9781522585312 (h/c) | ISBN 9781522585329 (s/c) | ISBN 9781522585336 (eISBN)
Subjects: LCSH: Microwave communication systems. | Photonics. | Microwave devices.
Classification: LCC TK5103.4833 .C66 2020 | DDC 621.36/5--dc23 LC record available at https://lccn.loc.gov/2018055424

This book is published in the IGI Global book series Advances in Computer and Electrical Engineering (ACEE) (ISSN: 2327-039X; eISSN: 2327-0403)

British Cataloguing in Publication Data
A Cataloguing in Publication record for this book is available from the British Library.

All work contributed to this book is new, previously-unpublished material.
The views expressed in this book are those of the authors, but not necessarily of the publisher.

For electronic access to this publication, please contact: eresources@igi-global.com.

Advances in Computer and Electrical Engineering (ACEE) Book Series

ISSN:2327-039X
EISSN:2327-0403

Editor-in-Chief: Srikanta Patnaik, SOA University, India

MISSION

The fields of computer engineering and electrical engineering encompass a broad range of interdisciplinary topics allowing for expansive research developments across multiple fields. Research in these areas continues to develop and become increasingly important as computer and electrical systems have become an integral part of everyday life.

The **Advances in Computer and Electrical Engineering (ACEE) Book Series** aims to publish research on diverse topics pertaining to computer engineering and electrical engineering. **ACEE** encourages scholarly discourse on the latest applications, tools, and methodologies being implemented in the field for the design and development of computer and electrical systems.

COVERAGE

- Qualitative Methods
- Programming
- VLSI Design
- Digital Electronics
- Computer Science
- Circuit Analysis
- Algorithms
- Electrical Power Conversion
- Analog Electronics
- Power Electronics

IGI Global is currently accepting manuscripts for publication within this series. To submit a proposal for a volume in this series, please contact our Acquisition Editors at Acquisitions@igi-global.com or visit: http://www.igi-global.com/publish/.

The Advances in Computer and Electrical Engineering (ACEE) Book Series (ISSN 2327-039X) is published by IGI Global, 701 E. Chocolate Avenue, Hershey, PA 17033-1240, USA, www.igi-global.com. This series is composed of titles available for purchase individually; each title is edited to be contextually exclusive from any other title within the series. For pricing and ordering information please visit http://www.igi-global.com/book-series/advances-computer-electrical-engineering/73675. Postmaster: Send all address changes to above address. ©© 2019 IGI Global. All rights, including translation in other languages reserved by the publisher. No part of this series may be reproduced or used in any form or by any means – graphics, electronic, or mechanical, including photocopying, recording, taping, or information and retrieval systems – without written permission from the publisher, except for non commercial, educational use, including classroom teaching purposes. The views expressed in this series are those of the authors, but not necessarily of IGI Global.

Titles in this Series

For a list of additional titles in this series, please visit:
https://www.igi-global.com/book-series/advances-computer-electrical-engineering/73675

Multi-Objective Stochastic Programming in Fuzzy Environments
Animesh Biswas (University of Kalyani, India) and Arnab Kumar De (Government College of Engineering and Textile Technology Serampore, India)
Engineering Science Reference • ©2019 • 420pp • H/C (ISBN: 9781522583011) • US $215.00

Renewable Energy and Power Supply Challenges for Rural Regions
Valeriy Kharchenko (Federal Scientific Agroengineering Center VIM, Russia) and Pandian Vasant (Universiti Teknologi PETRONAS, Malaysia)
Engineering Science Reference • ©2019 • 432pp • H/C (ISBN: 9781522591795) • US $205.00

Secure Cyber-Physical Systems for Smart Cities
Riaz Ahmed Shaikh (King Abdulaziz University, Saudi Arabia)
Engineering Science Reference • ©2019 • 280pp • H/C (ISBN: 9781522571896) • US $225.00

Handbook of Research on Smart Power System Operation and Control
Hassan Haes Alhelou (Tishreen University, Syria) and Ghassan Hayek (Tishreen University, Syria)
Engineering Science Reference • ©2019 • 489pp • H/C (ISBN: 9781522580300) • US $265.00

Cases on Modern Computer Systems in Aviation
Tetiana Shmelova (National Aviation University, Ukraine) Yuliya Sikirda (National Aviation University, Ukraine) Nina Rizun (Gdansk University of Technology, Poland) and Dmytro Kucherov (National Aviation University, Ukraine)
Engineering Science Reference • ©2019 • 488pp • H/C (ISBN: 9781522575887) • US $225.00

Harnessing the Internet of Everything (IoE) for Accelerated Innovation Opportunities
Pedro J.S. Cardoso (University of Algarve, Portugal) Jânio Monteiro (University of Algarve, Portugal) Jorge Semião (University of Algarve, Portugal) and João M.F. Rodrigues (University of Algarve, Portugal)
Engineering Science Reference • ©2019 • 383pp • H/C (ISBN: 9781522573326) • US $225.00

For an entire list of titles in this series, please visit:
https://www.igi-global.com/book-series/advances-computer-electrical-engineering/73675

IGI Global
DISSEMINATOR OF KNOWLEDGE

701 East Chocolate Avenue, Hershey, PA 17033, USA
Tel: 717-533-8845 x100 • Fax: 717-533-8661
E-Mail: cust@igi-global.com • www.igi-global.com

Siddhartha Bhattacharyya would like to dedicate this book to his father, Late Ajit Kumar Bhattacharyya; his mother, Late Hashi Bhattacharyya; his beloved wife, Rashni; his schoolmates Anshuman Haldar, Bishanka Sen, Chandan Saha, Pritish Debsarma, and Sasanka Sekhar Banerjee.

Arpan Deyasi would like to dedicate this book to R. Loudon, who first enlightened the physics of electromagnetic wave propagation through absorbing dielectrics.

Pampa Debnath would like to dedicate this book to her respected father, Mr. Parimal Debnath; mother, Mrs. Krishna Debnath, her beloved husband, Mr. Snehasis Roy; and kid, Master Deeptanshu Roy.

Nilanjan Dey would like to dedicate this book to his parents and wife for their encouragement, patience, and support throughout the project.

Editorial Advisory Board

Asit K. Datta, *University of Calcutta, India*
Tarun Kumar Gangopadhyay, *Central Glass and Ceramic Research Institute, India*
Goran Klepac, *University of Algebra, Croatia*
Pathik Kumbhaka, *NIT Durgapur, India*
S. P. Maity, *IIEST Shibpur, India*
Monojit Mitra, *IIEST Shibpur, India*
Sourangshu Mukhopadhyay, *University of Burdwan, India*
Biswanath Roy, *Jadavpur University, India*

List of Reviewers

Arup Bhattacharjee, *RCC Institute of Information Technology, India*
Nandan Bhattacharyya, *RCC Institute of Information Technology, India*
Swapan Bhattacharyya, *Dr. Sudhir Chandra Sur Degree Engineering College, India*
Krishnendu Chattopadhyay, *West Bengal University of Technology, India*
Prosenjit Roy Chowdhury, *Acharya Prafulla Chandra College, India*
Anirban Karmakar, *Tripura University, India*
Surajit Mandal, *B. P. Poddar Institute of Management and Technology, India*
Soumen Mukherjee, *RCC Institute of Information Technology, India*
Ayan Pradhan, *Acharya Prafulla Chandra College, India*
Jayanta Kumar Rakshit, *NIT Agartala, India*
Partha Pratim Sarkar, *University Institute of Technology, India*
Vaibhav Shukla, *Allenhouse Institute of Technology, India*

Table of Contents

Detailed Table of Contents

Microwave photonics is the arena of research in the 21st century due to ever-increasing ultra-large bandwidth and the meticulous availability of data with very low cost. In this context, conventional optoelectronic devices are replaced by novel photonic counterparts, both in transreceiver design as well as devices and systems. The major objective of this replacement is to reduce noise by means of lower scattering, where photons are only responsible for propagation of electromagnetic wave. With introduction of novel materials, low-loss communication system can now be designed at beyond THz range, mainly due to the physical realization of electromagnetic bandgap structure. This chapter is extended towards plasmonics with the intension of making sensors for beyond THz applications.

The microstrip antenna (MSA) consists of a dielectric substrate in between a metallic conducting patch and a ground plane. The most common forms of the MSA are the rectangular and circular patch MSAs. There are several microstrip antenna analysis methods. The most popular models are transmission-line model, cavity model, method of moments, FDTD method, and finite element method. The transmission-line model is the simplest of these methods, and it provides good

equivalent magnetic surface of monopole antenna. Impedance bandwidth of the above slot antenna is enhanced through various tuning stubs. In case of forklike tuning stub, the obtained bandwidth is about 900MHz, for hexagonal stub the available bandwidth is 1751MHz. Further improvement in bandwidth is proposed through rotation of hexagonal wide slot, results in wide bandwidth of 5165 MHz covering all the WLAN and WiMAX applications.

A novel compact ultra-wideband (UWB) multiple-input multiple-output (MIMO) slot antenna with band notch characteristics is presented for portable wireless UWB applications. The antenna comprises of co-planar waveguide feed (CPW) and two radiating monopoles oriented in orthogonal orientation for providing orthogonal radiation patterns. A Minkowski fractal parasitic stub along with a Minkowski fractal grounded stub has been placed at 45° between the monopoles to reduce the coupling between them, which in turn establishes high isolation between the radiators. An excellent band notch characteristic is obtained at 5.5 GHz by etching a modified E-shaped compact slot on the radiators. Results show that the designed antenna meets -10 dB impedance bandwidth and -17 dB isolation throughout the entire operating band (3.1 -12 GHz). Novelty of this design lies in improving isolation using fractal which occupies less space in compared to other isolation mechanisms in MIMO structures. The simulated and measured results depict that the proposed antenna is convenient for MIMO diversity systems.

The kapton polyimide material is considered to design conformal antenna with spiral square for radio frequency identification (RFID) and wireless local area network (WLAN) applications. In this chapter, the analysis and investigation has been carried out with spiral square techniques using coplanar waveguide (CPW) feed. The proposed antenna operates at 5.8 GHz with impedance bandwidth of 170 MHz (5.73 - 5.9 GHz) with return loss -25.6 dB and gain is 2.4 dBi. The proposed antenna has considered with different bending angles for investigating the conformal characteristics due to flexibility of the material. These results are presented for omni-directional radiation patterns.

Optics is considered a potential candidate for the realization of logic devices, digital optical systems for communication, and computation exploiting its super-fast speed. Optical logic gates also can act on the basis of frequency conversion process of some nonlinear materials. Further, in this chapter, the authors have mentioned the dibit representation technique for reducing bit error problem at the input and output terminals of all optical digital logic circuits and a control input for selecting particular logic operation. Here the authors have proposed frequency encoded all optical dibit-based integrated AND and OR logic gates with control input, where a single circuit acts as both AND logic gate and OR logic gate using the optical switches like reflected semiconductor optical amplifier and add/drop multiplexer.

The leading content provider companies like Google, Yahoo, and Amazon installed mega-data centers that contain hundreds of thousands of servers in very large scale. The current data center systems are organized in the form of the hierarchal tree structure based on bandwidth-limited electronic switches. Modern data center systems face a number of issues like high power consumption, limited bandwidth availability, server connectivity, energy and cost efficiency, traffic complexity, etc. One of the most feasible solution of these issues is the use of optical switching technologies in the core of data center systems. In this chapter a brief description about the modern data center system is presented, and some prominent optical packet switch architectures are also presented in this chapter with their pros and cons.

A uniaxial birefringent crystal lens with its optic axis perpendicular to the system axis and sandwiched between two properly oriented linear polarizers behaves as an isotropic lens with a radially varying complex mask on its pupil plane. The proposed system may be adapted for both apodization and enhanced resolution just by rotating one of the two linear polarizers even when it is illuminated with a polychromatic source of light. Hence, the system may find applications in the fields of spectroscopy and astronomy. In general, it behaves as a double focus lens. However, by varying the birefringent lens parameters, it is possible to obtain a noticeably large depth-of-focus compared to an identical isotropic lens. An optical imaging system with large depth-of-focus is a prerequisite for many fields of applications particularly for microscope imagery, medical imaging, as well as for automatic inspection in microelectronics industry.

This chapter is proposed with an approach to analyze reflectance as a function of negative index material thickness for different parameters under the surface plasmon condition and extended approach towards the field enhancement of electric field as function of incidence angle and transmittance as function of incidence angle has been analyzed. This chapter can reflect the good comparison between 3 layer medium and n layer medium model. Characteristic impedance of MIM surface plasmon structure is analytically calculated considering the effect of both Faraday inductance and kinetic inductance. Effect of metal layer thickness, insulator thickness, and electron density are tailored to observe the impedance variation with frequency. Wavelength dependence of characteristic impedance and quality factor of MIM (metal-insulator-metal) surface plasmon structure is analyzed. Structural parameters and damping ratio of the structure is tuned within allowable limit to analyze the variation after detailed analytical computation.

The realization of all-optical polarization switch and all-optical logic gates based on polarization-conversion on single silicon micro-ring resonator (MRR) is demonstrated. By adjusting the mode state of the input source as well as the pump light, the all-optical polarization switch, and hence, all-optical NOT, OR/NOR. AND-NAND logic gates are realized. The design is ultra-compact, ultrafast, and less optical power is required for all-optical polarization-conversion-based switch and logic gates, respectively. The MRR also shows outstanding performance as its Q (quality) factor is very high. The design is robust, simple, stable, easy-to-fabricate, and silicon-on-insulator (SOI) compatible. The structure is compatible for interconnects and capable for integrating in electronics as well as in plasmonics circuits.

Preface

Microwave photonics and information optics are the present trends of research in order to meet the ever-increasing demand of high bandwidth and precision of data along with ultrafast speed with low cost. In order to reduce noise at the communication trans-receivers, scattering in the devices needs to be decreased, which ignites the replacement of optoelectronic devices by photonic devices, where in the latter case, only photons are responsible for propagation of electromagnetic wave. This novel feature transformed the high-frequency communication scenario, and corresponding devices in trans-receiver sections and associated filters are modified with the intention of low-loss communication. Information transfer in the digital domain requires optical logic gate, and henceforth, a comprehensive discussion on this need-of-the-hour subject is very much needed for acquaintance with the undergraduate and graduate students.

The origin of contemporary developments may be considered the analysis of electromagnetic bandgap structure, and different novel photonic structures are therefore proposed for making efficient communication system in THz and beyond THz ranges. Microwave photonics in this context plays a critical role where integrated circuits, systems and applications of millimeter wave devices, novel plasmonic sensors, metamaterials, optical modulators and detectors requires extensive research study. Computational researches in these domains, either RF, or photonics, speaks in the language of Maxwell's equations. Developments of integrated silicon photonic circuits lead to the possibility of high bandwidth and high bit-rate transmission. Modern antennas are therefore invented to facilitate this state-of-the-art communication system, and is therefore included as one of the major subject area of this book.

This book would come to the benefits of several categories of students and researchers. At the undergraduate level, it will serve as references for selected topics, and also to find out contemporary research problems in this emerging area. The intended audience includes graduate students, who may consider it as reference, and may concentrate on few chapters as text for special/elective papers. At the doctoral level, apart from compulsory coursework, this book can be treated for identifying the direction of works, and thrusting areas in the domain of microwave photonics.

The enriched and wider spectrum for the proposed book may be a subject of interest of young faculties, existing research communities, new research aspirants from the domain of physics, electronic science and electrical/electronic engineering across the globe.

The book comprises 11 well-versed chapters reporting latest trends in research on photonic devices.

Microwave photonics is the arena of research in the 21st century due to the ever-increasing of ultra-large bandwidth and meticulous availability of data with very low cost. In this context, conventional optoelectronic devices are replaced by novel photonic counterparts, both in transreceiver design, as well as devices and systems. The major objective of this replacement is to reduce noise by means of lower scattering, where photons are only responsible for propagation of electromagnetic wave. Chapter 1 illustrates the advent of novel materials due to which low-loss communication system can now be designed at beyond THz range, possible mainly due to the physical realization of electromagnetic bandgap structure.

Chapter 2 deals with microstrip antenna (MSA), which consists of a dielectric substrate in between a metallic conducting patch and a ground plane. The chapter also focuses on the most common forms of MSA and the analysis methods ranging from Transmission-line model, Cavity model, Method of moments, FDTD method, Finite Element method. It also provides a comparative study of the different methods.

In unbounded media wave propagation is supposed to be unguided. In TV and Radio broadcasting, unbounded medium propagation of wave is required. Here transmission of information is destined for one and all who may be interested. Another way of transmitting information is by guided media. Guided media acts to direct the transmission of energy from Transmitter to receiver. Chapter 3 elucidates different forms of transmission lines with special reference to their implementation.

Microstrip patch antennas are printed antennas, finds suitability because of light weight, low volume, thin profile, dual frequency and dual polarization operation, compatible with MMIC. The objective of Chapter 4 is to exhibit the investigations on the Bandwidth Enhancement of Microstrip Antennas with special reference to Microstrip line fed wide slot antennas. Performances are realised and validated through experimental studies on the impedance properties by VNA and radiation properties by pattern measurement setup.

A novel compact Ultra-wideband (UWB) Multiple Input Multiple Output (MIMO) slot antenna with band notch characteristics is presented in Chapter 5 for portable wireless UWB applications. The antenna comprises of co-planar waveguide feed (CPW) and two radiating monopoles oriented in orthogonal orientation for providing orthogonal radiation patterns. A Minkowski fractal parasitic stub along with a Minkowski fractal grounded stub has been placed at 45° between the monopoles to reduce the coupling between them which intern establish high isolation between the radiators. An excellent band notch characteristic is obtained at 5.5 GHz by etching a modified E-shaped compact slot on the radiators. Results show that the designed antenna meets -10 dB impedance bandwidth and -17 dB isolation throughout the entire operating band (3.1 -12 GHz).

In Chapter 6, a compact patch antenna has been presented for wireless applications. A simple rectangular microstrip antenna with small and narrow X-shaped slots on radiating patch has presented to operate wide bandwidth frequency. The X-slot antenna is operated below 10dB return loss with single resonating frequency at 9.9GHz. The return loss is observed wide bandwidth of -35.48dB with 2.80GHz (8.49-11.30GHz) impedance bandwidth of 28.25% is obtained. The current distribution and radiation patterns of X-slots antenna have been presented in the results.

Nowadays, optics is considered as potential candidate for realization of logic devices, digital optical systems for communication and computation exploiting its super-fast speed. Optical logic gates also can act on the basis of frequency conversion process of some nonlinear materials. In Chapter 7, the authors have mentioned the dibit representation technique for reducing bit error problem at the input and output terminals of all optical digital logic circuits and a control input for selecting particular logic operation. Here the authors have proposed frequency encoded all optical dibit based integrated AND and OR logic gates with control input, where a single circuit acts as both AND logic gate and OR logic gate using the optical switches like reflected semiconductor optical amplifier and add/drop multiplexer.

The leading content provider companies like google, yahoo, amazon installed mega data centers which contain hundreds of thousands of servers in very large scale. The current data center systems are organized in the form of the hierarchal tree structure based on bandwidth limited electronic switches. Modern data center systems face number of issues like high power consumption, limited bandwidth availability, server connectivity, energy and cost efficiency, traffic complexity etc. The one of the most feasible solution of these issues is the use of optical switching technologies in the core of data center systems. In Chapter 8, a brief description about the modern data center system is presented. In addition, some prominent optical packet switch architectures are also presented in this chapter with their pros and cons.

A uniaxial birefringent crystal lens with its optic axis perpendicular to the system axis and sandwiched between two properly oriented linear polarizers, behaves as an isotropic lens with a radially varying complex mask on its pupil plane. Chapter 9 proposes a system which may be adapted for both apodization and enhanced resolution just by rotating one of the two linear polarizers even when it is illuminated with a polychromatic source of light. Hence, the system may find applications in the fields of spectroscopy and astronomy.

Chapter 10 focuses on an approach to analyze Reflectance as a function of negative index material thickness for different parameters under the surface plasmon condition and extended approach towards the field enhancement of electric field as function of incidence angle and transmittance as function of incidence angle has been analyzed. This chapter can reflect the good comparison between 3 layer medium and n layer medium model Characteristic impedance of MIM surface plasmon structure is analytically calculated considering the effect of both Faraday inductance and kinetic inductance.

The realization of all-optical polarization-switch and all-optical logic-gates based on polarization-conversion on single silicon micro-ring resonator (MRR) is demonstrated in Chapter 11. By adjusting the mode state of the input source as well as the pump light, the all-optical polarization switch and hence all-optical NOT, OR/ NOR. AND-NAND logic gates are realized. The design is ultra-compact, ultrafast and very less optical power is required for all-optical polarization-conversion based switch and logic gates respectively.

The objective of this volume is to exhibit a wider aspect of microwave engineering and photonic technology and its applications. The works are precisely focused considering the developments of devices/structures with different materials having different optical properties. This sets the present trend of research in microwave photonics. Coverage is proposed in such a way that it can ignite the budding researchers for further exploring the promising area so that new insightful contributions may be generated in near future. In that context, the present endeavor can fulfill its objective by creating potential impact on a larger set of audience. The future ideas can add valued contribution in this field.

Siddhartha Bhattacharyya
RCC Institute of Information Technology, India

Pampa Debnath
RCC Institute of Information Technology, India

Arpan Deyasi
RCC Institute of Information Technology, India

Nilanjan Dey
Techno India College of Technology, India
March 2019, Kolkata, India

Chapter 1
Applications of Electromagnetic Bandgap Structure in Microwave Photonics

Arpan Deyasi
RCC Institute of Information Technology, India

Pampa Debnath
RCC Institute of Information Technology, India

Siddhartha Bhattacharyya
https://orcid.org/0000-0003-0360-7919
RCC Institute of Information Technology, India

ABSTRACT

Microwave photonics is the arena of research in the 21st century due to ever-increasing ultra-large bandwidth and the meticulous availability of data with very low cost. In this context, conventional optoelectronic devices are replaced by novel photonic counterparts, both in transreceiver design as well as devices and systems. The major objective of this replacement is to reduce noise by means of lower scattering, where photons are only responsible for propagation of electromagnetic wave. With introduction of novel materials, low-loss communication system can now be designed at beyond THz range, mainly due to the physical realization of electromagnetic bandgap structure. This chapter is extended towards plasmonics with the intension of making sensors for beyond THz applications.

DOI: 10.4018/978-1-5225-8531-2.ch001

Copyright © 2019, IGI Global. Copying or distributing in print or electronic forms without written permission of IGI Global is prohibited.

INTRODUCTION

Twenty-first century is the epoch of technology, where evolution and disappearance of a new product takes within a very small time-interval. Due to ever-changing societal need, different new branches of engineering are sprouting day-by-day, may be termed as technological renaissance; where re-invention and re-engineering are the other sides of the same coin in the need of survival of civilization. Along with the change of industry requirement, need of the time is also to amend the curriculum in engineering discipline, where once signified and highlighted contributions are considered as basic and fundamentals after a few years, ultimately may be eliminated from the higher-level academics in order to include into lower-level syllabi so that later re-engineered concepts are incorporated to upgrade the academic framework. However, in such turbulence of technical innovations, a very few concepts are still remains equally acceptable in all forms of learners, researchers, scientists without which we can't even think of progress of Society. To pay homage to those great path-finders, who have paved the way a long ago even with the unavailability of sophisticated instruments and computational advancements, there path-breaking works is tribute in various forms after a sufficient long time-interval. But rarest of rare works are celebrated after hundred or hundred-fifty years of their first publication/announcements, owing to their novelty, consequence, connotation, implication and with possibility of making more fruitful exploration even after that. Year of 2015 was such a year in the history of human civilization when all parts of the scientific community across the world celebrate a particular noble contribution which may be treated as the peer driving force of scientific advancement even at twenty-first century. Yes, dear readers, we are talking about *Maxwell's Equation,* whose hundred-and-fifty years are celebrated worldwide in 2015, after the first publication in 1865 preceded by its first public announcement at Royal Society of London at 1863 (Sengupta & Sarkar, 2003; Selvan, 2007; Sarkar, 2006). Probably the direction of chariot for civilization is destined at that very moment, which is later extrapolated by Erwin Schrodinger.

Looking approximately twenty-five years behind through the window of 2018, one landmark in the experimental communication engineering was made at 900 MHz radio spectrum, when former Finnish prime minister made conversation with mayor at Tampere (Andreas, 2005). The communication has historical importance as it was the world's first interaction through 2G GSM network, built by Telenokia and Siemens; operated by Radiolinja. If we further look back twenty-five years, pioneering result was published in the Proceedings of IEE entitled as "Dielectric-fibre Surface Waveguide for Optical Frequencies" (Kao & Hockham, 1986); for which Charles Kao obtained Nobel prize in 2009. The study reveals the suitability of using glass-fiber to implement optical communication; and is regarded as "The basis of today's optical fiber communication". Fifty years behind the discovery of optical fiber, a four-way international call was made between Newyork, San Fransisco, Washington and Jekyll Island (Riordan & Hoddeson, 1998). The first phrase travels from Newyork to San Fransisco "Ahoy, Ahoy, Mr. Watson, are you there? Do you hear me?" The distance between Newyork and San Fransisco is 3400 miles approximately; and Alexander Graham Bell made the first transcontinental phone call.

The above-mentioned three groundbreaking achievements are the fruits of the pioneering theoretical work, first published in 1865, one hundred and fifty years ago; with the title "A dynamical Theory of Electromagnetic Field" (Maxwell, 1865). The great unification of physics (it combines electricity, magnetism and optics) was invented by James Clerk Maxwell. Arguably he is called "The Father of Electrical Engineering", who explained electromagnetic phenomenon by 20 scalar equations, later reduced into 4 vector equations by OLiver Heavyside, and experimentally verified by Heinrich Hertz. Maxwell theoretically established that light is an electromagnetic wave. He pointed out that light is an undulation in the same medium caused by electric and magnetic phenomena According to him, "What, then, is light according to the electromagnetic theory? It consists of alternate and opposite rapidly recurring transverse magnetic disturbances, accompanied with electric displacements, the direction of the electric displacement being at the right angles to the magnetic disturbance, and both at right angles to the direction of the ray.............." (Maxwell, 1865). He computed relation between the electric displacement, true conduction and the total current compounded of both; the relation between the lines of magnetic force and the inductive coefficients of a circuit; relation between the strength of a current and its magnetic effects; value of electromotive force in a body, as arising from the motion of the body in the field, the alteration of the field itself, and the variation of electric potential from one part of the field to another; relation between electric displacement and electric current, and the electromotive force which produces it; the relation between the amount of free electricity at any point, and the electric displacements in the neighborhood; derived the coefficients of induction between two circuits. Later showed how a circuit containing both capacitance and inductance would respond when connected to generators containing alternating currents of different frequencies. He developed the phenomenon of electrical resonance in parallel to acoustic resonance. He proved that there must be electromagnetic waves; whose speed he calculated would be identical to the speed of light.

In 2018, when we are so advanced with latest communication gadgets, when our daily life is surrounded by fourth generation system, when wired and wireless communication controls our every footsteps, when our decision is virtually controlled by world wide web; it becomes duty of an electrical engineer to pay homage to the great scientist; and this chapter tries to show the deepest respect of the forefounder of modern civilization. Learning becomes passion if one tries to know the history of science and its chronological development, and it is the right time to revere the work of J. C. Maxwell on the celebration of 150 years of his famous lecture at Royal Society of London, globally celebrated as International Year of Light.

As civilization progresses with the advent of technology based on the fundamental equations, new concepts are being developed to improve the communication system, and for that purpose, various novel structures are successfully proposed. One of the latest candidate in this series is the electromagnetic bandgap structure, may be called as the analogous of semiconductor/dielectric in electronics. In the next few pages, we will describe the detailed physics of this structure, and its different modifications proposed by eminent workers in this field (Verma & Daya, 2011; Yang & Samii, 2009; Yang, Fan, Chen, She, & Feng, 2005; Islam & Alam, 2013; Lee, Yeo, & Mittra, 2003).

ELECTROMAGNETIC BAND GAP STRUCTURE

Electromagnetic bandgap materials are one of the most rapidly advancing materials in the electromagnetic arena (Bhavarthe, Rathod, & Reddy, 2018; Huang & Wu, 2017). They have ability to persuade the propagation of electromagnetic waves to a level that was not possible earlier. Due to their unique properties, EBG materials are very popular in scientific society. Generally, EBG structures are defined as artificial periodic structures that avert or assist the propagation of electromagnetic waves in a specified band of frequencies for all incident angles and all polarization states. They are also known as high impedance surface due to their ability to suppress the surface wave at certain operational frequencies. In recent years, there has been rapid increase in utilization of Electromagnetic bandgap (EBG) structures in electromagnetic and antenna community (Yang, Fan, & Feng, 2005; Hajlaoui, 2018; Santos-Díaz, Galaz-Larios, Ramírez-García, 2017; Elwi, 2017).

The EBG terminology is based on the total internal reflection, phenomenon of Photonic crystal in optics, which is realized by periodic structures. EBG Structures are popularly known as photonic crystals that are artificially synthesized crystals which control light completely. The EBG structure is originated from the two papers published by Eli Yablonovitch (Yablonovitch, 1987) in 1987. In the 1980's Yoblonovitch stated that "this PBG, produced by periodic variation in the refractive index of the structure, can be very useful as it can be used to eliminate the spontaneous emission of photons at certain frequency bands". Because of their ability of suppressing surface waves, they are also renamed as high impedance surface. Due to plethora of opportunities, EBG structures now-a-days produce wide variety of design alternatives in the area of microwave photonics.

The structure is also called as photonic bandgap structure also, when it is applied for optical device design, or when photons are the only responsible carriers for output instead of combination of electrons and photons. In the next part, we are discussing detailed of photonic bandgap, and it's importance.

PHOTONIC BAND GAP: IT'S IMPORTANCE

Photonic Band Gap materials, also known as photonic crystals, are materials which have a band gap due to a periodicity in the materials dielectric properties (Andreani et al., 2003). The band gap in photonic crystals represents the forbidden energy range where wave behaving photons can't be transmitted through the material (D'Orazio, De Palo, De Sario, Petruzzelli, & Prudenzano, 2003). By imitating the periodicity of photonic crystals one can tailor the specific band gap of a structure by defining a pattern with repeating regions, typically holes in a square or hexagonal arrangement, which alternates between materials with high and low dielectric constant.

Figure 1. Brillouin zone in 1D photonic crystal

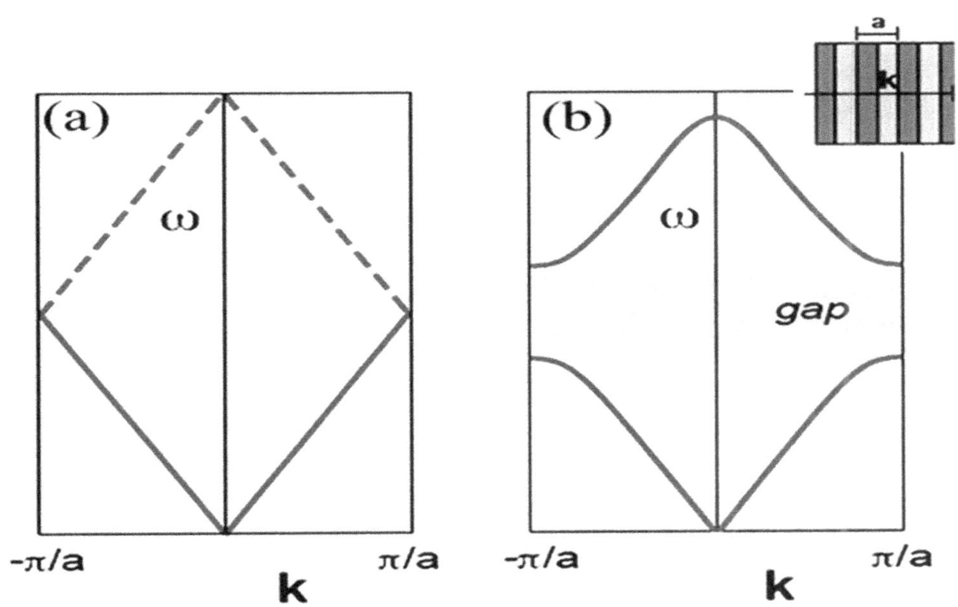

In semiconductor physics, first Brillouin zone is one of the most important concepts for formulation of band structure where a uniquely defined primitive cell in reciprocal space is named as 'Brillouin zone'. The boundaries of this cell are given by planes related to points on the reciprocal lattice. For computation purpose in order to define the zone boundaries, similar method applicable for Wigner–Seitz cell in the Bravais lattice is utilized. The importance of the Brillouin zone lies from the physical behavior of Bloch wave description of waves in a periodic medium, in which it is found that the solutions can be completely characterized by their behavior in a single Brillouin zone.

PHOTONIC CRYSTAL

Photonics is a technology of generating and harnessing light and other forms of radiant energy whose quantum unit is photon. Recent advance in photonics allows precision manipulation and detection of the properties of light, resulting in dramatic improvements in the performance of the existing technologies, and which can efficiently be used in designing device for optical communication (D'Orazio, De Palo, De Sario, Petruzzelli, & Prudenzano, 2003) and information processing (Russell, 2006) with superior performance. In 1887, the famous English Scientist Lord Rayleigh made a novel experiment consisting of periodic multi-layer dielectric stacks, which surprisingly exhibit a photonic band-gap along the direction of stack material variation, and is along one dimension only. Exactly 100 years after, a new group of workers are attracted after the breakthrough research of Yablonovitch (1987) and John on periodic structure behavior at optical frequency, where the refractive index variation is observed along more than one dimension. These are termed in modern days as photonic crystals.

In order to artificially create photonic crystal as the optical microstructure, we have to make controlled periodic growth of metal-dielectric/dielectric materials with alternating regions of various dielectric constants (it may be noted in this context that only one type of material is obtained) where localization of propagating electromagnetic wave can only be obtained by particularly choosing refractive index and thickness variations. The variation of refractive indices of the constituent materials may be looked as the inverse variation of the bandgaps of the materials, and henceforth, it may be considered as multiple quantum well (Maity, Chottopadhyay, Banerjee, & Deyasi, 1400-1405) with considerable magnitude of conduction band discontinuity. This property is utilized to make electromagnetic bandgap where propagation of electromagnetic wave is restricted in certain wavelengths and is allowed to propagate in other spectrum. Therefore, it may be realized as efficient band pass/reject filter for photonic integrated circuit (Arafin & Coldren, 2018), optical transmitter (Altug & Vučković, 2005), optical receiver (Nozaki et al., 2016), sensor (Liu & Salemink, 2012), photonic crystal fiber (Limpert et al., 2003), quantum information processing (Azuma, 2008) applications. This novel microstructure already replaced conventional optical fiber for efficient communication. Photonic crystal fiber is a revolutionary concept, and now realizable due to the advancement in microelectronic technologies. For different communication applications, designing photonic crystal fibers and other notable devices is very important, and thus role of material composition (Gao, Chen, Qiu, Lu, & Huang, 2011) plays a crucial role in this context. Photonic quantum well systems are also studied in recent past to make a comparative analysis between low-dimensional electronic and photonic systems, and resonant tunneling is characterized for this structure.

Photonic crystal is one of the most promising platforms for optical information processing due to the ability of exhibiting arbitrarily different dispersions for the propagation of electromagnetic waves. Photonic crystals without a cPBG can be designed to obtain super collimators and super-lenses. Based on the same effect, two photons that impinge a photonic crystal with the same angle but a slightly different energy, may find EFSs with a very different curvature. As a consequence their propagation angles would be very dissimilar. This is known as super-prism effect and could be applied to the fabrication of small integrated multiplexers. Fabrication of integrated circuits in which information carriers were photons instead of electrons is one of the most expected applications for photonic crystals. In PCFs, photonic crystal with photonic bandgap is constructed to prevent light propagation in certain directions with a certain range of wavelengths. Contrary to normal fiber optics, PCFs use total internal reflection or light confinement in hollow core methods to propagate light. Light propagation in PCFs is far superior to standard fiber, which uses constant lower refractive index cladding. Yablonovitch predicted that certain

Figure 2. Photonic crystal in 1D, 2D, 3D forms

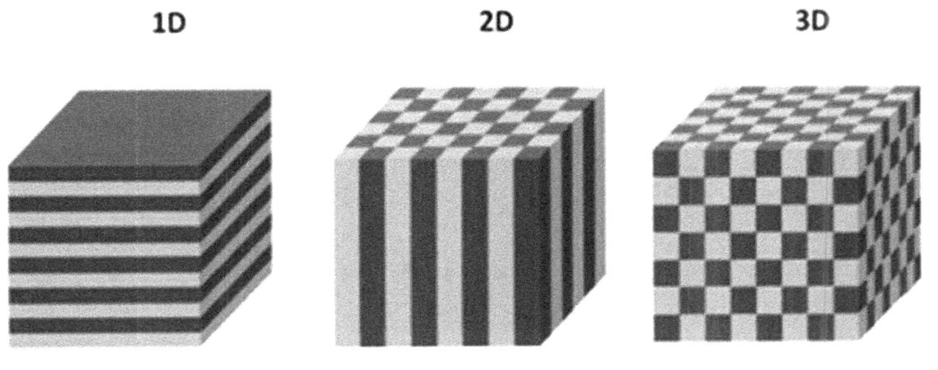

three-dimensionally periodic photonic crystal structures contained complete photonic bandgap. The presence of band-gaps in photonic crystal meant that for certain frequency range there were no propagating modes through the material. This means that if used as a cladding material, the signal would not be able to propagate into and through the cladding. This prediction opened the field of fiber optic manufacturing to the possibility of producing a core-cladding interface that produces total internal reflection.

Photonic crystals are available in 1D, 2D, 3D structure. After initial breakthrough by Yablonovitvh, one-dimensional photonic crystal receives wide-spread attention. This is already used in thin-film optics, where refractive index of different glasses and mirrors are determined and tuned by making layers on the structure with variable thicknesses. 2D and 3D structures are already investigated. These works show novel optical properties in terms of both fundamental and applied researches. From commercial perspectives, 2D structures are more favorable than the existing 1D structure. One rel example in support of the statement is the overwhelming acceptance of photonic crystal fiber, which basically replaces the conventional optical fibers. These PCF's are made by 2D structures. These 2D structures effectively confines light with completely diverse individuality compared to conventional optical fiber, and these fibers are widely used for applications in nonlinear devices and guiding exotic wavelengths. In this context, it is better to mention that 3D structures are yet to find proper commercial applications till the development of 3D photonic IC, and hence less research was carried out on it till now. However, it is noteworthy to mention that these devices offer additional features such as optical nonlinearity, which are extremely essential for the operation of optical transistors used in optical computers.

Figure 3. Microstructured fiber

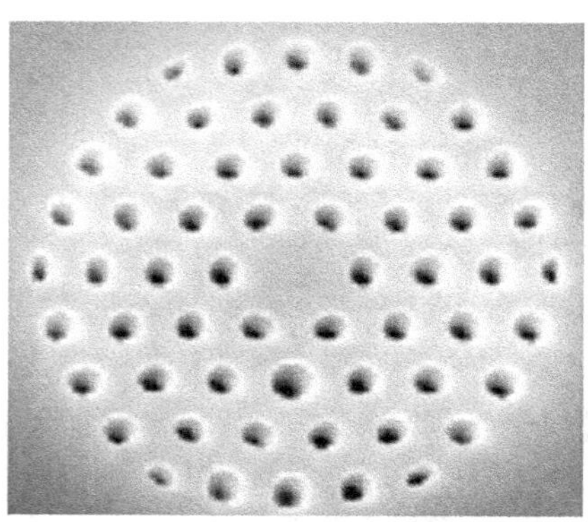

Photonic metamaterials, at submicron scale, maneuver light or any electromagnetic wave (it depends on the internal and material parameters) at optical frequencies. Till 2019, sub-wavelength structures designed at beyond THz frequencies depicted a few, questionable, results only at visible wavelengths. In this context, frequency selective surfaces should be mentioned as they have provided similar optical properties of subwavelength structured metamaterials as exhibited not only by photonic crystals, but also by diffraction gratings. Dielectric mirrors with various coatings on the front surfaces display similarities to those metamaterials. But these structures, as mentioned by experts of material science and also form the domain of photonics, considered distinct so far than commercial subwavelength structures, owing to the fact that their features are prearranged for the wavelength at which they operate and thus cannot be estimated as harmonized substance.

Photonic Multiple Quantum Well

In this section, we will deal with the merging of quantum mechanics with photonic counterpart through a beautiful analogy. To begin with, we start with the basic concept of quantum well. A quantum well is a thin dimensional layer sandwiched between two other layers (may be of same type, or may be of different types) where normally, the bandgap of the inter layer is smaller than the covering layers. This modified band structure helps to confine quasi-particles (known as electrons or holes) inside the well, and its dimension is perpendicular to the layer surface. But a great look reveals that the movement in the other dimensions (where quantum confinements are not taken place) is not restricted. Obviously, it is a quantum effect. Owing to the confinement, fundamental electronic properties of the structure is modified. In this context, analysis/comparative study may begin with density of states function. Depending on the geometry of the quantum well profile, density of states is modified keeping the nature of variation same. Latest works are reported for complex device structures where DOS basically gives the idea of getting current from the device.

On the other hand, let's begin with the introduction of photonic crystal in the domain of optical engineering. It is a new class of structures, which can be depicted as optically active devices, or optically resonant devices. It has attracted tremendous attention after the works of Yablonovitch (1987). Inside these structures, propagating electromagnetic wave or light, interacts with the materials due to their periodic change of dielectric constant, and these interactions produces resonance in the presence of internal excitations of constituent materials. This ultimately results strong frequency dispersion of ingredient dielectric constants (Ivchenko, Voronov, Erementchouk, Deych, & Lisyansky, 2004) an extreme example of such type of interactions produce optical lattice, where resonant elements, which are already well-localized, are sporadically disseminated throughout the medium with predefined (mathematically) uniform dielectric permittivity. At the beginning of the introduction of optical lattice, concepts are generated that the structures are formed by cold atoms. But with the advent of quantum structures in late 1980s, this concept is extensively applied to multiple quantum well structure (MQW) with some special kind. This is considered as a semiconductor analog of one-dimensional optical lattice (Vallini, Gu, Kats, Fainman, & Frateschi, 2013). MQW is an intermittent multilayer configuration built of two or more (first case, it is asymmetric, otherwise becomes asymmetric) semiconductor materials. One of the most used example in this world is the combination of instance, GaAs and $Al_xGa_{1-x}As$, in which electrons and holes are separately confined in thinner layers of a material with a smaller band gap (quantum wells, smaller w.r.t the other materials) alienated by wider bandgap layers (quantum barriers).

Now we will join these two concepts. The MQW structures are termed as photonic crystal or may be termed, if we look the initial input from a different angle. In quantum mechanics, from analysis point-of-view, we consider as energy input from any one side of the quantum barriers from the external world. On the other words, we consider that the input of electromagnetic wave in the quantum structure as energy flux. Schrödinger equation is solved based on this concept. If we just change our perspective of energy input, we will find a different scenario. We will consider the input property as field with polarization, and the entire mathematical analysis becomes different. Because we have to replace the concept of bandgap by corresponding refractive indices of the materials, along with the dimensions of the individual layers. Henceforth, the fundamental governing equation now becomes

Figure 4. Photonic multiple quantum well

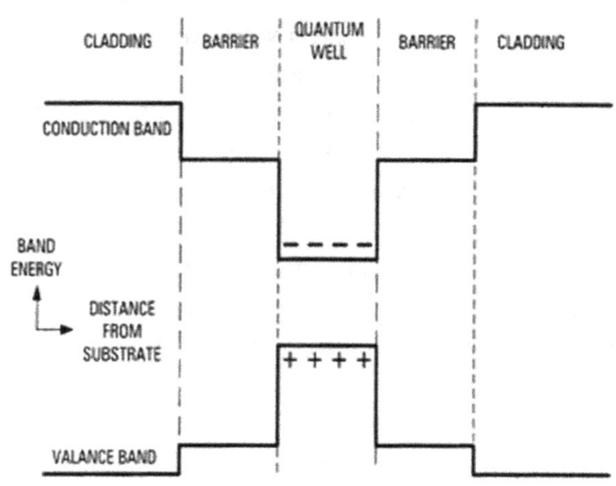

the Bragg's law, and ultimately guiding of e.m wave is realized. Therefore, they may be considered as the opposite sides of the same coin, different due to the analytical perspectives. However, the concept of photonic crystal is also applicable for micron structures, which is not possible for quantum analysis. Quantum well devices feature heterostructures where constituent materials are of group III-V and II-VI materials. Hence, in the term "PMQW", we consider nanometric photonic crystals where compositions are made by semiconductor materials. Here starts the journey.

What Is Metamaterial?

The word 'meta' means 'beyond' and 'metamaterial' refers to 'beyond conventional materials'. The matematerials are completely man-made and consists of some special properties that are not found in nature. Metamaterials can allow electric field component as well as magnetic field component of light to be coupled, and this is one of the main property of materials. Another property is negative refractive index, which is the most fundamental characteristics of light propagation in material. The metamaterials with negative refractive index can lead to be a superlens, which is capable of imaging objects. It can also give a fine structure, which has much smaller wavelength than the light. Another applications of metamaterials include antennas are- optical nano circuits, optical nano lithography and meta coatings that can make a object invisible. Negative index materials are the most exiting application of metamaterials, which has a huge use in optical technologies. The notion of a negative refractive index is one of such case which provides a unique opportunity for researcher to consider and revise the interpretation of very basic law.

Role of Metamaterial in Photonic Crystal

Artificial materials i.e., metamaterials have same properties that are mare pronounced than natural materials. Negative index of refraction is one of the properties of metamaterial and that's why it is called negative index material or left handed material. Negative index materials have electro dynamic characteristics. The electrodynamics of the materials with ε and μ is different from the materials with positive ε and μ. There are possible three cases: 1. There are no differences, 2. negative values of ε and μ are in principle impossible because of its conflicts with basic principles, 3. Negative values of ε and μ is possible as electrodynamics of such materials differ from positive ε and μ. The first metamaterials are consisted of copper rings and straight wires. The size of those elements and the distance between them are smaller than the wavelength. The lattice constants of a normal material is much bellow then the wavelength of visible light, that's why the electric field and magnetic field response of such materials can be described by ε and μ.

The light propagation in the photonic crystals is not an average effect of atoms in common crystals and this light propagation in any photonic crystal is the result of Bragg Diffraction at each atom. So, periodical structure is very important for photonic crystal. ε and μ cannot describe the light propagation and light refraction at the boundary, i.e., light waves in a photonic crystal is considered as Bloch Waves. There are many applications of these negative index materials – those are, microwave applications, branch live couplers, ring couplers. Currently the negative index material fabrication is much more demanding than the negative index material design and optimization. Though the slabs of this material is bigger compared with the wavelength, that cannot be considered as optical super lenses, but then negative index material slabs can open a new opportunity for overcoming the diffraction limit.

Classification of Metamaterial

The response of a system to the presence of Electromagnetic field is determined by the properties of the materials involved. These properties define the macroscopic parameters permittivity ε and permeability μ of materials. On the basis of permittivity ε and permeability μ, the metamaterials are classified in following four groups.

1. **Double Positive (DPS) Material:** The materials which have both permittivity & permeability greater than zero ($\varepsilon > 0, \mu > 0$) are called as double positive (DPS) materials. Most occurring media (e.g. dielectrics) fall under this designation.

Figure 5. Classification of Metamaterials

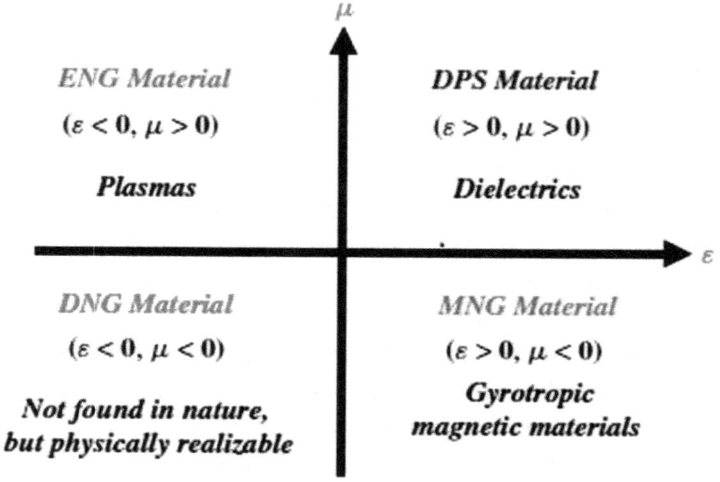

2. **Epsilon Negative (ENG) Material:** If a material has permittivity less than zero and permeability greater than zero ($\varepsilon < 0, \mu > 0$) it is called as epsilon negative (ENG) material. In certain frequency regimes, many plasmas exhibit these characteristics.
3. **Mu Negative(MNG) Material:** If a material has permittivity greater than zero & permeability less than zero ($\varepsilon > 0, \mu < 0$) it is called as mu negative (MNG) material. In certain frequency regimes, some gyro tropic material exhibits these characteristics.
4. **Double Negative (DNG) Material:** If a material has permittivity & permeability less than zero ($\varepsilon < 0, \mu < 0$) it is termed as double negative (DNG) material. This class of materials can only been produced artificially.

APPLICATION OF METAMATERIAL

Metamaterials are under consideration for many applications. They are-

1. **Antennas:** Metamaterial antennas are a class of antennas that use metamaterials to improve performance. Demonstrations showed that metamaterials could enhance an antenna's radiated power. Materials that can attain negative permeability allow for properties such as small antenna size, high directivity and tunable frequency.
2. **Absorber:** A metamaterial absorber manipulates the loss components of metamaterials' permittivity and magnetic permeability, to absorb large amounts of electromagnetic radiation. This is a useful feature for solar photovoltaic applications. Loss components are also relevant in applications of negative refractive index (photonic metamaterials, antenna systems) or transformation optics (metamaterial cloaking, celestial mechanics), but often are not utilized in these applications.
3. **Superlens:** A superlens is a two or three-dimensional device that uses metamaterials, usually with negative refraction properties, to achieve resolution beyond the diffraction limit (ideally, infinite resolution). Such a behaviour is enabled by the capability of double-negative materials to yield negative phase velocity. The diffraction limit is inherent in conventional optical devices or lenses.
4. **Light and Sound Filtering:** Metamaterials textured with nanoscale wrinkles could control sound or light signals, such as changing a material's color or improving ultrasound resolution. Uses include nondestructive material testing, medical diagnostics and sound suppression. The materials can be made through

a high-precision, multi-layer deposition process. The thickness of each layer can be controlled within a fraction of a wavelength. The material is then compressed, creating precise wrinkles whose spacing can cause scattering of selected frequencies.

METAMATERIAL IN PHOTONIC CRYSTAL

Light is the ultimate means of sending information to and from the interior structure of materials _ it packages data in a signal of zero mass and unmatched speed. However, light is, in a sense, `one-handed' when interacting with atoms of conventional materials. This is because from the two field components of light - electric and magnetic - only the electric `hand' efficiently probes the atoms of a material, whereas the magnetic component remains relatively unused because the interaction of atoms with the magnetic-field component of light is normally weak.

Metamaterials, that is, artificial materials with rationally designed properties, can allow both field components of light to be coupled to meta-atoms, enabling entirely new optical properties and exciting applications with such `two-handed' light. Among the fascinating properties is a negative refractive index. The refractive index Metamaterials with negative refraction may lead to the development of a superlens capable of imaging objects and fine structures that are much smaller than the wavelength of light.

Photonic metamaterials, nanometer scale, manipulate light at optical frequencies. To date, sub-wavelength structures have shown only a few, questionable, results at visible wavelengths. Photonic crystals and frequency-selective surfaces such as diffraction gratings, dielectric mirrors and optical coatings exhibit similarities to subwavelength structured metamaterials. However, these are usually considered distinct from subwavelength structures, as their features are structured for the wavelength at which they function and thus cannot be approximated as a homogeneous material. However, material structures such as photonic crystals are effective in the visible light spectrum. The middle of the visible spectrum has a wavelength of approximately 560 nm (for sunlight). Photonic crystal structures are generally half this size or smaller, that is <280 nm.

Metamaterial becomes subject of research in the last few years due to its capability of improving SNR at desired communication spectra (Xiong, Hong, & Peng, 2012; Hwang, Liu, & Chin, 2009). More precisely, array of metamaterial can be looked as one-dimensional photonic crystal which can effectively improve the signal strength at the optical communication window. This property can effectively be utilized for controlled electromagnetic wave propagation, with the possibility to exhibit desired transmittivity at the central wavelength. DNG material can be considered as a suitable choice for this purpose, replacing conventional material system (Robinson & Nakkeeran, 2011), or semiconductor heterostructure (Maity, Chottopadhyay, Banerjee, & Deyasi, 2013). But most of the works available in literature are dealt with materials with positive refractive indices, whereas the present paper deals with the works carried out with negative r.i material system.

Rudziński (2007) analyzed electromagnetic wave propagation inside 1D photonic crystal for oblique incidence. Resonant tunneling inside 1D crystal was theoretically (Xu et al., 2007) and experimentally (Lin, Hietala, Lyo, & Zaslavsky, 1996) verified at lower wavelength region. Gao (Gao, Chen, Qiu, Lu, & Huang, 2011) calculated the effect of refractive index on transmission spectra for bandpass filter design. The structure is already considered as a suitable candidate for photodetector (Reininger et al., 2012) and integrated circuit applications.

Pioneering work in the field of DNG material based optical communication system was initiated a few years ago by Povinelli (Povinelli, Johnson, & Joannopoulos, 2003) in 2003. He showed that electromagnetic metamaterial has synthetic structure that is not found in naturally occurring materials or composites. Photonic crystals have magnetic resonances that can lead to a negative effect magnetic permeability. If the used defect inside PC is point defect, then it resembles a magnetic moment, which can be incorporated into a finite crystal so that the structure becomes a magnetic emitter. This can be designed to operate even at optical frequencies. Later Caglayan (Caglayan, Bulu, Loncar, & Ozbay, 2009) experimentally observed subwavelength localization using metamaterial based cavities. Veselago (Veselago, Braginsky, Shklover, & Hafner, 2006) made a detailed review work on SNG and DNG materials related with Bragg diffraction principle. The light propagation in the photonic crystals is not an average effect of atoms in common crystals and this light propagation in any photonic crystal is the result of Bragg Diffraction at each atom. So, periodical structure is very important for photonic crystal. ε and μ cannot describe the light propagation and light refraction at the boundary, i.e., light waves in a photonic crystal is considered as Bloch Waves. Baba (Baba, Matsumoto, & Asatsuma, 2008) also analyzed negative refraction in photonic materials. Photonic crystals are composed of dielectric that exhibit negative refraction based on anomalous dispersion arising from multi-dimentional periodicity. These types of crystals have application to a compact demultiplexer and a parallel optical coupler. Propagating wave characteristics is also analyzed (Awasthi, Mishra, Malaviya, & Ojha, 2009) for designing antenna, optical nano-circuit, superlens. It can allow electric field component as well as magnetic field component of light to be coupled, and this is one of the main properties of materials. When a light wave is passing through these materials, the wave become p-polarized or s-polarized. When the transmission spectrum of p-polarized wave is moving away, it shows that the gap between the width and depth reduces. When the incident angle is greater than 50°, the gap for p-polarized light is eliminated.

Recently one type of crystal is introduced, which has tunable light wave propagation that is its refractive indices can be changed, when its direction become changed and this crystal is called liquid crystal (Karaomerlioglu, Mamedov, & Ozbay, 2014). The tunable light wave propagation in 2D hole type photonic crystal is investigated by Liu, infiltrated with nematic liquid crystal. Using hybrid conventional wave guide Liu and Chen proposed a tunable field sensitive polarizer and photonic crystal structure with nematic liquid crystals. The tunable photonic crystal wave guide coupler with nematic liquid crystals etc. According to DNG materials, SNG materials and infinite materials, the photonic crystals are made up of synthetic, periodic, dielectric materials which can exhibit negative refraction behavior which are determined by the characteristics of the band structures. El-Khozondar (El-Khozondar, El-Khozondar, & Shabat, 2014) proposed the waveguide structure. It is a new type of photonic structure is introduced which consists of an alternative lamellae of metamaterials and dielectrics. This structure works as a wave guide in the visible region, i.e., near about 450 nm and 750 nm. This structure also allows transmitting the short IR light.

Photonic crystal structures using metamaterials can be used as sensors, optical modulators and semiconductor lasers are based on the existence of surface waves (Namdar, Entezar, Rahimi, & Tajalli, 2010). Surface waves are nonradiative modes which are propagating through the interface with amplitudes in every bounding medium. In periodic systems modes are localized at a surface, which are known as Tamm states. Surface waves (SWs) have some potential to become alternatives and also have some advantages- 1st, due to the scaling nature of dielectric PCs, it can exist virtually and 2nd, low dielectric loss can lead to sharp resonant coupling between the incoming light and SWs. Effects of lossy materials are analyzed by researchers recently (Shi, Chen, Han, & Soukoulis, 2013). The frequency dependence of the phase angle of the Bloch impedance is influenced by the ratio of the electric to the magnetic damping coefficient of the DNG metataterials. When Bloch impedance become maximum then the reflection will be strongest and absorption will be minimum and when Bloch impedance become maximum then the reflection will be strongest and absorption will be minimum, and when Bloch impedance become minimum then reflection will be weakest and absorption will be maximum.

The metamaterial inspired micro structure that is RPCs are invariant under translations, by implementing radially dependent constitutive parameter functions in multilayered systems. The fundamental characteristics of the micro structure can derive from the periodicity of the radial anisotropic constitutive parameters, for a 2D configuration (Xu et al., 2007). Depending on the number of constitutive parameters and layers implemented, a practical design of a microwave microstructure is perform.

MICROWAVE PHOTONICS

Microwave photonics is the arena of research in the twenty-first century due to the ever-increasing of ultra-large bandwidth and meticulous availability of data with very low cost. In this context, conventional optoelectronic devices are replaced by novel photonic counterparts, both in transreceiver design, as well as devices and systems. The major objective of this replacement is to reduce noise by means of lower scattering, where photons are only responsible for propagation of electromagnetic wave. With introduction of novel materials, low-loss communication system can now be designed at beyond THz range (Díaz, Galaz-Larios, & Ramírez-García, 2017; Dong & Liu, 2016), possible mainly due to the physical realization of electromagnetic bandgap structure. Work is extended towards plasmonics (Yokota et al., 2015) with the intension of making sensors for beyond THz applications.

Solution of Maxwell's equation subject to appropriate boundary conditions gets the prime importance for today's communication system development, be in unbounded or bounded media. Different forms of periodic structures are already experimentally realized, and their interaction with electromagnetic wave produces different novel phenomena with remarkable features, which are not generally available. Precisely, these structures exhibit pass and stop bands, and different bandgap are identified, analogous to semiconductor structures. These properties are explored in design of filters, frequency-selective surfaces, gratings, photonic crystals, antennas, waveguides etc, which may be grouped under the banner of "electromagnetic Bandgap Structures". Modern antennas are based on various types of EBG structure (Luberto & Fano, 2015). Computational electromagnetic plays a key role in this context for theoretical prediction of state-of-the-art device performances.

The concept of Microstrip antenna has been proposed first by Deschamps in 1953, though Munson and Howell developed practical antennas in the year of 1970. Various benefits of microstrip antennas such as light weight, low cost, compatible with high technology integrated circuits and ease of fabrication have led to expand applications and to develop the research as a distinguish entity in the area of high frequency antenna. With an ever-increasing demand for wireless communications systems, the requirements of low profile miniaturized antenna has brought microstrip antenna to the leading edge. Inspite of several advantages they suffers from narrow bandwidth, low gain and excitation of surface waves. In present scenario miniaturized antennas have high demand in mobile communication systems. Thus for practical applications, reduction in size along with enhancement of gain and bandwidth become the key design issues for MSA.

In present scenario wireless communication system is one of the fast rising technologies. The demand for higher data transmission rates has moved this technology towards higher frequency region. The Radio Telegraphy demonstrated by Guglielmo

Marconi in 1897 has been considered as the establishment of wireless revolution. Different technologies have been used to implement wireless communication systems. Infrared technology has been realized by transmitting infrared light. For efficient performance and fast data rate Light of sight (LOS) transmission is required. This simple secured and cheap technology is suitable for short distance wireless communications. For long distance communication Radio Frequency (RF) technology has been employed. Mobile communication, Satellite communication are the most popular examples of RF technology.

The communication media that are used for transmission of information can be considered as bounded or unbounded media. Bounded media or guided media provides a physical path through which signals are transmitted which includes twisted pair transmission line, coaxial cable, waveguide and optical fiber. Twisted pair is most commonly used and cheaper, light weight. Coaxial cable has been used for long distance telephone line. The feature of high bandwidth, higher noise immunity and transmission of digital signal at very high rate makes coaxial cable more attractive for bounded transmission. Waveguide is used to transfer electromagnetic energy. Low loss, high quality factor and high power handling capability are beneficial property of waveguide. Waveguide are used in photonic integrated circuits, optical fiber communication and optical inter Light weight, wide bandwidth, and low attenuation of signal and immune to electromagnetic interference are the main characteristic features of optical fiber.

EBG structures can also be fabricated using artificially negative refractive index materials, termed as metamaterials. These materials can be used to reduce the spectrum width of antenna operating region in order to enhance SNR, and also utilized to design photonic crystal. The later can be considered as the building block of twenty-first century. It is also realized very recently that metamaterial based lenses may be used to view live viruses.

Probably the most fascinating outcome of EBG structure is the photonic crystal (PhC). Devices and components based on PhC can replace their bulk optoelectronic counterparts. It is also available in nature, and it can now-a-days used from low-loss waveguiding to produce inverse Doppler Effect. Recent published literatures suggested that it can produce on-chip quantum optical applications (Liu et al., 2018).

It is presently applied to generate surface plasmon resonance (SPR) in novel periodic structures, and that also can be locally controlled. This property can effective be utilized to design biosensors when coupled to photonic crystal fiber (PCF). This is used as a very sensitive detector with reduced fabrication cost.

REFERENCES

Altug, H., & Vučković, J. (2005). Photonic Crystal Nanocavity Array Laser. *Optics Express, 13*(22), 8819–8828. doi:10.1364/OPEX.13.008819 PMID:19498914

Andreani, L. C., Agio, M., Bajoni, D., Belotti, M., Galli, M., Guizzetti, G., ... Vecchi, G. (2003). Optical Properties and Photonic Mode Dispersion in Two-Dimensional and Waveguide- Embedded Photonic Crystals. *Synthetic Metals, 139*(3), 695–700. doi:10.1016/S0379-6779(03)00330-8

Andreas, M. (2005). *Wireless Communications*. Wiley-IEEE Press.

Arafin, S., & Coldren, L. A. (2018). Advanced InP Photonic Integrated Circuits for Communication and Sensing. *IEEE Journal of Selected Topics in Quantum Electronics, 24*(1), 1–12. doi:10.1109/JSTQE.2017.2754583

Awasthi, S. K., Mishra, A., Malaviya, U., & Ojha, S. P. (2009). Wave Propagation in a One-Dimensional Photonic Crystal with Metamaterial. *Solid State Communications, 149*(33-34), 1379–1383. doi:10.1016/j.ssc.2009.05.017

Azuma, H. (2008). Quantum Computation with Kerr-Nonlinear Photonic Crystals. *Journal of Physics. D, Applied Physics, 41*(025102).

Baba, T., Matsumoto, T., & Asatsuma, T. (2008). Negative Refraction in Photonic Crystals. *Advances in Science and Technology, 55*, 91–100. doi:10.4028/www.scientific.net/AST.55.91

Bhavarthe, P. P., Rathod, S. S., & Reddy, K. T. V. (2018). Mutual coupling reduction in patch antenna using Electromagnetic Band Gap (EBG) structure for IoT application. *International Conference on Communication information and Computing Technology*. 10.1109/ICCICT.2018.8325867

Caglayan, H., Bulu, I., Loncar, M., & Ozbay, E. (2009). Experimental Observation of Subwavelength Localization using Metamaterial-Based Cavities. *Optics Letters, 34*(1), 88–90. doi:10.1364/OL.34.000088 PMID:19109649

D'Orazio, A., De Palo, V., De Sario, M., Petruzzelli, V., & Prudenzano, F. (2003). Finite Difference Time Domain Modeling of Light Amplification in Active Photonic Bandgap Structures. *Progress in Electromagnetics Research, 39*, 299–339. doi:10.2528/PIER02112501

Díaz, A. M. S., Galaz-Larios, M. C., & Ramírez-García, E. (2017). Analysis and design of electromagnetic band gap structures with application in planar antennas. *42nd International Conference on Infrared, Millimeter, and Terahertz Waves*. 10.1109/IRMMW-THz.2017.8067190

Díaz-Rubio, A., Carbonell, J., Torrent, D., Cervera, F., Kirleis, M. A., Piqué, A., & Sánchez-Dehesa, J. (2012). Radial Photonic Crystals for Microwave Operation. *The Sixth International Congress on Advanced Electromagnetic Materials in Microwaves and Optics*, 393-395.

Dong, Y., & Liu, P. (2016). Applications of graphene in terahertz electromagnetic band-gap structure designs. *Asia-Pacific International Symposium on Electromagnetic Compatibility*. 10.1109/APEMC.2016.7522966

El-Khozondar, H. J., El-Khozondar, R. J., & Shabat, M. M. (2014). Metamaterial-dielectric Photonics Crystal Waveguide Structure. *Optics*, *4*(1-2), 1–4.

Elwi, T. A. (2017). Electromagnetic band gap structures based on ultra wideband microstrip antenna. *Microwave and Optical Technology Letters*, *59*(4), 827–834. doi:10.1002/mop.30397

Gao, Y., Chen, H., Qiu, H., Lu, Q., & Huang, C. (2011). Transmission Spectra Characteristics of 1D Photonic Crystals with Complex Dielectric Constant. *Rare Metals*, *30*(2), 150–154. doi:10.100712598-011-0215-7

Gao, Y., Chen, H., Qiu, H., Lu, Q., & Huang, C. (2011). Transmission Spectra Characteristics of 1D Photonic Crystals with Complex Dielectric Constant. *Rare Metals*, *30*(2), 150–154. doi:10.100712598-011-0215-7

Hajlaoui, E. A. (2018). New triple band electromagnetic band gap microstrip patch antenna with two shaped parasitic elements. *Journal of Computational Electronics*, *17*(1), 452–457. doi:10.100710825-017-1100-x

Huang, W., & Wu, Z. (2017). Study on band gap characteristics of EBG structures with interdigital and spiral ground plane. *AIP Conference Proceedings*, 1890.

Hwang, R. B., Liu, H. W., & Chin, C. Y. (2009). A metamaterial-based E-plane Horn Antenna. *Progress in Electromagnetics Research*, *93*, 275–289. doi:10.2528/PIER09050606

Islam, M. T., & Alam, M. S. (2013). Compact EBG structure for alleviating mutual coupling between patch antenna array elements. *Progress in Electromagnetics Research*, *137*, 425–438. doi:10.2528/PIER12121205

Ivchenko, E. L., Voronov, M. M., Erementchouk, M. V., Deych, L. I., & Lisyansky, A. A. (2004). Multiple-quantum-well-based photonic crystals with simple and compound elementary supercells. *Physical Review B: Condensed Matter and Materials Physics*, *70*(19), 195106. doi:10.1103/PhysRevB.70.195106

Kao, K. C., & Hockham, G. A. (1986). Dielectric-fibre surface waveguides for optical frequencies. *IEE Proceedings. Part J, Optoelectronics*, *133*(3), 191–198. doi:10.1049/ip-j.1986.0030

Karaomerlioglu, F., Mamedov, A. M., & Ozbay, E. (2014). Optical Properties of Metamaterial-based Devices Modulated by a Liquid Crystal. *Applied Physics. A, Materials Science & Processing*, *117*(2), 611–619. doi:10.100700339-014-8711-x

Lee, Y., Yeo, J., & Mittra, R. (2003). Investigation of electromagnetic bandgap (EBG) structures for antenna pattern control. *Proceedings of the IEEE International Antennas and Propagation Symposium*, 2, 1115-1118.

Limpert, J., Schreiber, T., Nolte, S., Zellmer, H., Tunnermann, T., Iliew, R., ... Jakobsen, C. (2003). High Power Air-Clad Large-Mode-Area Photonic Crystal Fiber Laser. *Optics Express*, *11*(7), 818–823. doi:10.1364/OE.11.000818 PMID:19461794

Lin, S. Y., Hietala, V. M., Lyo, S. K., & Zaslavsky, A. (1996). Photonic Bandgap Quantum Well and Quantum Box Structures: A high-Q Resonant Cavity. *Applied Physics Letters*, *68*(23), 3233–3235. doi:10.1063/1.116558

Liu, F. (2018, July). High Purcell factor generation of indistinguishable on-chip single photons. *Nature Nanotechnology*. PMID:30013218

Liu, Y., & Salemink, H. W. M. (2012). Photonic Crystal-based All-Optical On-Chip Sensor. *Optics Express*, *20*(18), 19912–19920. doi:10.1364/OE.20.019912 PMID:23037043

Luberto, M., & Fano, W. G. (2015). Microstrip antenna design using EBG (Electromagnetic Band Gap) structures at 2.4GHz. *XVI Workshop on Information Processing and Control*. 10.1109/RPIC.2015.7497097

Maity, Chottopadhyay, Banerjee, & Deyasi. (n.d.). Novel band-pass filter design using photonic multiple quantum well structure with p-polarized incident wave at 1550 nm. *Journal of Electron Devices, 17*, 1400-1405.

Maity, A., Chottopadhyay, B., Banerjee, U., & Deyasi, A. (2013). Novel Band-Pass Filter Design using Photonic Multiple Quantum Well Structure with p-Polarized Incident Wave at 1550 nm. *Journal of Electron Devices*, *17*, 1400–1405.

Maxwell, J. C. (1865). A dynamical theory of the electromagnetic field. *Philosophical Transactions of the Royal Society of London*, *155*, 459–512. doi:10.1098/rstl.1865.0008

Namdar, A., Entezar, S. R., Rahimi, H., & Tajalli, H. (2010). Backward TAMM States In 1D Single-Negative Metamaterial Photonic Crystals. *Progress In Electromagnetics Research Letters*, *13*, 149–159. doi:10.2528/PIERL09121305

Nozaki, K., Matsuo, S., Fujii, T., Takeda, K., Ono, M., Shakoor, A., ... Notomi, M. (2016). Photonic-Crystal Nano-Photodetector with Ultrasmall Capacitance for On-Chip Light-To-Voltage Conversion without an Amplifier. *Optica*, *3*(5), 483–492. doi:10.1364/OPTICA.3.000483

Povinelli, M. L., Johnson, S. G., Joannopoulos, J. D., & Pendry, J. B. (2003). Toward Photonic-Crystal Metamaterials: Creating Magnetic Emitters in Photonic Crystals. *Applied Physics Letters*, *82*(7), 1069–1071. doi:10.1063/1.1544428

Reininger, P., Kalchmair, S., Gansch, R., Andrews, A. M., Detz, H., Zederbauer, T., ... Strasser, G. (2012). Optimized Photonic Crystal Design for Quantum Well Infrared Photodetectors. *Proceedings of the Society for Photo-Instrumentation Engineers*, *8425*, 84250A. doi:10.1117/12.923045

Riordan, M., & Hoddeson, L. (1998). *Crystal Fire: The Birth of the Information Age* (1st ed.). New York: Sloan Technology Series.

Robinson, S., & Nakkeeran, R. (2011). Photonic Crystal Ring Resonator based Add-Drop Filter using Hexagonal Rods for CWDM Systems. *Optoelectronics Letters*, *7*(3), 164–166. doi:10.100711801-011-0172-2

Rudziński, A. (2007). Analytic Expressions for Electromagnetic Field Envelopes in a 1D Photonic Crystal. *Acta Physica Polonica A*, *111*(3), 323–333. doi:10.12693/APhysPolA.111.323

Russell, P. S. J. (2006). Photonic-Crystal Fibers. *Journal of Lightwave Technology*, *24*(12), 472–474. doi:10.1109/JLT.2006.885258

Santos-Díaz, A. M., Galaz-Larios, M. C., & Ramírez-García, E. (2017). Analysis and design of electromagnetic band gap structures with application in planar antennas. *42nd International Conference on Infrared, Millimeter, and Terahertz Waves.*

Sarkar, T. K. (2006). *"Who Was James Clerk Maxwell and What Is/Was His Electromagnetic Theory", Seminar paper on IEEE AP/MTT HK Joint Chapter*. City University of Hong Kong.

Selvan, K. T. (2007). Engineering Education: Presentation of Maxwell's Equations in Historical Perspective and the Likely Desirable Outcomes. *IEEE Antennas & Propagation Magazine*, *49*(5), 155–160. doi:10.1109/MAP.2007.4395329

Sengupta, D. L., & Sarkar, T. K. (2003). Maxwell, Hertz, the Maxwellians, and the Early History of Electromagnetic Waves. *IEEE Antennas & Propagation Magazine*, *45*(2), 13–19. doi:10.1109/MAP.2003.1203114

Shi, F., Chen, Y., Han, P., & Soukoulis, C. M. (2013). Investigation of One-Dimensional Photonic Bandgap Structures containing Lossy Double-Negative Metamaterials through the Bloch Impedance. *Journal of the Optical Society of America. B, Optical Physics*, *30*(6), 1473. doi:10.1364/JOSAB.30.001473

Vallini, F., Gu, Q., Kats, M., Fainman, Y., & Frateschi, N. C. (2013). Carrier saturation in multiple quantum well metallo-dielectric semiconductor nanolaser: Is bulk material a better choice for gain media? *Optics Express*, *21*(22), 25985–25998. doi:10.1364/OE.21.025985 PMID:24216824

Verma, R., & Daya, K. S. (2011). Effect of Forbidden Bands of Electromagnetic Bandgap Engineered Ground Plane on the Response of Half Wave Length Linear Microwave Resonator. *Journal of Applied Physics*, 109.

Veselago, V., Braginsky, L., Shklover, V., & Hafner, C. (2006) Negative Refractive Index Materials. *Journal of Computational and Theoretical Nanoscience*, *3*(2), 1–30. doi:10.1166/jctn.2006.3000

Xiong, H., Hong, J. S., & Peng, Y. H. (2012). Impedance Bandwidth and Gain Improvement for Microstrip Antenna using Metamaterials. *Radioenginnering*, *21*, 993–998.

Xu, X., Chen, H., Xiong, Z., Jin, A., Gu, C., Cheng, B., & Zhang, D. (2007). Fabrication of Photonic Crystals on Several Kinds of Semiconductor Materials by using Focused-Ion Beam Method. *Thin Solid Films*, *515*(22), 8297–8300. doi:10.1016/j.tsf.2007.02.061

Yablonovitch, E. (1987). Inhibited Spontaneous Emission in Solid-State Physics and Electronics. *Physical Review Letters*, *58*(20), 2059–2061. doi:10.1103/PhysRevLett.58.2059 PMID:10034639

Yang, L., Fan, M., Chen, F., She, J., & Feng, Z. (2005). A Novel Compact Electromagnetic–Bandgap (EBG) Structure and Its Applications for Microwave Circuits. *IEEE Transactions on Microwave Theory and Techniques*, *53*(1), 183–190. doi:10.1109/TMTT.2004.839322

Yang & Samii. (2009). *Electromagnetic Band Gap Structures in Antenna Engineering*. Cambridge University Press.

Yang, Fan, & Feng. (2005). A spiral electromagnetic bandgap (EBG) structure and its application in microstrip antenna arrays. *Asia-Pacific Microwave Conference Proceedings*. doi:10.1063/1.5005254

Yokota, H. (2015). *Control of localized surface plasmon resonance energy in monolayer structures of gold and silver nanoparticles* (Vol. 40). Physical Chemistry Chemical Physics.

Chapter 2
Microstrip Antenna

Nandan Bhattacharyya
RCC Institute of Information Technology, India

Jawad Yaseen Siddiqui
Calcutta University, India

ABSTRACT

The microstrip antenna (MSA) consists of a dielectric substrate in between a metallic conducting patch and a ground plane. The most common forms of the MSA are the rectangular and circular patch MSAs. There are several microstrip antenna analysis methods. The most popular models are transmission-line model, cavity model, method of moments, FDTD method, and finite element method. The transmission-line model is the simplest of these methods, and it provides good physical insight but is less accurate. The cavity model is more accurate compared to the transmission-line model, but cavity model is more complex. Though cavity model gives good physical insight, it is rather difficult to model coupling. The full-wave models (which include primarily integral equations/moment method) are very accurate, very versatile, but they are the most complex models and usually give less physical insight. This chapter explores the microstrip antenna.

INTRODUCTION

Microstrip Antenna

In microstrip antenna (MSA) there is a dielectric in between a top metallic patch and a bottom ground plane. Input to MSA is generally provided using a coaxial probe or by using a microstrip edge feed. Usually the top metallic patch is shaped rectangular and circular. However, depending on application requirement patch shapes other than rectangular and circular are used to achieve desired MSA performance.

DOI: 10.4018/978-1-5225-8531-2.ch002

Copyright © 2019, IGI Global. Copying or distributing in print or electronic forms without written permission of IGI Global is prohibited.

Advantages: low profile, simple and inexpensive to manufacture, conformable, mechanically robust, and very versatile.

Disadvantages: MSApower handling capacity is low, having low efficiency, high quality factor, narrow bandwidth, lacks polarization purity, spurious radiation.

There are several microstrip antenna analysis methods. The most accepted models are:

1. Transmission line model
2. Cavity model
3. Method of moments
4. FDTD method
5. Finite Element method

The transmission line model is the simplest of these methods. Both transmission line model and cavity model furnishes good physical insight but it is difficult to model coupling. Compared to the transmission-line model, cavity model is more accurate but is more complex. Whereas full-wave models including primarily integral equations, Method of Moments are more accurate and versatile at the cost of increased complexity in modeling and less physical insight. These models could be used to analyze regular or arbitrary shaped element, stacked elements and arrays of elements.

Transmission Line Model

Transmission line model is the most reasonable choice for the study of microstrip antennas as the physical structure of microstrip antennas have been derived from microstrip transmission lines. This transmission line model was the first technique employed to investigate a rectangular microstrip antenna by Munson (1974).

Transmission Line Model for Rectangular Microstrip Antennas

Rectangular microstrip antenna comprises of a dielectric in between a conducting patch at the top and a conducting ground plane at the bottom. The antenna feeding is done by using a coaxial probe or by using a microstrip transmission line. Figure 1 illustrates a rectangular microstrip antenna with microstrip transmission line feeding.

The patch is characterised by the length L, the width W, the thickness t, the conductivity σ. The dielectric substrate has a relative permittivity ε_r, and a loss tangent δ, length L_{sub}, a width W_{sub}, and a thickness h. The conducting ground plane has length L_{sub}, width W_{sub} and a thickness t.

Figure 1. Rectangular microstrip antenna

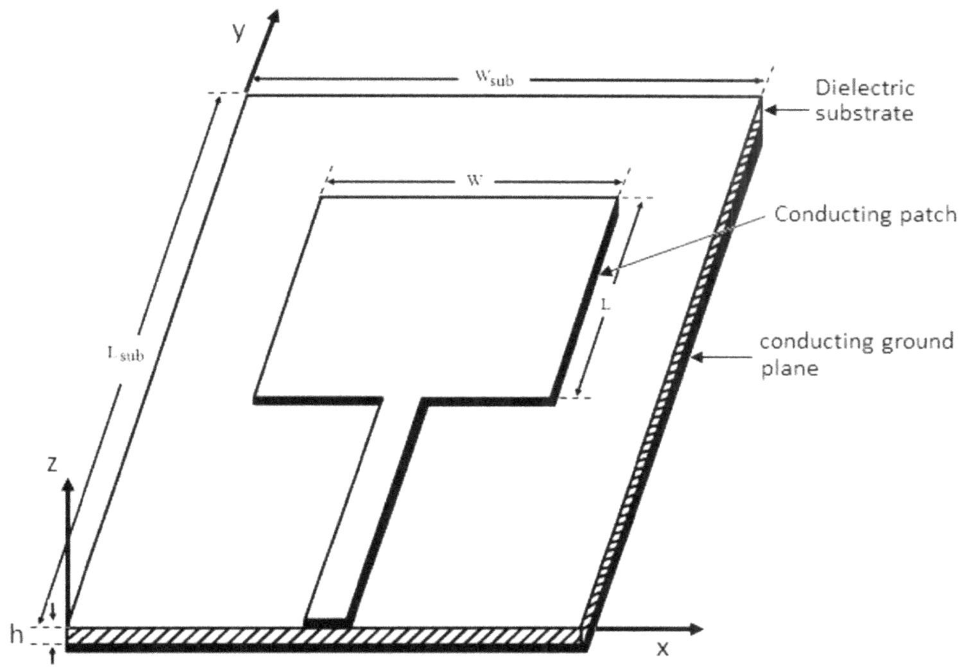

As shown in Figure 2, the fields of the microstrip antenna could be separated into an interior region and an exterior region (Carver & Coffey, 1979) in the analytical methods. The interior region of the microstrip antenna is modeled as a section of transmission line. The patch size and substrate parameters determines the characteristic impedance, Z_0, and propagation constant, β, for the line.

Figure 2. Microstrip antenna fields division: Interior and exterior region

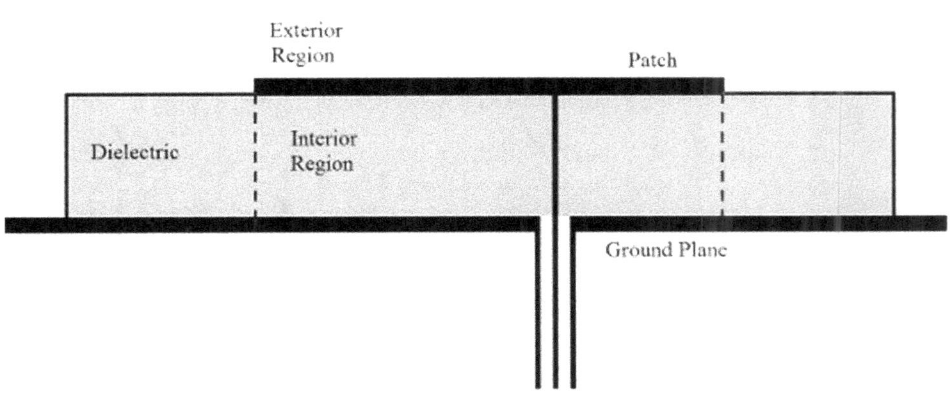

The volume enclosed by the top patch conductor, the projection of the patch periphery onto the ground plane and the portion of the ground plane underneath the patch is considered as the interior region. The fields in this region can be modeled as a transmission line section or a cavity giving rise to the designations *transmission line model* and *cavity model*. The exterior region is the remaining space comprising the rest of the ground plane, the rest of the dielectric, and air at the upperside of the patch conducting surface.

The transmission line represented by the conducting patch is open circuited at the end. Therefore the current at the end must be zero and voltage must be maximum. For a half wavelength long patch, the voltage, current and impedance along the patch length is as shown in figure 3. In this model it is assumed that the voltage, current and impedance remain constant along the transverse direction i.e. along the width. Microstrip antenna surface current distribution is depicted in figure 4.

From the voltage distribution shown in figure 3, it follows that the electric field along the length of patch (L = λg/2) is as shown in figure 5(a).

From figure 5 it is evident that the fringing electric fields from the two ends add along y direction while they cancels along z direction. This resultant y directed

Figure 3. Voltage (V), current (I) and impedance (Z) distribution along the patch length

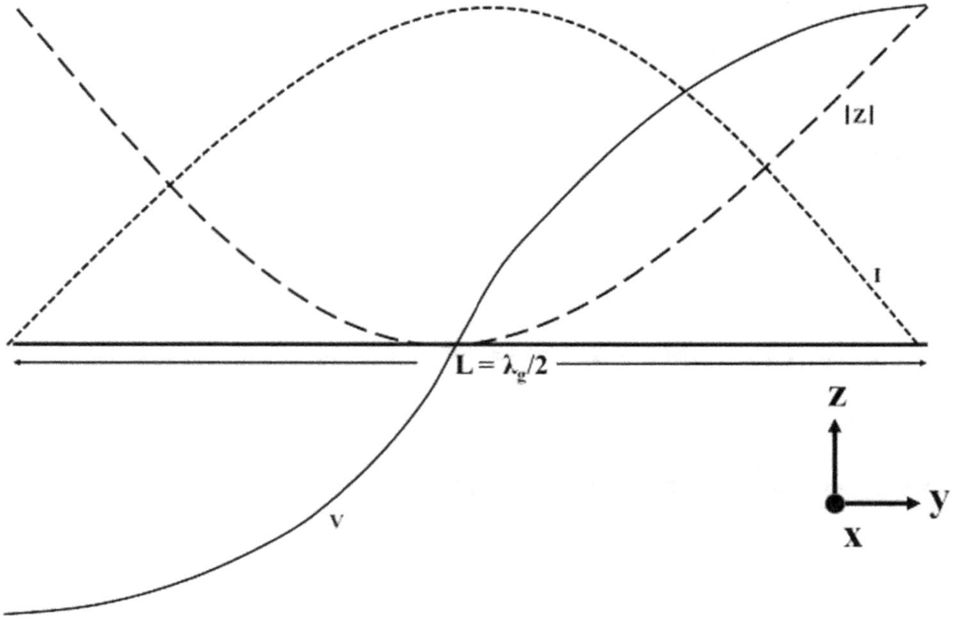

Figure 4. Current distribution on the patch's surface

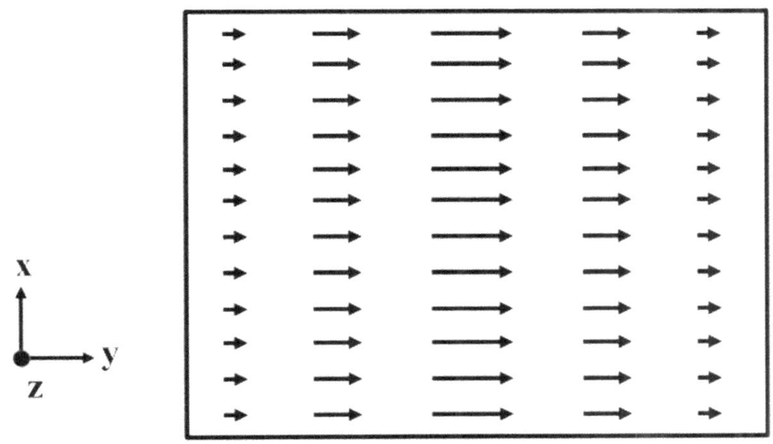

Figure 5. Electric field variation along the length of the patch

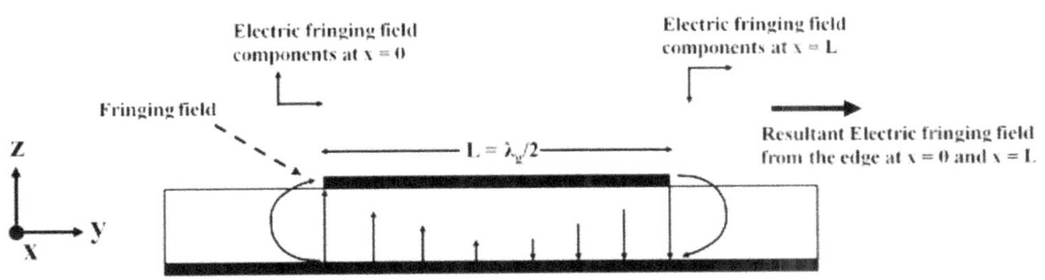

field contributes towards radiated field from the patch. Frequencies for which the patch length is not half wavelength long the resultant field along the y direction will obviously be less than the field for half wavelength long patch. That is why the length of the top metal patch L is chosen to be one half of the guided wavelength. The two edges of the patch at y =0 and L are therefore the radiating edge for the patch.

As there is no transverse voltage variation, voltage along the width (x = 0 to W) of the patch at any distance along length is constant. Therefore the fields due to the two edges (at x =0 and W) cancel each other as shown in figure 6. Therefore the patch edges at x =0 and W acts as nonradiating edges.

The source of radiation is the electric field tangential to the patch surface. The two open ends of the patch contributes towards radiation. The field from these two radiating edges can be comprehended as field from two radiating apertures. Accordingly the field due to a patch antenna is approximated by an array of two slot antennas separated by a distance L. These slots which are open ended terminations

Figure 6. Electric fringing field from the patch edges

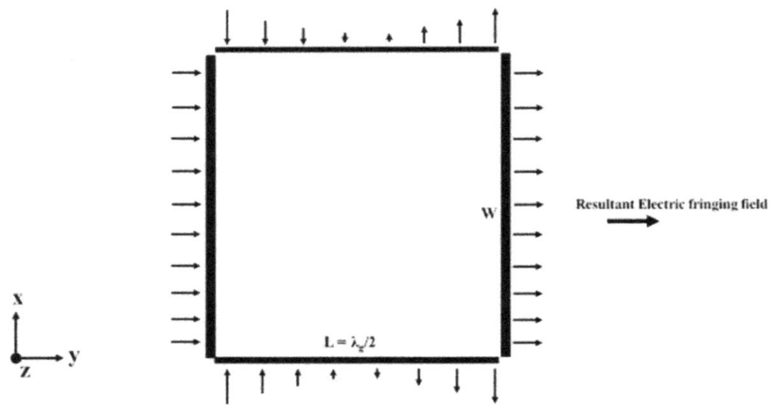

could be represented by parallel admittances $Y_s = Gs + jBs$. Here, *Gs* is the conductance related with the radiated power from the edge or wall, and *Bs* is the susceptance which accounts for the energy stored in the fringing field adjacent to the edge.

From the impedance profile over the patch surface shown in figure 3, it is apparent that it is possible to find a point inside the patch surface where the impedance matching with the feeding line is easier. In microstrip line feeding technique, input to the antenna is provided either by means of a 50 ohm microstrip transmission line

Figure 7. Microstrip patch antenna feeding technique:(a) coaxial feed, (b) microstrip feed

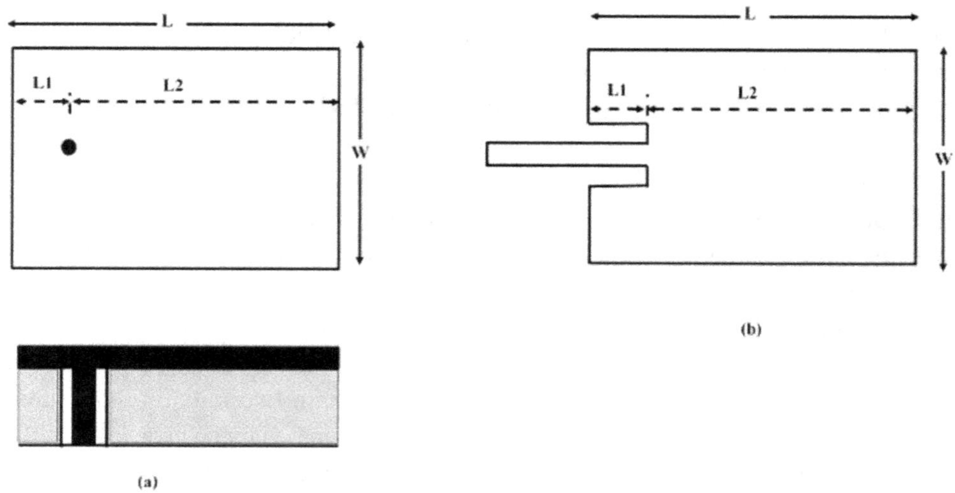

connected with the radiating edge of the antenna or tapped inside the patch surface at a point x = L1 for better impedance matching. In coaxial feeding a coaxial connection has to be made at a point x = L1 on the patch surface to achieve 50 Ohm impedance matching. The microstrip antenna feeding techniques have been shown in figure 7.

Based on these identifications, rectangular patch antenna and it's equivalent transmission line model is shown in Figure 8.

Input Resistance: The edge admittances could be transformed to attain the antenna input admittance at the feed point. Thus from the equivalent model illustrated in Figure 8, the input admittance is obtained as

Figure 8. Rectangular microstrip antenna and it's equivalent transmission line model

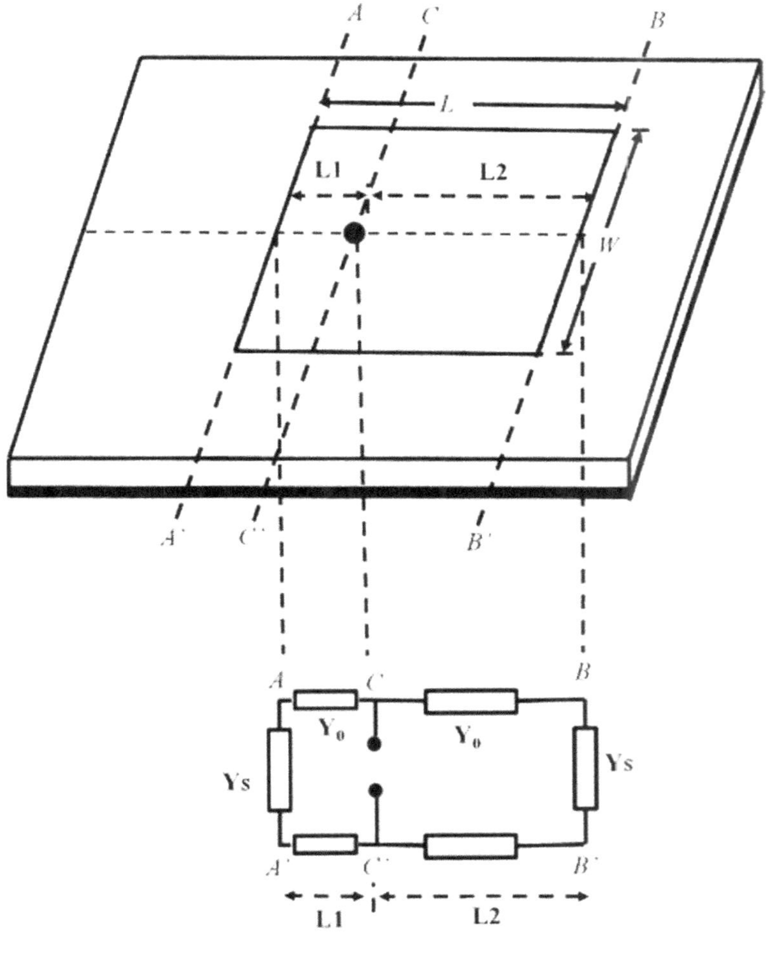

$$Y_{in} = Y_0 \left(\frac{Y_0 + jY_s \tan(\beta L_1)}{Y_s + jY_0 \tan(\beta L_1)} + \frac{Y_0 + jY_s \tan(\beta L_2)}{Y_s + jY_0 \tan(\beta L_2)} \right), \qquad L_1 + L_2 = L$$

where Y_0 is the characteristic admittance of the patch fed at $x = L1$.

Mutual conductance between the radiating edges was introduced in this model by Derneryd (1976, 1978).

$$G_m = \frac{1}{60\pi^2} \int_0^{\pi/2} Sin^2 \left(k_0 \frac{W}{2} cos\theta \right) tan^2\theta \, Sin\theta \, J_0 \left(k_0 L \, Sin\theta \right) d\theta$$

Modifying the edge admittance Ys to include Gm,

$$Y_s = G_s - G_m + jB_s$$

For resonant patch, $\beta\ (L1 + L2) \approx \pi$, the input resistance at resonance is given by Derneryd (1978).

$$R_{in} = \frac{1}{2G} \left\{ cos^2 (\beta L_1) + \frac{G^2 + B_s^2}{Y_0^2} sin^2 (\beta L_1) - \frac{B_s}{Y_0} \sin(2\beta L_1) \right\}$$

$$\approx \frac{1}{2G} cos^2 (\beta L_1) \qquad since G, B_s \ll Y_0$$

where $G = Gs - Gm$.

The fields at the patch edges undergo fringing where some of the field travels in air and some in substrate due to finite patch dimensions. Fraction of field that lies outside the substrate is called fringing field. To account for fringing field an effective relative dielectric constant has been introduced.

The fact that the field lines do not stop at the patch end and extends beyond the patch end could be modeled by an electrical lengthening of the patch by an amount of Δl.

One estimate of Δl is given by Hammersted (1975),

$$\frac{\Delta L}{h} = 0.412 \frac{\left(\varepsilon_{r_{eff}} + 0.3\right)\left(\frac{W}{h} + 0.264\right)}{\left(\varepsilon_{r_{eff}} - 0.258\right)\left(\frac{W}{h} + 0.8\right)}$$

where for W/h >1,

$$\varepsilon_{r_{eff}} = \frac{\varepsilon_r + 1}{2} + \frac{\varepsilon_r - 1}{2}\left[1 + 12\frac{h}{W}\right]^{-1/2}$$

$$L_{eff} = L + 2\Delta L$$

$$L_{eff} = \frac{\lambda_0}{2} = \frac{c}{2\sqrt{\varepsilon_{r_{eff}}} f_r}$$

$$f_r = \frac{\lambda_0}{2} = \frac{c}{2L_{eff}\sqrt{\varepsilon_{r_{eff}}}}$$

Various improvement in this transmission line model has been reported by Pues & Capelle (1984), Lier (1982).

Cavity Model

The fields within the dielectric of a microstrip antenna could be found by modeling it as a cavity (Lo, 1979), (Lo &Richards, 1981), (Richards, 1981)with electric conductors above and under it and magnetic conducting wall along the border of the patch. The cavity model allows accurate prediction of electric and magnetic field beneath the patch. But this approximate model leads to reactive input impedance.

A distribution of charge on the upper and bottom surface of the patch and on the ground plane is created due to attraction and repulsion when the patch antenna is excited. Attraction takes place between the opposite charges on the bottom surface of patch and on the surface of ground plane. Thus due to this attractive mechanism a charge concentration underneath the patch surface is created. Again due to repulsion between like charges in the bottom surface of the patch, some charges from the bottom surface has been pushed to the top surface. Movement of charges due to attraction and repulsion creates two charge densities Jt and Jb in the top and bottom layer. Distribution of charge and current density creation in a microstrip antenna is shown in figure 9.

Now if the substrate height is small compared to patch width, which is normally the case, the attractive mechanism dominates. Thus charge concentration and surface current is more at the bottom surface of the patch. In the limiting case of height to width ratio approaching zero, there would be no current on the top surface of patch. This in turn prohibits the tangential magnetic field component along the patch edge. Thus the four side walls of the cavity could be modeled as magnetic conducting surface. For practically realizable small substrate height to patch width ratio the tangential component of magnetic field at the ends would not vanish. Still they will be small enough to approximate the side wall as magnetically conducting. Figure 10 shows the microstrip antenna cavity model with electric and magnetic wall.

If the boundary of the cavity and the material inside it is modeled as lossless, the input impedance becomes purely reactive and will not incorporate the radiation phenomenon. A loss mechanism has been introduced to account for the radiation. To make the cavity structure lossy, so that it represents an antenna, an effective loss tangent δ_{eff} has been introduced. The loss tangent is chosen as the reciprocal of the antenna quality factor Q, to represent the loss from the cavity due to radiation. The field inside the cavity forms standing wave.

Figure 9. Distribution of charge and current density in a microstrip patch

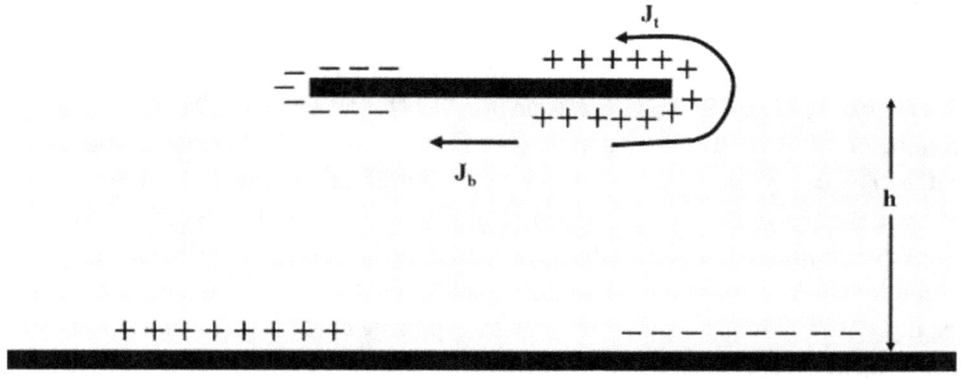

Figure 10. Cavity model with electric and magnetic wall

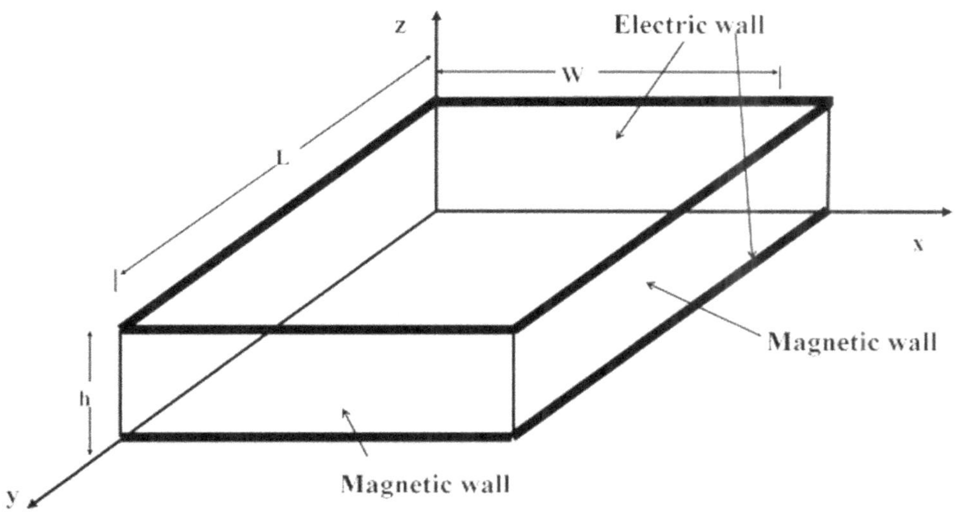

To find the field configurations field variation along the height is considered constant as height is small compared to wavelength.

The wave numbers are given by,

$$K_x = \left(\frac{m\pi}{h}\right), \ m = 0, 1, 2.....$$

$$K_y = \left(\frac{n\pi}{L}\right), \ n = 0, 1, 2.....$$

$$K_z = \left(\frac{p\pi}{W}\right), \ p = 0, 1, 2.....$$

$$m = n = p \neq 0$$

m, n, p represents number of half cycles along x, y and z directions respectively.

The cavity resonant frequencies are,

$$(f_r)_{mnp} = \frac{1}{2\pi\sqrt{\mu\varepsilon}}\sqrt{\left(\frac{m\pi}{h}\right)^2 + \left(\frac{n\pi}{L}\right)^2 + \left(\frac{p\pi}{W}\right)^2}$$

Generally for microstrip patch antennas,

substrate thickness (h)<<patch width (W) and

substrate thickness (h)<<patch length (L)

When patch length (L)>patch width (W)>substrate thickness (h), the dominant mode i.e. the lowest frequency mode is TM_{010}. The resonant frequency of dominant mode is given by,

$$(f_r)_{010} = \frac{1}{2L\sqrt{\mu\varepsilon}} = \frac{v_0}{2L\sqrt{\varepsilon_r}}$$

If patch length (L)>patch width (W)> L/2 >substrate thickness (h), the next higher order mode is TM100 and corresponding resonant frequency is

$$(f_r)_{001} = \frac{1}{2W\sqrt{\mu\varepsilon}} = \frac{v_0}{2W\sqrt{\varepsilon_r}}$$

If patch length (L)/2 >patch width (W)>substrate thickness (h), the second order mode is TM_{020} and the corresponding resonant frequency is

$$(f_r)_{020} = \frac{1}{2L\sqrt{\mu\varepsilon}} = \frac{v_0}{L\sqrt{\varepsilon_r}}$$

If patch width (W)/2 >patch length (L)>substrate thickness (h), the second order mode is TM_{200}.

Figure 11. Electric field distribution for different cavity modes

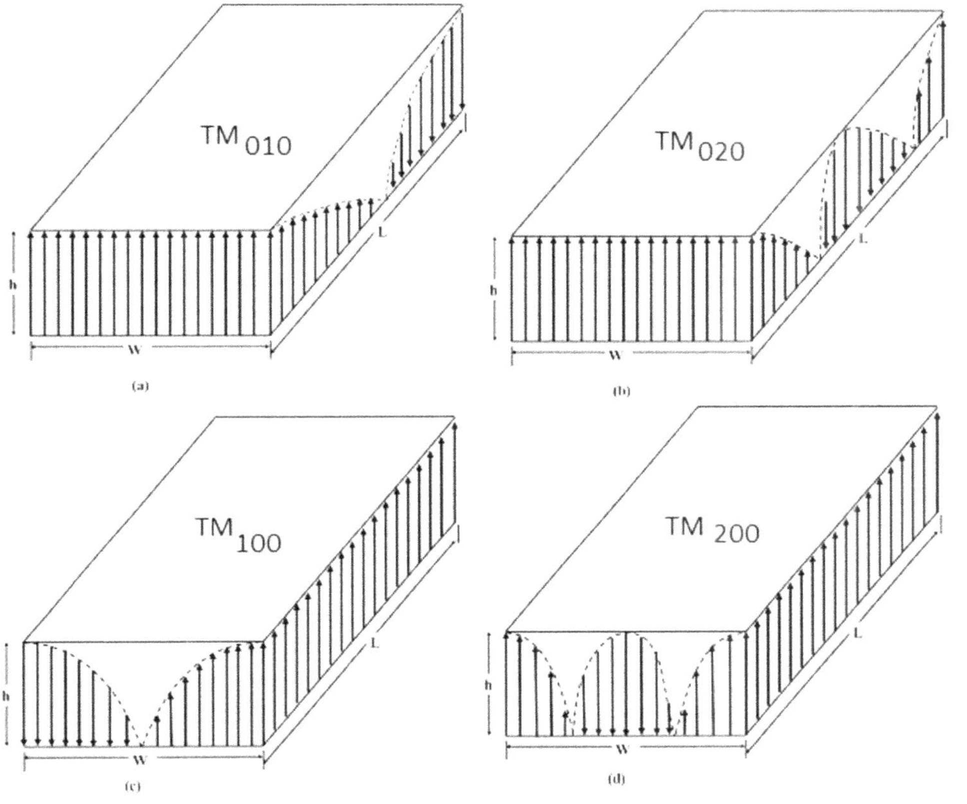

The distribution of electric field for the modes TM_{010}, TM_{020}, TM_{100}, TM_{200} is shown in figure 11.

In transmission line model transverse field variation is assumed to be zero. Energy is assumed to be propagating only in longitudinal direction. Though the dominant mode (TM_{010}) is mainly contributing, other higher modes also affect the performance.

REFERENCES

Carver, K. R., & Coffey, E. L. (1979). *Theoretical Investigation of the Microstrip Antenna*. Technical Report PT-00929, Physical Science Laboratory, New Mexico State University, Las Cruces, NM.

Derneryd, A. (1976). Linearly Polarized Microstrip Antennas. *IEEE Transactions on Antennas and Propagation*, *24*(6), 846–851. doi:10.1109/TAP.1976.1141445

Derneryd, A. (1978). A Theoretical Investigation of the Rectangular Microstrip Antenna Element. *IEEE Transactions on Antennas and Propagation*, *26*(4), 532–535. doi:10.1109/TAP.1978.1141890

Hammersted, E. O. (1975). Equations for microstrip Circuit Design. *5th European Microwave Conference*.

Lier, E. (1982). Improved Formulas for Input Impedance of Coax-Fed Microstrip Patch Antennas. *IEE Proc.*, *129*(H).

Lo, Y. T. (1979). Theory and Experiment on Microstrip Antennas. *IEEE Transactions on Antennas and Propagation*, *AP-27*.

Lo, Y. T., & Richards, W. F. (1981). Perturbation Approach to Design of Circularly Polarized Microstrip Antennas. *Electronics Letters*, *17*(11), 383. doi:10.1049/el:19810269

Munson, R. (1974). Conformal Microstrip Antennas and Microstrip Phased Arrays. *IEEE Transactions on Antennas and Propagation*, *22*(1), 74–78. doi:10.1109/TAP.1974.1140723

Pues, H., & Capelle, A. V. D. (1984). Accurate Transmission Line Model for the Rectangular MicrostripAntenna. *IEE Proc.*, *131*(H).

Richards, W. F. (1981). An Improved Theory for Microstrip Antennas and Applications. *IEEE Transactions on Antennas and Propagation*, *AP-29*.

Chapter 3
Transmission Line and Its Implementation

Pampa Debnath
RCC Institute of Information Technology, India

Arpan Deyasi
RCC Institute of Information Technology, India

ABSTRACT

In unbounded media, wave propagation is supposed to be unguided. The existence of uniform plane wave is considered to be all through the space. Electromagnetic energy related with the wave stretched over a broad area. In TV and radio broadcasting, unbounded medium propagation of the wave is required. Here transmission of information is destined for one and all who may be interested. Another way of transmitting information is by guided media. Guided media acts to direct the transmission of energy from transmitter to receiver. Transmission lines are usually used in low frequency power distribution and in high frequency communications as well as in the ethernet and internet in computer networks. Two or more parallel conductors may be used to construct a transmission line, which connects source to a load. Typical transmission lines consist of coaxial line, waveguide, microstrip line, coplanar waveguide, etc. In this chapter, problems related with transmission lines are solved with the help of EM field theory and electric circuit theory.

DOI: 10.4018/978-1-5225-8531-2.ch003

Copyright © 2019, IGI Global. Copying or distributing in print or electronic forms without written permission of IGI Global is prohibited.

INTRODUCTION

From last few decades high frequency communication technology has direct impact in our day to day life. More lithe and liberated organizations have been generated by Globalization where communication technology, the most important tool has been realized to make them different. Nowadays Swift development of wireless communications compared to the past eased with knowledge and coordination amongst people. People are now enjoying Wi-Fi network almost everywhere letting users to make connection on their smart phones, tablets and swapping data to a significant rate. Most of the wireless communication operates within the frequency range of 800 MHz to 3GHz. Good propagation characteristics; low attenuation and large wavelength have been observed for such frequency range. However continuous rise of the number of users per coverage area, shifted the main attention of the operators on capacity rather than area of coverage. Therefore the key objective of wireless communication is to achieve high capacity. Lack of accessible spectrum is additional problem faced by operators and the enlargement of 4G technologies depend on the accessibility of spectrum. For these purpose, the interest of using higher frequencies particularly in millimeter wave band has been grown. Due to atmospheric absorption the propagation has been attenuated at millimeter wave frequencies. This attenuation is not desirable while looking for an elongated coverage range and is fruitful when function for short distance to comprehend interference between several applications with unlicensed spectrum.

The expansion of planar transmission lines help to develop miniaturized of microwave circuits. These transmission lines are light weight with low profile characteristics [Collin,R.E 1992, Collin,R.E 1960, Liboff, R,L, and Dalmon G,C 1985, Marcuvitz, N 1986]. The characteristic of line impedance has been controlled by line geometry. Thin film technology has been used to fabricate the transmission line in one step. Several transmission lines exist that are used for different Microwave frequency range and for different applications.

COAXIAL LINE

The most extensively used TEM transmission line is the coaxial cable [Collin,R.E 1960, Liboff, R.L, and Dalmon G,C 1985] represented in Figure 1. It comprises of two conductors with inner conductor of radius a and outer conductor of radius b. The gap between two conductors is filled with dielectric material such as Teflon.

Figure 1. Configuration of coaxial line

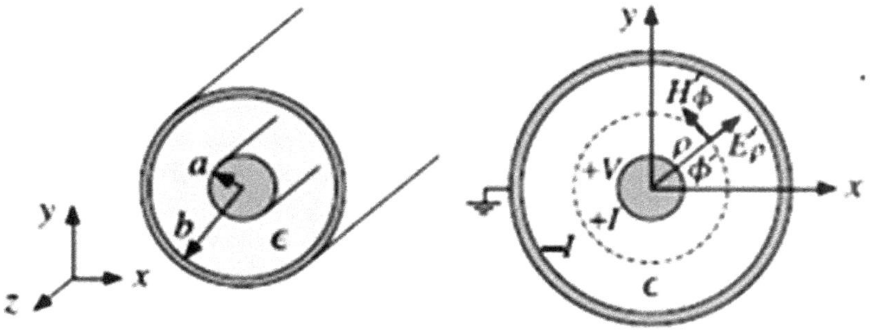

The problems related to electrostatic can be explained expediently in cylindrical coordinate ρ, ϕ′. The Laplace equation has been satisfied by potential φ′(ρ, ϕ).

The corresponding problems related to electrostatic can be explained expediently in cylindrical coordinates ρ, φ′. The Laplace's equation can be satisfied by potential ϕ′(ρ, φ):

$$\nabla_t^2 \varphi' = \frac{1}{\rho}\frac{\partial}{\partial \rho}\left(\rho \frac{\partial \varphi'}{\partial \rho}\right) + \frac{1}{\rho^2}\frac{\partial^2 \varphi'}{\partial^2 \varnothing} = 0$$

The potential is independent on the azimuthal angle φ′ due to cylindrical symmetry. Therefore

$$\frac{1}{\rho}\frac{\partial}{\partial \rho}\left(\rho \frac{\partial \varphi'}{\partial \rho}\right) = 0 => \rho \frac{\partial \varphi'}{\partial \rho} = B => \varphi'\left(\rho\right) = A + Bln\rho$$

where two constants of integration are A and B. Considering the condition that the outer conductor is grounded and the voltage of the inner conductor is held to be V′. The constants A=-Bln b and B=-V′ln(b/a), resulting the potential:

$$\varphi'\left(\rho\right) = \frac{V'}{\ln\left(b / a\right)} \ln\left(b / a\right)$$

It seems that radial component is associated with the electric field Eρ′ and azimuthal component is related with magnetic field Hφ′.

$$E_{\rho}' = \frac{V'}{\ln\left(\frac{b}{a}\right)}, \qquad H_{\varphi}' = \frac{V}{n\ln\left(b/a\right)}\frac{1}{\rho}$$

To achieve the current integrate H_{ϕ}' around the inner conductor:

$$I = \int_{0}^{2\pi} H'_{\varphi}\,\rho d\varphi' = \int_{0}^{2\pi} \frac{V'}{n\ln\left(\frac{b}{a}\right)}\frac{1}{\rho}\rho d\varphi' = \frac{2\pi V'}{n\ln\left(\frac{b}{a}\right)}$$

Therefore the inductance and capacitance per unit length can be written as:

$$Z = \frac{n}{2\pi}\ln\left(\frac{b}{a}\right), L_{1} = \frac{\mu}{2\pi}\ln\left(\frac{b}{a}\right), \; C_{1} = \frac{2\pi\epsilon}{\ln\left(b/a\right)}$$

By solving above equations one can state the magnetic field in the form of:

$$H'_{\varnothing} = \frac{1}{2\pi\rho}$$

Using Ampere's law around a closed circular wire of radius 'ρ' surrounded by the conductor kept inside, the same expression can be obtained. The transmitted power P_t can be stated either in terms of voltage V′ or in terms of the highest value of electric field which take place at ρ=a,

$$P_{t} = \frac{1}{2Z}\left|V'\right|^{2} = \frac{\pi\left|V'\right|^{2}}{n\ln\left(\frac{b}{a}\right)} = \frac{1}{n}\left|E_{a}\right|^{2}\left(\pi a^{2}\right)\ln\left(b/a\right)$$

WAVEGUIDE

Introduction

Field distribution of transmission line structures for theoretical investigation has been carried out by a number of investigators in long decade ago at several frequency ranges appropriate to the particular applications [Marcuvitz, N 1986, Jiao, C. Q 2011]. With the advancement of communication skill specifically at microwave and millimeter wave frequencies for accessibility of higher bandwidth, study of transmission lines are carried out towards higher frequency range [Simon, W, Werthen, M. and Wolff, I 1998, Zhao, D, Ding, Y. G, Wang, Y. and Ruan, C. J 2010].

Thus precise study of waveguide configurations is incredibly significant, which depends on its structural form, as well as on the characteristics of the medium. Derivations of theoretical equations for resistive profiles and for propagation characteristics have been done earlier but one basic area of research which has to be considered is the deviation of the simulated contour from the analytical studies for higher order modes [Zhao D et.al. 2010]. Electric and magnetic fields distributions are also having key significance concerning the deliberation of waveguides as basic integrated circuits components.

For calculating the electromagnetic performance of a configuration, HFSS is an interactive software package. Using HFSS, one can calculate fundamental characteristic of electromagnetic field, feature of port impedances and propagation constants, the resonances or eigen modes, S-parameters in generalized form and renormalized S-parameters to exact port impedances of a configuration. HFSS is an extraordinary form of tool exclusively modeled for extracting modal factors by device simulation including active and passive as well as antennas with specific structures, material characteristics at required frequency range using finite element method (FEM) [Lee, C. S., et.al. 1985]. It is necessary for implementing high speed and high frequency components used in contemporary electronic devices.

Thus precise outcomes acquired from HFSS software may be employed for modeling of electromagnetic configurations, and outcomes of simulation are reasonably supportive before fabricating of practical components. For designing trans-receiver in communication systems, the awareness of field variations and modal distribution [HFSS 2013, Southworth, G C 1980], corrugated rectangular waveguide configuration can be analyzed using appropriate material [Marcueitz, N 1951].

The characteristics of rectangular waveguide such as characteristic impedance, guided wavelength and propagation constant have been simulated using HFSS software for primary four modes and simulated data as well as results obtained from mathematical relationship are compared. Theoretical analysis has been carried out from C to Ku band. To study the basic characteristics of rectangular waveguide as well as to approximate the accuracy of HFSS based structure in association with rectangular electromagnetic configuration at various modes, this analysis has become very helpful (Das. A.et.al. 2012).

Mathematical Analysis

Consider a hollow waveguide rectangular in shape filled by air with conducting walls as shown in Figure 2. Material with various permittivity and permeability may be considered as dielectric instead of air. The propagation constant for TE_{mn} of a rectangular waveguide has been given by:

$$\beta = \omega\sqrt{\mu\varepsilon}\sqrt{1-(\omega_c / \omega)^2}$$

where ω is the frequency of propagation, ω_c is the cutoff frequency, μ, ε is the permeability and the permittivity of the medium respectively. The propagation wave length which is known as guided wavelength is given by:

$$\lambda_g = \frac{\lambda}{\sqrt{1-(\lambda / \lambda_c)^2}}$$

where λ_c is the cutoff wavelength and λ is the unbounded dielectric wavelength.

Characteristic impedance is specified as

$$Z = \frac{\eta}{\sqrt{1-(\omega_c / \omega)^2}}$$

where η is the intrinsic impedance of the unbounded medium.

Figure 2. Schematic diagram of rectangular waveguide

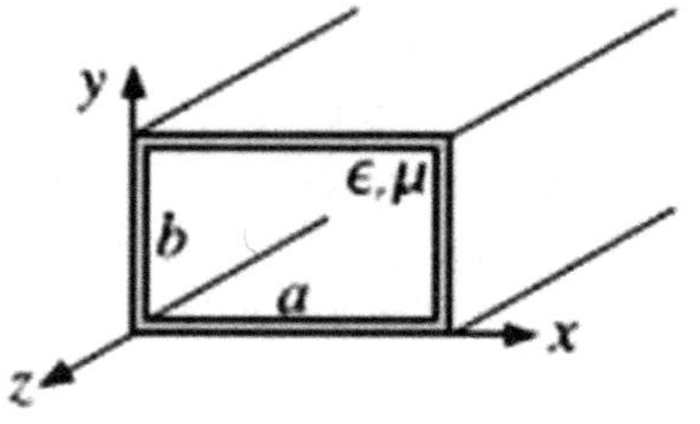

Profile Analysis

In this section comparative studies of characteristic profile for basic first modes of rectangular waveguide have been carried out. It has been noted from the characteristics profile of propagation constant that simulated result has been departed from computed theoretical result for higher order modes (TE_{11}/TM_{11}). It can be found out for each mode that incremental rate of propagation constant reduces with increasing frequency. It has been shown in Figure 3.

The profile for guided wavelength has been shown in Figure 4 where with increase in frequency, propagation constant reduces for a particular mode. For very high frequency profile is almost constant. In higher order modes divergence has also been observed. It may be stated that the contour of TE_{11} and TM_{11} has been same.

The impedance reduces with increasing frequency for a particular mode as shown in Figure 5. For a specific TE mode, a considerable amount of variation has been noted between these contours which are very significant for modal analysis in different applications of electromagnetic field.

HFSS software has been used for the analysis of field for different modes. Variations of E & H field inside the waveguide for first four TE modes have been shown in Figure 6(a) to Figure 6(d). It may be found out that for TE_{10} and TE_{20} mode, electric fields are vertically oriented whether horizontally oriented electric fields have been

Figure 3. Comparative study of Propagation constant contour of rectangular waveguide at C, X and Ku bands (Das. A.et.al. 2012)

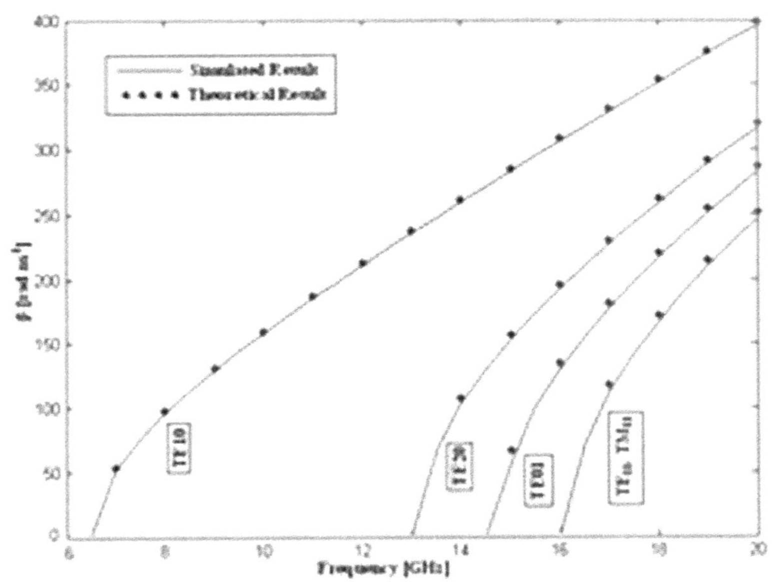

Figure 4. Variation of Guided wavelength contour of rectangular waveguide at C to Ku bands (Das. A.et.al. 2012)

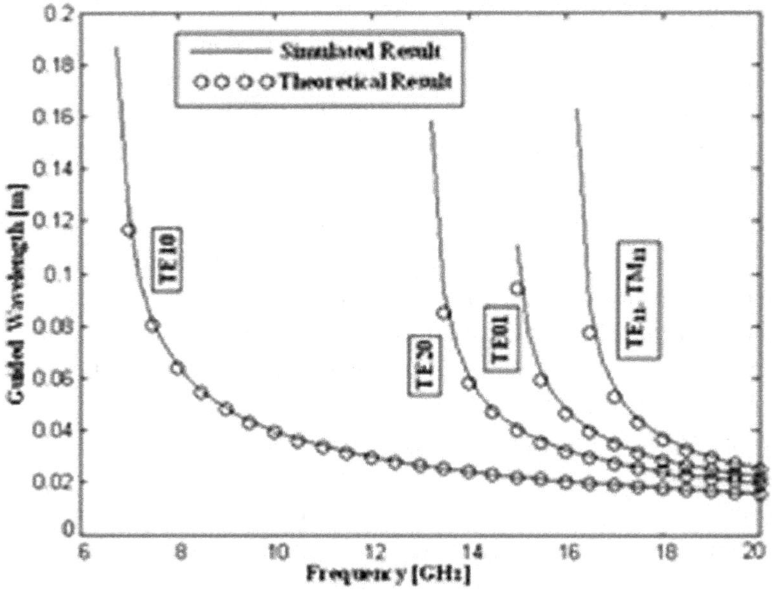

Figure 5. Variation of Impedance contour of rectangular waveguide at C to Ku bands (Das. A.et.al. 2012)

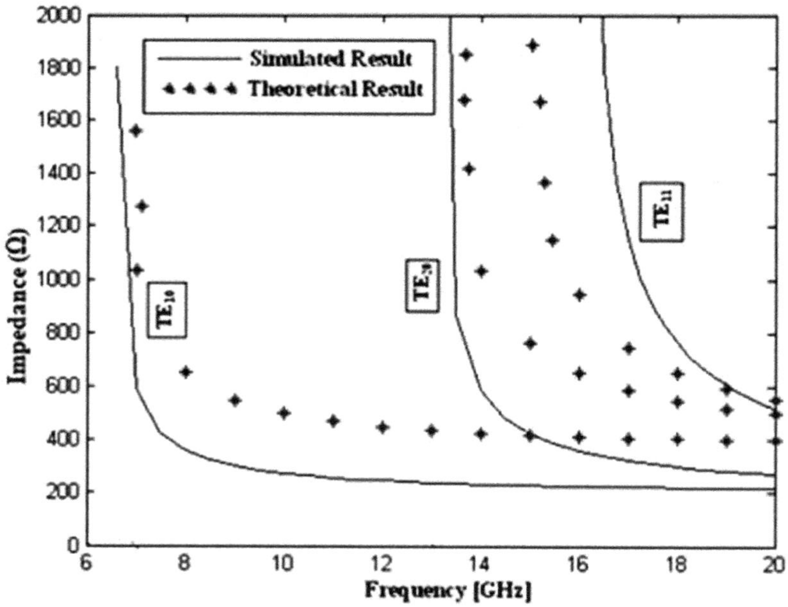

Figure 6a. Field variation for TE_{10} mode (Das. A.et.al. 2012)

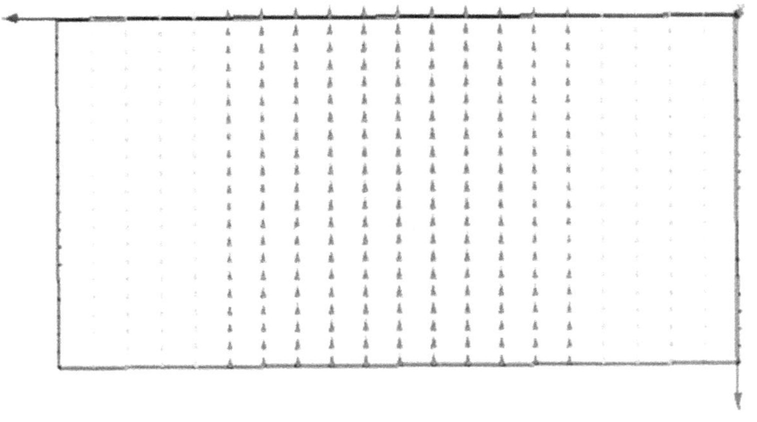

Figure 6b. Field variation for TE_{20} mode (Das. A.et.al. 2012)

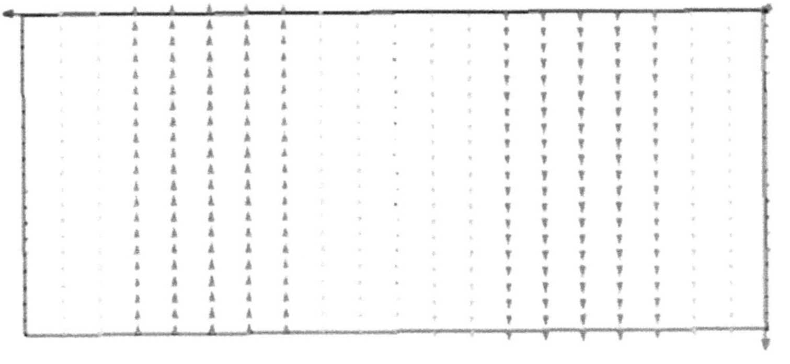

Figure 6c. Field variation for TE_{01} mode (Das. A.et.al. 2012)

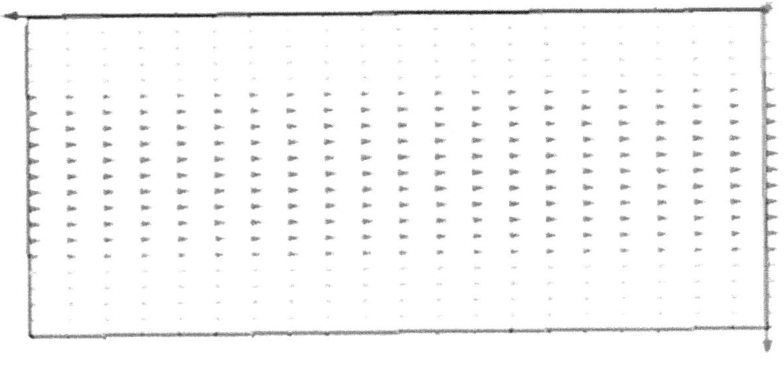

Figure 6d. Field variation for TE_{11} mode (Das. A.et.al. 2012)

observed for TE_{01} mode. Both horizontal and vertical distribution can be observed for TE_{11} mode at various spatial part of the waveguide. Published simulated profiles available in different literatures have displayed similar characteristics [Marcueitz, N 1951, Eshrah, I. A., et. al. 2005, Pozar, D. M 1998] as shown in Figure 6.

Electric and magnetic field variations as shown in Figure 6a, Figure 6b, Figure 6c, and Figure 6d inside the structure under consideration (waveguide) is one of the critical factors when various field applications of electromagnetic radiation is concerned. It can be stated that maximum intensity has been observed at the mid region but variation in non uniform nature has been observed towards the ends and has been presented in Figure 7 and Figure 8. Also it has been found out from 3D

Figure 7. Electric field variation inside rectangular waveguide (Das. A.et.al. 2012)

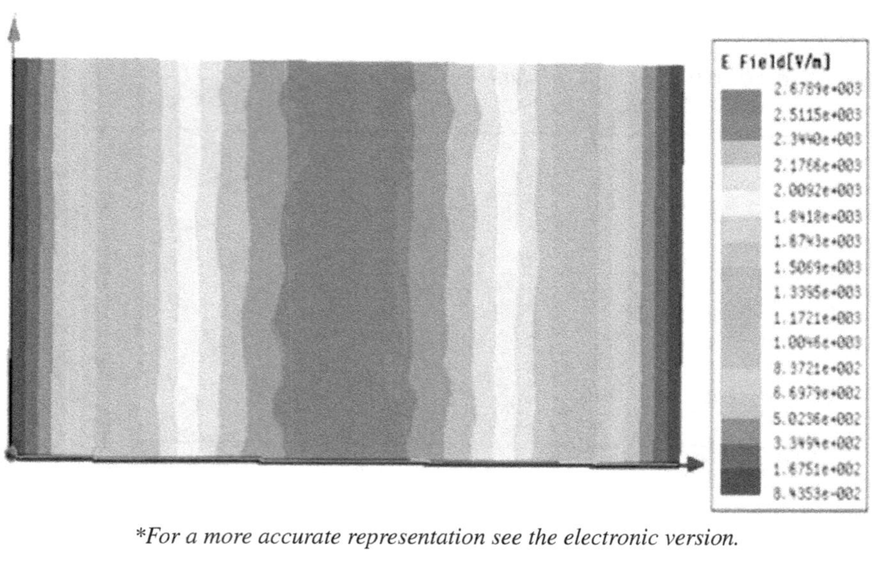

**For a more accurate representation see the electronic version.*

Figure 8. Magnetic field variation inside rectangular waveguide (Das. A.et.al. 2012)

**For a more accurate representation see the electronic version.*

graph that the variation of intensity is not the step function rather diminishes in exponential manner [Das, A. and Das, S. K 2009].

Numerical simulation using Finite element method has been done to study the features of rectangular waveguide. Simulations have been carried out at C, X and Ku band i.e. lower microwave frequencies, because it is significant for laboratory purpose and to perform simple experiments using microwave test bench. First four modes have been considered for analytical purpose to approximate the little divergence from the theoretical outcomes. Distribution of fields along with intensity profile has been simulated to recognize the properties. The above analytical approach is necessary to develop a new area as of Corrugated RW as left-handed metamaterial transmission lines.

MICROSTRIP LINE

Microstrip line comprises of a thin conductor on a low loss dielectric substrate above the ground plane. As the dimension of microwave devices is very tiny, the process of signal input to these devices and pull out power from them employs microstrip lines.

The configuration given in Figure 10 shows a typical cross section of a micro strip line. Owing to absence of ground plane and dielectric substrate above the strip, the electric field lines partially follow the lower dielectric substrate and stays partially in air. Therefore the mode of propagation is not pure TEM but quasi TEM

Figure 9. Microstrip line

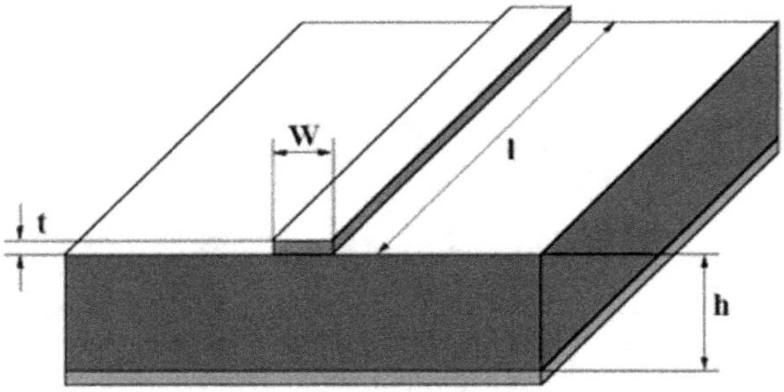

Figure 10. Field configuration

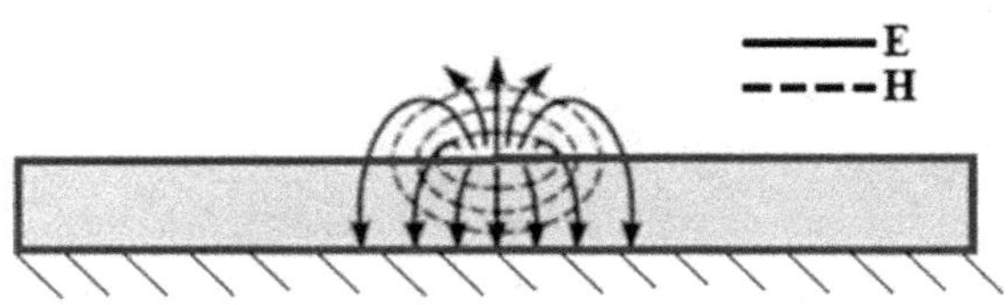

[Edwards, T 1992, Gupta, K, C., *et. al.*1979]. In microstrip line the radiation loss is proportional to the square of frequency. The field can be mostly restricted inside the dielectric by using high dielectric material which reduces the radiation loss of the open configuration.

$$\varepsilon_{effe} = \frac{\varepsilon_{re}+1}{2} + \frac{\varepsilon_{re}-1}{2}\left[\left(1+\frac{12h}{w}\right)^{-\frac{1}{2}} + 0.04\left(1-\frac{w}{h}\right)^{2}\right]; \qquad w/h \le 1$$

$$\varepsilon_{effe} = \frac{\varepsilon_{re}+1}{2} + \frac{\varepsilon_{re}-1}{2}\left[\left(1+\frac{12h}{w}\right)^{-\frac{1}{2}}\right]; \qquad w/h \gg 1$$

where ε_{re} is the relative dielectric constant of the substrate.

The characteristics impedance of microstrip lines can be written as

$$Z_o' = \frac{60}{\sqrt{\varepsilon_{effe}}}\ln\left[\frac{8h}{w}+\frac{w}{4h}\right] ohm, \qquad w/h \le 1$$

$$Z'_o = \frac{376.7}{\sqrt{\varepsilon_{effe}}[\frac{w}{h} + 2.4 + 0.667\ln\left(\frac{w}{h} + 1.444\right)} ohm; for\, w / h > 1$$

$$Z'_o = \frac{376.7}{\sqrt{\varepsilon_{effe}}} \frac{h}{w} ohm; \frac{w}{h} \gg 1$$

The guided wavelength for the quasi-TEM mode can be given as

$$\lambda'_g = \lambda_o \sqrt{\varepsilon_{effe}}$$

Two types of losses can be realized for the substrate material with non-magnetic property:

1. Dielectric loss which occurs in the substrate
2. Ohmic loss due to finite conductivity of strip conductor and ground plane.

If α_{ct} and α_{dt} are the ohmic attenuation constant and dielectric attenuation constant, total attenuation constant can be stated as $\alpha_t = \alpha_{ct} + \alpha_{dt}$ where

$$\alpha_{dt} = \frac{\sigma_d}{2} \sqrt{\frac{\mu}{\varepsilon}} == 27.3 \frac{\varepsilon_{effe} - 1}{\varepsilon_{effe} - 1} \frac{\varepsilon_r}{\varepsilon_{effe}} \frac{tan\delta}{\lambda_g'} dB / m$$

where dielectric substrate conductivity $= \sigma_d$ and dielectric loss tangent $= \tan\delta = \sigma_{dt}/\omega\varepsilon$.

At microwave frequencies the key attenuation factor happens owing to the finite conductivity of strip for low loss dielectric substrate. Due to the current on the strip conductor, it gives ohmic losses.

Considering uniform distribution of current in the area of $-w/2<x<w/2$, ohmic attenuation constant can be given as

$$\alpha_{ct} \approx \frac{8.686}{Z_o' w} \sqrt{\frac{\pi f \mu}{\sigma}} \quad dB / m$$

The most commonly used substrate material for microstrip lines are PTFE/Teflon, RT Duroid 5880, sapphire and alumina etc. Most widely used material is alumina for frequencies up to 20GHz.

COPLANAR WAVEGUIDE

The coplanar waveguide was first introduced in 1969 by C.P.Wen for Gyromagnetic microwave device application with nonreciprocal characteristic. It has a capability to produce magnetic field with elliptical polarization in the slots.

Figure 11 shows the structural view of Coplanar waveguide. It consists of a Centre conductor and two Ground plane placed on a same surface of a dielectric slab. Two

Table 1.

Characteristics features	Coaxial Line	Waveguide	Microstrip Line
Preferred Mode	TEM	TE_{10}	Quasi-TEM
Other Mode	TM, TE	TM, TE	TM, TE
Power Capacity	Medium	High	Low
Dispersion	None	Medium	Low
Physical Size	Large	Real Large	Small
Bandwidth	High	Low	High
Loss	Medium	Low	High
Component Integration	Hard	Hard	Easy
Fabrication Ease	Medium	Medium	Real Easy

Figure 11. The structural view of coplanar waveguide

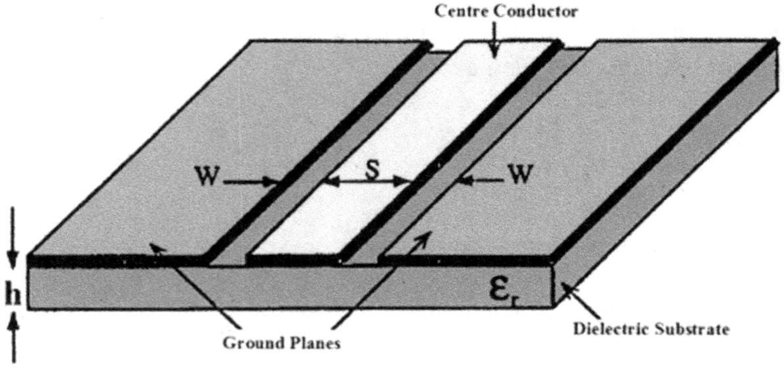

Dominant modes of propagation have been considered for coplanar line [Bellatoni J., Compton R and Levy H 1989, Iskander M and Lind T Dec. 1989]. One is the quasi-TEM mode and other is non TEM mode. Quasi-TEM mode frequently called the odd mode where the fields are 180 degree out of phase in two slots. The non TEM mode is often known as even mode where the fields in the two slots are in phase. Odd mode is most required in microwave circuits because of its low radiation properties. The unwanted even mode can be cancelled out either by preserving the symmetry of the structure or using bond wires to join the ground plane of CPW. Coaxial line, microstrip line or slot line can be used to excite Coplanar waveguide.

One of the main benefits of CPW is that all conductors are situated on the same plane and therefore via holes have not been required which make it easy to connect active/passive elements. Line impedance and phase velocity are less dependent on the height of the substrate than on the slot width to center conductor width (aspect ratio).

The basic merits of planar transmission lines are

1. Miniaturized size and light in weight
2. Improved reliability
3. Mounted components can be accessed easily
4. Characteristics impedance can be optimized by controlling the dimension in a single plane
5. Passive circuits are possible to design easily in one plane
6. Cost effective

There are several demerits exist in these transmission lines. The capability or handling low power due to small in size is one of the disadvantages of this line. Radiation losses are also another problem from open structures. By using high dielectric constant material, field lines can be confined more in dielectric substrate, therefore radiation losses can be minimized. The design of few active devices is also restricted due to low quality factor obtained from simple microstrip conformations.

REFERENCES

Bellatoni, J., Compton, R., & Levy, H. (1989). A New W-Band Coplanar Waveguide Test Fixture. *IEEE MTT-S International Microwave Symposium Digest*, 1203-1204. 10.1109/MWSYM.1989.38940

Collin, R. E. (1960). *Field Theory of Guided waves*. McGraw-Hill Book Co.

Collin, R. E. (1966). Foundation of Microwave Engineering (2nd ed.). McGraw-Hill Book Co.

Collin, R.E. (1992). *Foundation of Microwave Engg*. McGraw-Hill Inc. International.

Das, A., & Das, S. K. (2009). *Microwave Engineering*. McGraw Hill.

Das, A., Sarkar, M., Das, R., Das, D., & Debnath, P. (2012, March). Numerical Characterization of Rectangular Wave Guide in Free space using HFSS Software at lower microwave Frequencies. *IJETAE*, *2*(3), 60–64.

Edwards, T. (1992). *Foundation for Microstrip circuit Design* (2nd ed.). John Wiley and Sons.

Eshrah, I. A., Kishk, A. A., Yakovlev, A. B., & Glisson, A. W. (2005). Modal Analysis of Corrugated Rectangular Waveguides Supporting Left-Hand Propagation. *IEEE AP-Society International Symposium Digest*. 10.1109/APS.2005.1551408

Gupta, K. C., Ramesh, G., & Bahl, I. J. (1979). Microstrip Lines and Slot lines. Artech House, Inc.

High Frequency Structure Simulator based on the Finite Element Method. (2013). Ansoft Corporation.

Iskander, M., & Lind, T. (1989). Electromagnetic Coupling of Coplanar Waveguides and Microstrip Lines to Highly Lossy Dielectric Media. *IEEE MTT*, 1910-1917.

Jiao, C. Q. (2011). Selective Suppression of Electromagnetic Modes in a Rectangular Waveguide by using Distributed Wall Losses. *Progress in Electromagnetics Research Letters*, *22*, 119–128. doi:10.2528/PIERL11013102

Lee, C. S., Lee, S. W., & Chuang, S. L. (1985). Plot of Modal Field Distribution in Rectangular and Circular Waveguides. *IEEE Transactions on Microwave Theory and Techniques*, *33*(3), 271–274. doi:10.1109/TMTT.1985.1132998

Liboff, R. L., & Dalmon, G. C. (1985). *Transmission Lines, Waveguides and Smith Chart*. New York: McMillan Publishing Co.

Marcueitz, N. (1951). *Waveguide Handbook, Rad. Lab. Series* (Vol. 10). New York: McGraw Hill Book Co.

Marcuvitz, N. (1986). *Waveguide Hand Book. Peter Peregrinus Ltd.*

Pozar, D. M. (1998). *Microwave Engineering* (2nd ed.). New York: John Wiley &sons Inc.

Simon, W., Werthen, M., & Wolff, I. (1998). A Novel Coplanar Transmission Line to Rectangular Waveguide Transition. *IEEE MTT-Society International Symposium Digest*, 257-260. 10.1109/MWSYM.1998.689369

Southworth, G. C. (1980). *Principles and Application of Waveguide Transmission*. D Van Nostrand Co.

Zhao, D., Ding, Y. G., Wang, Y., & Ruan, C. J. (2010). Linear Analysis of a Rectangular Waveguide Cyclotron Maser with a Sheet Electron Beam. *Physics of Plasmas*, *17*(11), 113110. doi:10.1063/1.3514596

Chapter 4
Investigations on the Microstripline-Fed Wide-Slot Antennas for Wideband Applications

Krishnendu Chattopadhyay
MCKV Institute of Engineering, India

Sekhar Ranjan Bhadra Chaudhuri
Indian Institute of Engineering Science and Technology Shibpur, India

ABSTRACT

Microstrip patch antennas are printed antennas that find suitability because they are lightweight, low volume, thin in profile, dual frequency, and dual polarization operation, and compatible with MMIC. The objective of chapter is to exhibit the investigations on the bandwidth enhancement of microstrip antennas with special reference to microstrip-line-fed wide-slot antennas. Performances are realized and validated through experimental studies on the impedance properties by VNA and radiation properties by pattern measurement setup. An innovative method for the design of hexagonal wide-slot antenna has been proposed considering it as an equivalent magnetic surface of monopole antenna. Impedance bandwidth of the above slot antenna is enhanced through various tuning stubs. In case of forklike tuning stub, the obtained bandwidth is about 900MHz, for hexagonal stub the available bandwidth is 1751MHz. Further improvement in bandwidth is proposed through rotation of hexagonal wide slot, results in wide bandwidth of 5165 MHz covering all the WLAN and WiMAX applications.

DOI: 10.4018/978-1-5225-8531-2.ch004

Copyright © 2019, IGI Global. Copying or distributing in print or electronic forms without written permission of IGI Global is prohibited.

INTRODUCTION

Slot antennas are most common at microwave frequencies. Any slot has its complimentary form in wires or strips so that the pattern and impedance data of these forms can be used to predict the patterns and impedances of the corresponding slots. The theory is based largely on the generalization and extension of Babinet's principle. In its basic form a λ/2 slot is cut on a flat metal sheet. The width of the slot is much smaller than one wave length. The currents are not confined to the edges of the slot but spread over the sheet. The slot antenna is a very efficient radiator and radiation occurs equally from both sides of the sheet. Slot antennas are very practical in space applications because it can be flush- mounted on the surface of the aircraft or spacecraft. The main advantages of the radiating slots are of wider bandwidth, less interaction via surface waves, better isolation and negligible radiation from feed network. The slot antennas may be square, rectangle, circular, elliptical or any other configurations. The concept of microstrip slot antenna has evolved from slot antenna excited by a strip line. The strip and slot combinations offer an additional degree of freedom in the design of microstrip antennas (Yoshimura, 1972).

Printed wide slot antennas are acceptable now a days because of their wide impedance bandwidths. In addition to that, the printed planar antennas are compatible with active devices or microwave monolithic integrated circuits. In recent years, growing research activities on microstrip line fed printed wide slot antennas has been noticed (Dastranj *et al*,2010; Sung, 2011; Sung 2012; Chattopadhyay *et al*, 2013). The fundamental research on microstrip line fed slot antenna was first proposed by Y. Yoshimura in 1972. The technology was further extended to wide radiating slot by T.K. Sarkar *et al* in the year 1993, involved exhaustive analysis of electric field and input impedance (Sarkar *et al*, 1993). The wide bandwidth feature of slot antennas with stub of appropriate shape had been reported by many researchers. A square slot fed by a 50Ω microstrip line with forklike tuning stub at an operating frequency around 2GHz was found to exhibit a 1.5:1 VSWR bandwidth of 1GHz (Sze *et al*, 2001). Wideband Impedance matching over frequency range of 4255MHz was achieved by incorporating suitable tuning of three resonance frequencies in a rectangular slot antenna (Jang, 2001). Wide impedance bandwidths of 120%(1.82GHz-7.23GHz) and 110%(2.42GHz-8.48GHz) were obtained by Y.F. Liu *et al* in the experimental study with an arc shape slot and triangular slot respectively(Liu *et al*, 2004). Another excellent design of microstrip line fed printed wide slot antenna with a rotated square slot has been proposed where the impedance bandwidth of 2257MHz has been reported(Jan *et al*, 2005).

In this chapter a detailed investigation has been carried out on bandwidth enhancement of hexagonal wide slot antenna fed by microstrip line loaded with various geometries of tuning stub. Two such tuning stub geometries are considered for the same. One is forklike shape and the other is hexagonal shape with a straight section. In another study, rotation of hexagonal wide slot has been carried out towards improvement of bandwidth. Impedance and radiation characteristics of all the configurations stated above have also been demonstrated in the foregoing sections.

Physical Concept of Bandwidth Enhancement in Wide-Slot Antennas

A wide slot is a slot with an aspect ratio(width to height ratio) significantly smaller than that of the usual narrow slots, some times quite close to unity. For this wide slot configuration, two orthogonal resonant modes with similar slot radiation patterns and having same polarization planes are merged to get significant bandwidth enhancement (Kahrizi, Sarkar, & Maricevic, 1993). The optimization of coupling with suitable feeding structure is used to excite the dominant mode and different higher order modes which are responsible for large impedance bandwidth. To excite different higher order modes and maintain suitable coupling between feed structure and the radiating wide slot, the feed line is loaded by different shapes of tuning stub. The adjustment of different dimensions of tuning stub also provides suitable reactance matching to get wider impedance bandwidth. The coupling level is adjusted by observing the coupling loop of input impedance plot on smith chart for a particular antenna structure.

Design Aspect of Hexagonal Slot Antenna

Dimension of the Hexagonal Slot

It is noticed that the dimension of the slot antenna is directly related to the lower edge frequency of impedance bandwidth. The design concept has been developed from the theory of planar monopole antennas and applied to slot antenna design as complementary structure of monopole. For elliptical disc monopole, an empirical formula for estimating the lower edge frequency (f_L) of VSWR=2 bandwidth is derived based on the equivalence of planar configuration area to that of cylindrical wire of height L (which is same as the planar disc height), with equivalent radius r as shown in (Agrawall, Kumar, & Ray, 1998)

$$f_L = \frac{30 \times 0.24}{L + r} \tag{1}$$

Figure 1. Configuration of Microstrip Line fed Hexagonal slot with suitable dimensions

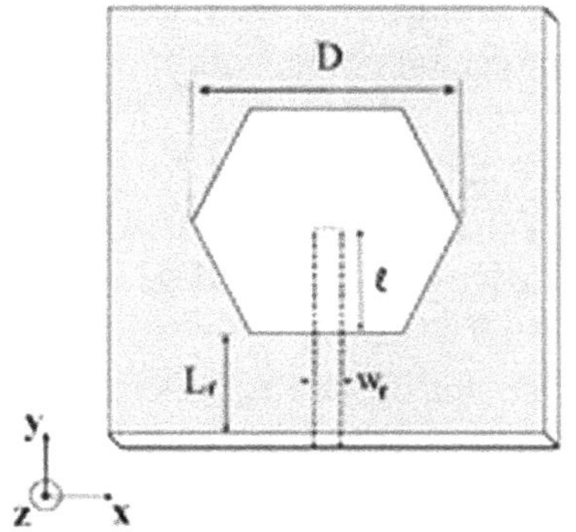

where f_L is in GHz, L and r in centimeter. The equivalent radius r is given by equating the corresponding surface areas i.e $2\pi rL = \pi ab$. Slot is an equivalent magnetic surface, where equation 3.1 has been empirically modified as reported earlier (Li, Liang, & Chen, 2006)

$$f_L = \frac{30 \times C}{L + r} \tag{2}$$

Here L and r are expressed in centimeter and f_L is in GHz. For the hexagonal slot under consideration as shown in Figure 1, L=D and r=0.103D. C is the element factor, equal to 0.379 for hexagonal slot according to above relation between r and D. The equivalent radius r is given by $2\pi rL = (3\sqrt{3}/2)(D/2)^2$,equating the areas of cylindrical wire monopole and the hexagonal surface of radius D/2, to any vertex. A comparative study between calculated and measured value of f_L for different dimensions of hexagonal slot length D has been carried out within the frequency range of 1.353GHz to 3.8GHz and it has been found that the measured frequencies according to the simple formulation agrees within $\pm 8\%$.

As the equivalent radius r stated above is not having significant dimension with respect to the slot dimension D, another approach of finding out lower edge frequency of impedance bandwidth (f_L) is the polynomial curve fitting which has been studied here. A polynomial equation to calculate the lower edge frequency of bandwidth (f_L) for a given dimension of slot D is given by

$$Y = 0.009x^4 - 0.2059x^3 + 1.7742x^2 - 7.1335x + 13.41 \tag{3}$$

where x is the variable for a given slot dimension D in centimeter and Y is the design frequency f_L in GHz, corresponding to the lower edge of bandwidth. The polynomial equation exhibits more accurate outcome, within $\pm$ 1.8% variation.

Studies on Bandwidth Enhancement of a Microstrip Line Fed Hexagonal Wide Slot Antenna Having Fork-Like Tuning Stub (Chattopadhyay et al., 2012)

In this work, a hexagonal wide slot, excited by a microstrip line with extended forklike tuning stub is the proposed configuration for wideband operation. A hexagonal wide slot fed by a simple 50Ω microstrip line having a straight section tuning stub as an extension of feed line is considered as reference antenna with respect to which the proposed antenna with fork like tuning stub is compared. The reference wide slot antenna has been designed utilizing the polynomial design equation presented in section 3. The substrate parameters i.e the substrate material relative permittivity, thickness and loss tangent are the same for both the reference and proposed configuration. Improvement in performance has been established by 1.5:1 and 2:1 VSWR bandwidth in order to get satisfactory antenna matching performance in terms of both simulation and experimental studies. Radiation pattern study is also exhibited to show the operation of proposed configuration as a good radiator as well as receiving element. The simulation study has been carried out by MoM based IE3D simulation software. The details of the proposed antenna simulation and experimental results are presented and discussed below.

Antenna Configuration

The reference antenna has been shown in Figure 2(a), printed on an FR4 substrate of thickness 1.6mm with relative permittivity of 4.4. The diameter (D) of the reference hexagonal slot is 53.7mm. All the parameters of the reference antenna are same as the proposed antenna, except the tuning stub configuration. The tuning stub of the reference antenna is a straight extension of feed line with a length (ℓ) of 20mm.

The geometry of the hexagonal wide slot proposed antenna with fork like tuning stub is shown in Figure 2(b). The proposed antenna has been printed on similar FR4 substrate having thickness of 1.6mm and relative permittivity of 4.4. The diameter (D) of the hexagonal slot is 53.7mm. The wide slot is fed by a 50Ω microstrip line, having width (W_f) of 3mm containing a fork like tuning stub that is printed on opposite side of the FR4 substrate. The tuning stub has been placed symmetrically with respect to the centre line along y-axis of the wide slot. The fork like tuning

Figure 2a. Configuration of microstrip line fed wide slot antenna with a simple tuning stub (reference antenna)

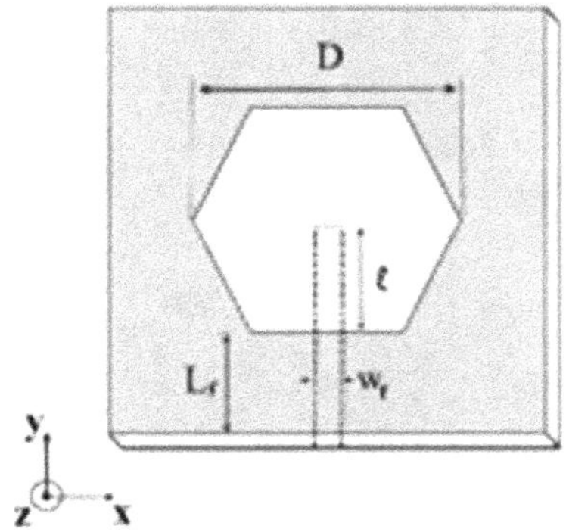

Figure 2b. Configuration of microstrip line fed wide slot antenna with fork-like tuning stub (proposed antenna)

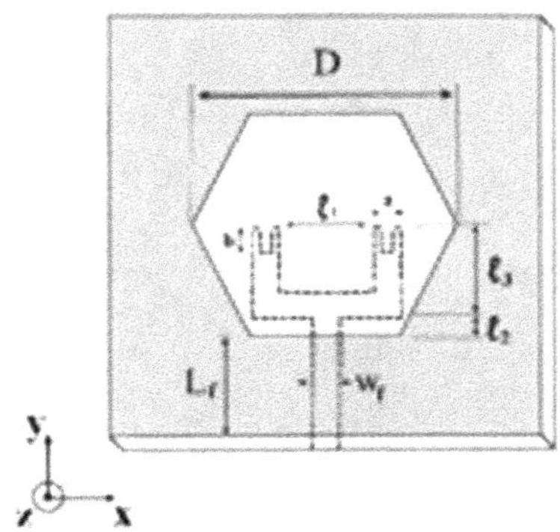

stub has a straight section of length (ℓ_2) and two branch sections of equal length (ℓ_3). There is a spacing (ℓ_1) between the inner edges of the two branch sections. For simplicity the width of all the sections are kept same and equal to the width of 50Ω microstrip feed line.

By selecting proper dimensions of ℓ_1, ℓ_2 and ℓ_3 of fork like tuning stub, impedance matching over an enlarged bandwidth has been obtained. Antenna1, Antenna2, Antenna3 and Antenna4 has been conceived for studies on different dimensions of ℓ_1, ℓ_2 and ℓ_3.. The Table 1 exhibits the different dimensions of the above antenna configurations where Antenna2 is considered as the proposed configuration. The photography of the slot side and feed side of the prototype has been shown in Figure 3(a) and 3(b).

Figure 3a. Slot side view of the prototype (Chattopadhyay et al, 2012)

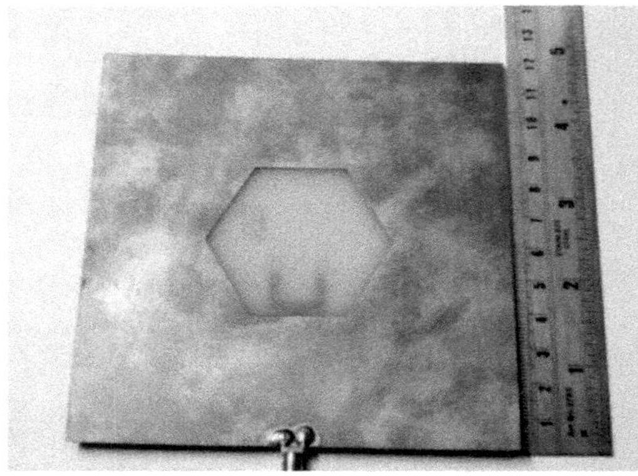

Figure 3b. Feed side view of the prototype (Chattopadhyay et al, 2012)

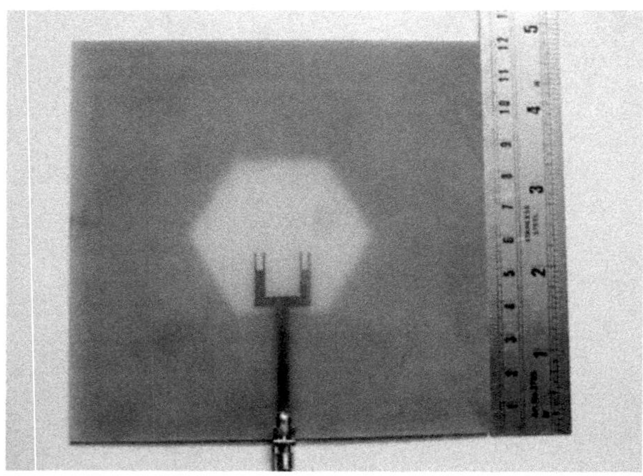

Table 1. Design parameters of different Antennas including proposed antenna (Antenna2) (Chattopadhyay et al, 2012)

	ℓ_1 (mm)	ℓ_2 (mm)	ℓ_3 (mm)
Antenna1	4	2	18
Antenna2 (Proposed)	10	2	16
Antenna3	12	2	16
Antenna4	10	0	19
D=53.7mm, a=2mm, b=5mm, W_f =3mm, L_f =36.7mm			

Discussion on Obtained Results

In this section, impedance, radiation and gain characteristics of the proposed antenna has been presented. Initially the results of simulation study are exhibited. The simulated VSWR responses of different configurations i.e. Antenna1, Antenna2 (proposed antenna), Antenna3 and Antenna4 has been compared with respect to the reference antenna and exhibited in Figure 4.

The simulation results for 1.5:1 and 2:1 VSWR bandwidth indicating higher (f_H) and lower (f_L) band edge frequency values are shown in Table 2.

The two standard reference levels of VSWR values i.e 1.5:1 and 2:1 are considered corresponding to the level of expected radiated power. To calculate the expected percentage of input power to be radiated, the formula $1 - \rho^2$ will be considered, where ρ is the magnitude of input reflection coefficient. When VSWR =2, the radiated

Figure 4. Simulated VSWR responses of different configurations (Chattopadhyay et al, 2012)

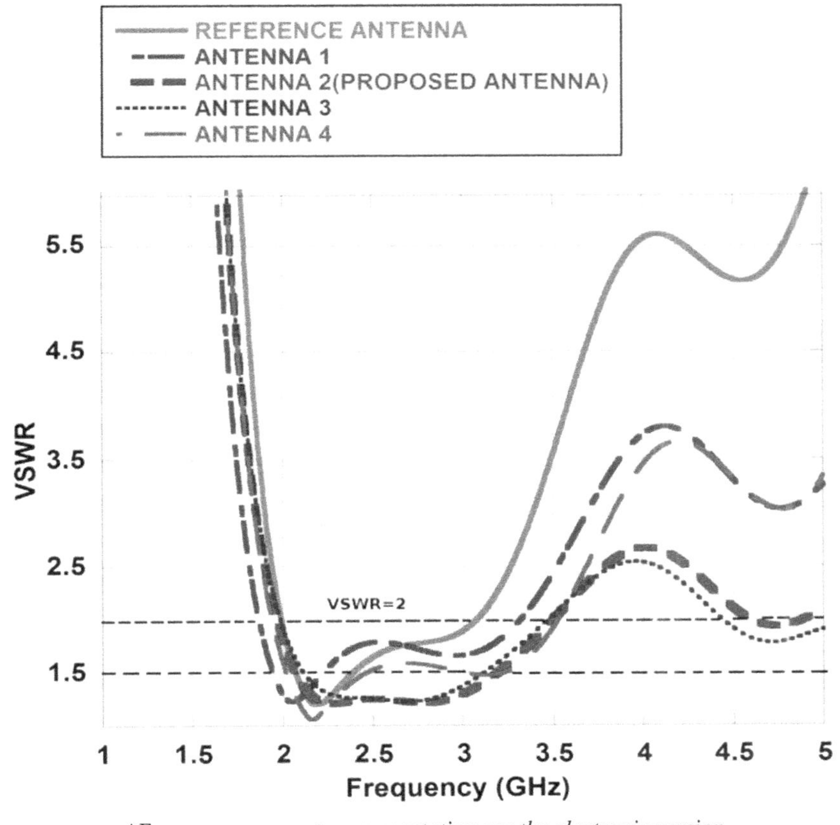

**For a more accurate representation see the electronic version.*

Table 2. Comparison between Proposed Antenna (Antenna2) and Reference Antenna corresponding to 1.5:1 and 2:1 VSWR Bandwidth and f L and f H (Chattopadhyay et al, 2012)

Antenna Types	f_L (MHz)	f_H (MHz)	VSWR Bandwidth (MHz) for VSWR=1.5
Reference Antenna	2069	2377	308
Antenna1	1938	2225	287
Antenna2 (Proposed)	2068	3206	1138
Antenna3	2105	3149	1044
Antenna4	2021	2465	444
Antenna Types	**f_L (MHz)**	**f_H (MHz)**	**VSWR Bandwidth (MHz) for VSWR=2**
Reference Antenna	1950	2420	470
Antenna1	1868	3303	1435
Antenna2 (Proposed)	1959	3516	1557
Antenna3	1983	3463	1480
Antenna4	1955	3522	1567

power will be 88% of the input power whereas for reference VSWR level 1.5, the radiated power will be 96% of the total input power to the antenna. In this context, VSWR=1.5 level is considered for high power radiation when needed.

From the obtained simulation results this has been observed that Antenna1, Antenna2 (proposed configuration), Antenna3 and Antenna4 exhibits enhanced bandwidth than the reference antenna in view of 2:1 VSWR. If 1.5:1 VSWR bandwidth is considered, the Antenna2, Antenna3 and Antenna4 are having larger bandwidth than the reference antenna. From the study between Antenna2 and Antenna 4 it is found that the inclusion of a small straight section (ℓ_2) in fork like tuning stub, results in significant improvement in VSWR bandwidth. The Antenna2 exhibits much higher 1.5:1 VSWR bandwidth i.e 1138MHz with respect to Antenna4 having 1.5:1 VSWR bandwidth 444MHz. But in view of 2:1 VSWR bandwidth, both are exhibiting almost similar performances. In this study, Antenna2 has been considered as proposed antenna, comparing both 1.5:1 and 2:1 VSWR bandwidth as it will become useful for radiation of more power over a wider bandwidth than antenna4 analyzed by 1.5:1 VSWR bandwidth. The Antenna2 will be an efficient design for wideband operation with high power. This has been observed that for 2:1 VSWR bandwidth; the Antenna2 and Antenna4 (where there is the absence of straight section ℓ_2) are exhibiting almost similar performances having a fabrication tolerance.

To explain the phenomenon of bandwidth enhancement, the impedance plots on smith chart for different configurations and magnetic current distribution of the proposed antenna has been exhibited in Figure 5(a-d) and Figure 6 (a-d) respectively.

With reference to Figure 5(a-d) it is observed that with the change in dimension of the fork like tuning stub the size of the coupling loop varies. For Antenna2 tightest coupling loop within VSWR=1.5 circle has been observed around the centre of the smith chart. This result leads to a largely enhanced impedance bandwidth.

The magnetic current distribution plots shown in Figure 6(a-d), exhibit the excitation of different coupled resonant modes corresponding to different path lengths that has been excited by the geometry of fork like tuning stub and hexagonal nature of wide slot. This has also been found that proper tuning among different excited modes is also responsible for wider impedance bandwidth.

From the comparison between the results of Antenna1 to Antenna4, the conclusion can be drawn that there exists an optimal spacing (ℓ_1) between two branch sections of fork like tuning stub. For Antenna2, the selected spacing is about one fifth of the slot diameter (D). From the comparison between Antenna2 and Antenna4 results it has been found that larger 1.5:1 VSWR bandwidth can be obtained by placing a straight section ℓ_2 of fork like tuning stub under the slot on opposite side of the substrate. The length ℓ_3 of the two branch sections along with grooved top edge is

Figure 5a. Impedance plot of reference antenna (Chattopadhyay et al, 2012)

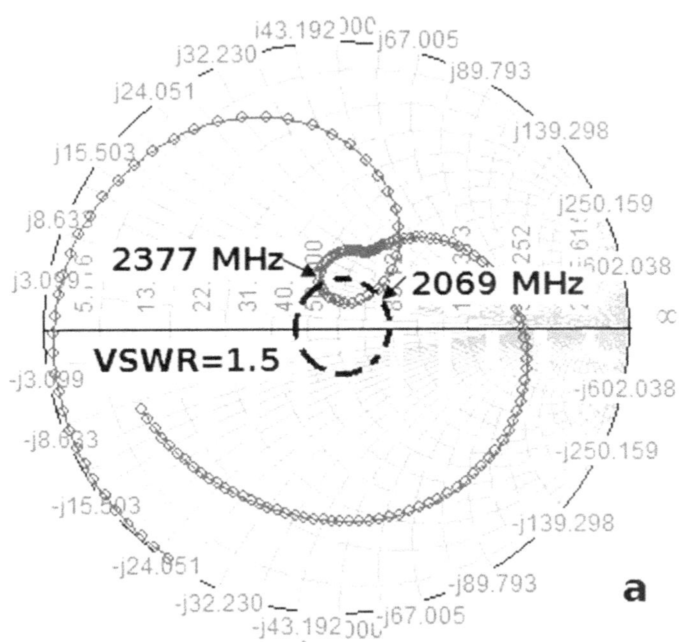

Figure 5b. Impedance plot of Antenna2(proposed antenna) (Chattopadhyay et al, 2012)

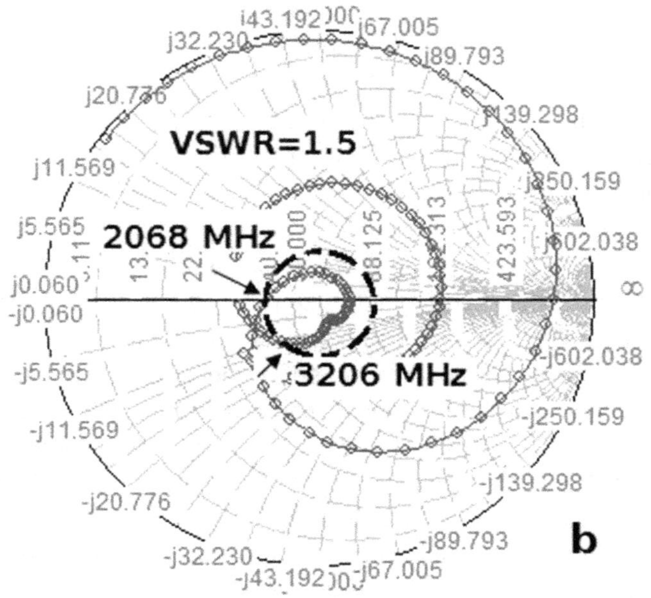

Figure 5c. Impedance plot of Antenna3 (Chattopadhyay et al, 2012)

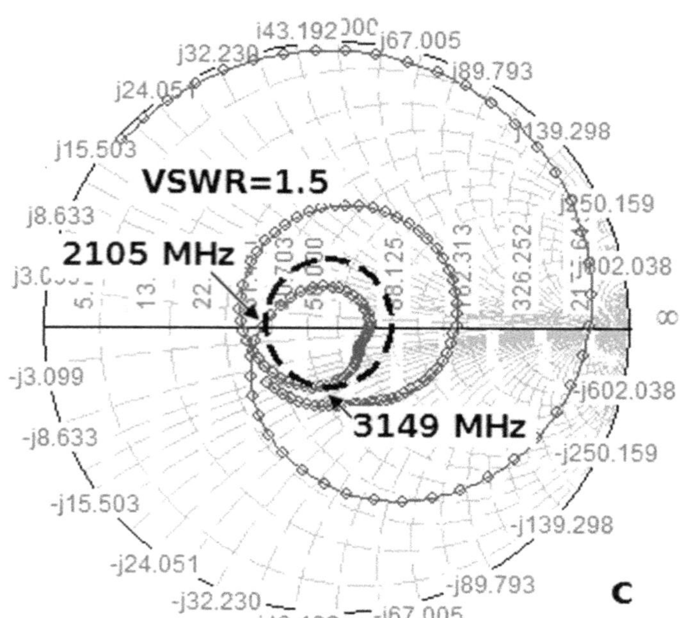

Figure 5d. Impedance plot of Antenna4 (Chattopadhyay et al, 2012)

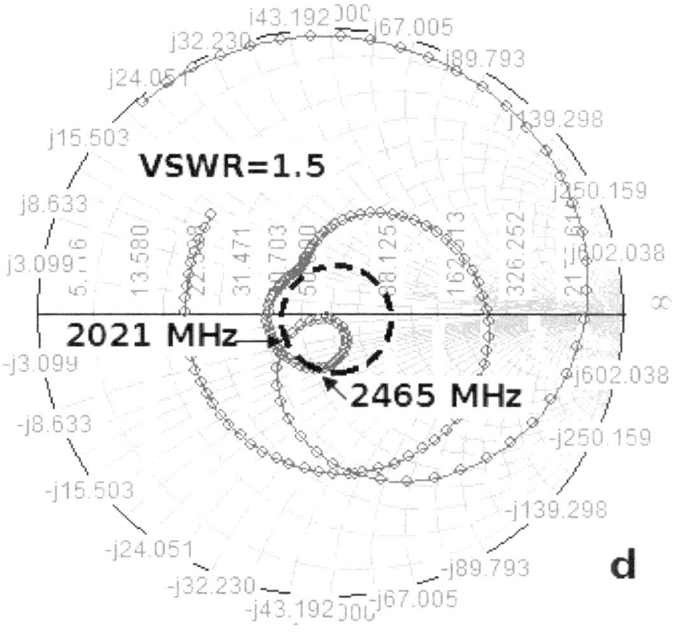

Figure 6a. Magnetic current distribution of the proposed antenna at 2100MHz (Chattopadhay et al, 2012)

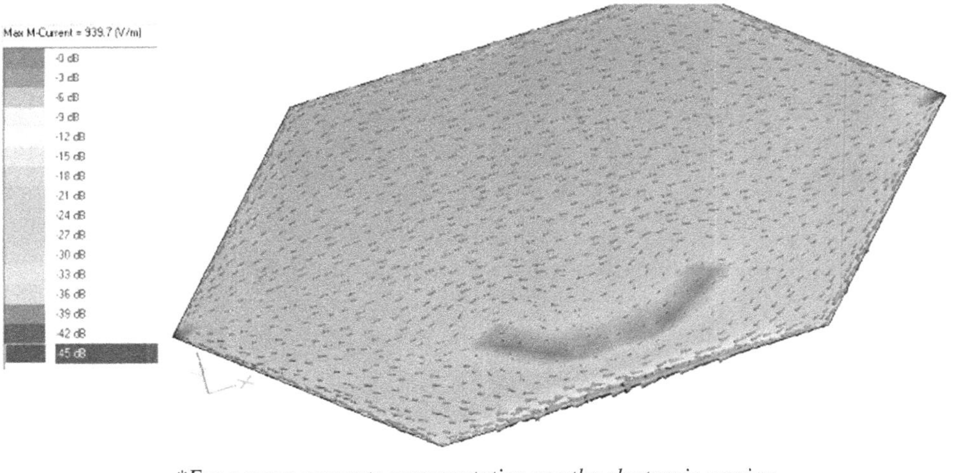

**For a more accurate representation see the electronic version.*

Figure 6b. Magnetic current distribution of the proposed antenna at 2500MHz (Chattopadhay et al, 2012)

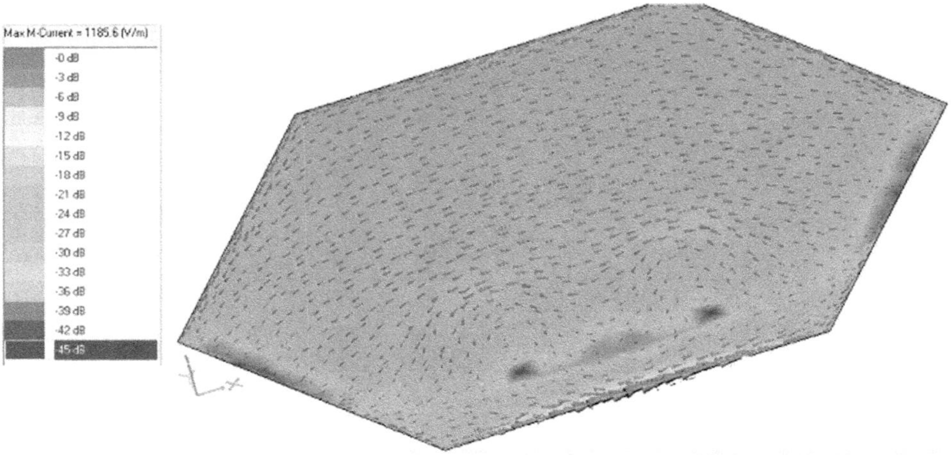

**For a more accurate representation see the electronic version.*

Figure 6c. Magnetic current distribution of the proposed antenna at 2800MHz (Chattopadhay et al, 2012)

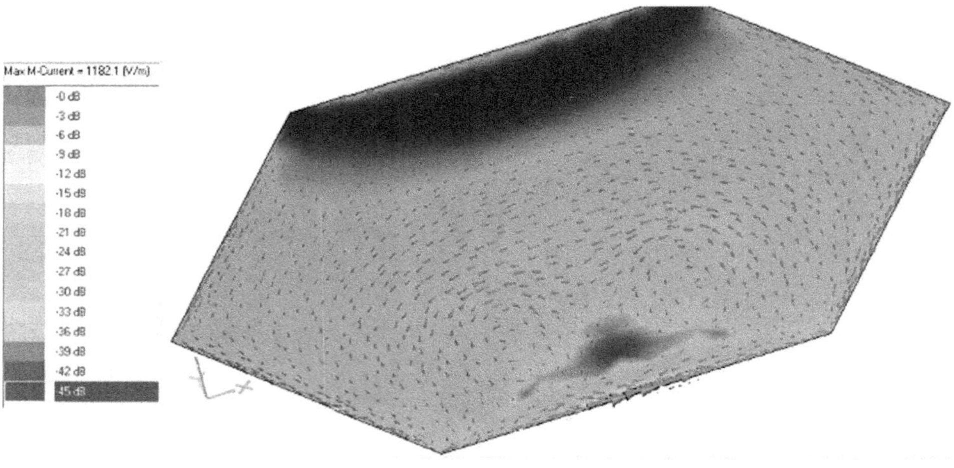

**For a more accurate representation see the electronic version.*

an important parameter to achieve easy fine tuning of the imaginary part of input impedance.

The experimental study has been done for the proposed antenna (Antenna2). The VSWR response of the fabricated antenna has been studied by Agilent Technologies PNA-L Network Analyzer (Model No: N5230A, 10MHz-20GHz). The simulated and measured VSWR responses of the proposed antenna have been exhibited in

Figure 6d. Magnetic current distribution of the proposed antenna at 3200MHz (Chattopadhay et al, 2012)

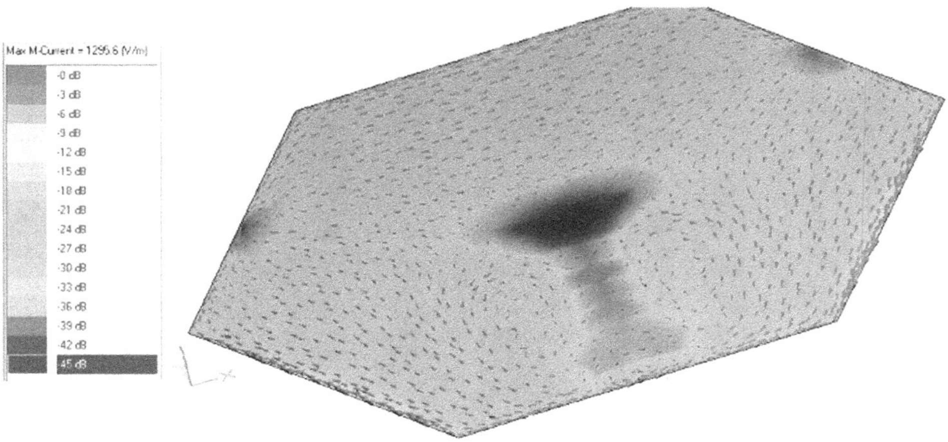

**For a more accurate representation see the electronic version.*

Figure 7 for 1.5:1 and 2:1 VSWR bandwidth. From the measured VSWR response this has been observed that the available bandwidth is 900MHz (2.095GHz-2.995GHz) for 1.5:1 VSWR and 1340MHz(2.010GHz-3.350GHz) for 2:1 VSWR. The experimentally available bandwidth is smaller than the simulated result. This may be due to tolerance in loss tangent value used in simulation and practically available substrate along with fabrication tolerances.

The radiation patterns of the proposed antenna prototype are experimentally measured and plotted in Figure 8 (a-c) for frequencies 2230MHz, 2500MHz and 2900MHz.

From the plotted radiation patterns this has been found that the wide slot antenna is a bidirectional radiator and almost similar radiation patterns are available in both sides. From the depicted patterns it has been observed that for the entire 1.5:1 VSWR bandwidth, the radiation patterns of the proposed antenna maintains same polarization and similar broadside radiation characteristics. This has been further noted that the cross-polarization levels are higher in H-plane(x-z plane) rather than E-plane(y-z plane). It is also observed that the cross-polarization level is increasing with the increase in frequency.

The study on variation of peak gain of the proposed antenna has been exhibited in Figure 9. The variation in peak gain is about 1.75dBi. The available peak gain is 3.025dBi within the frequency range corresponding to 1.5:1 VSWR although the peak gain is 3.15 dBi considering 2:1 VSWR bandwidth.

Figure 7. Comparison between simulated and measured VSWR response of the proposed antenna (Antenna2) (Chattopadhyay et al, 2012)

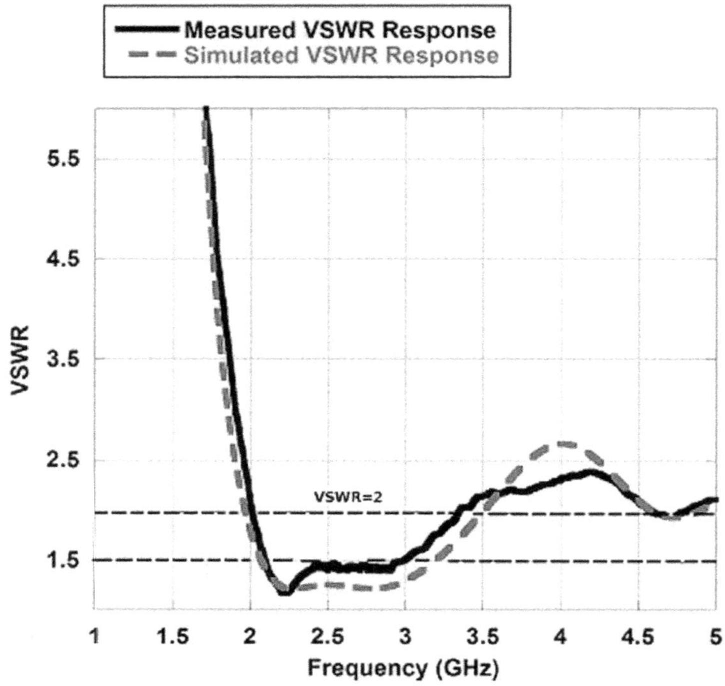

Figure 8a. E-plane and H-plane radiation patterns at 2230MHz (Chattopadhyay et al, 2012)

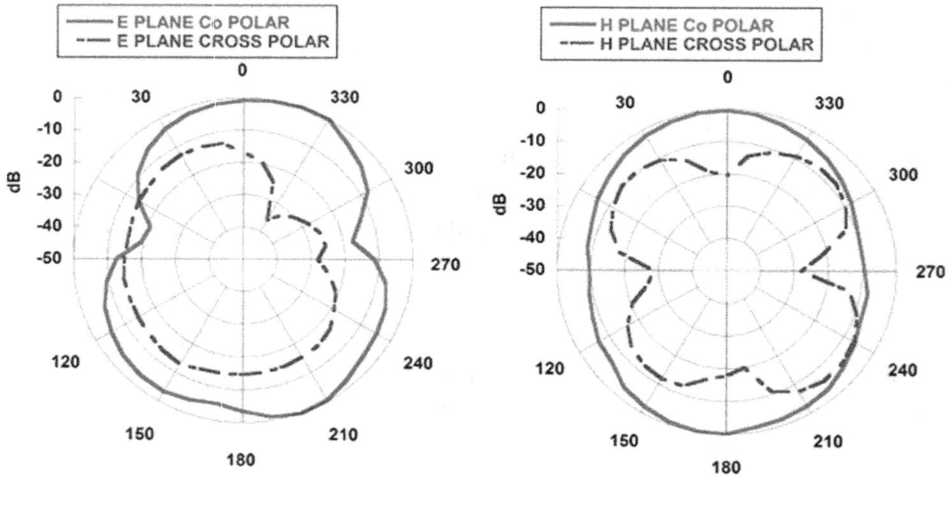

Figure 8b. E-plane and H-plane radiation patterns at 2500MHz (Chattopadhyay et al, 2012)

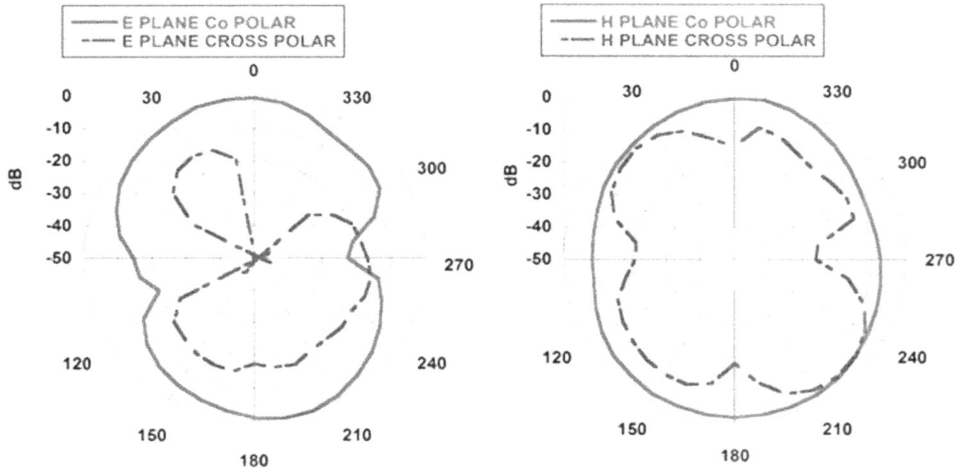

Figure 8c. E-plane and H-plane radiation patterns at 2900MHz (Chattopadhyay et al, 2012)

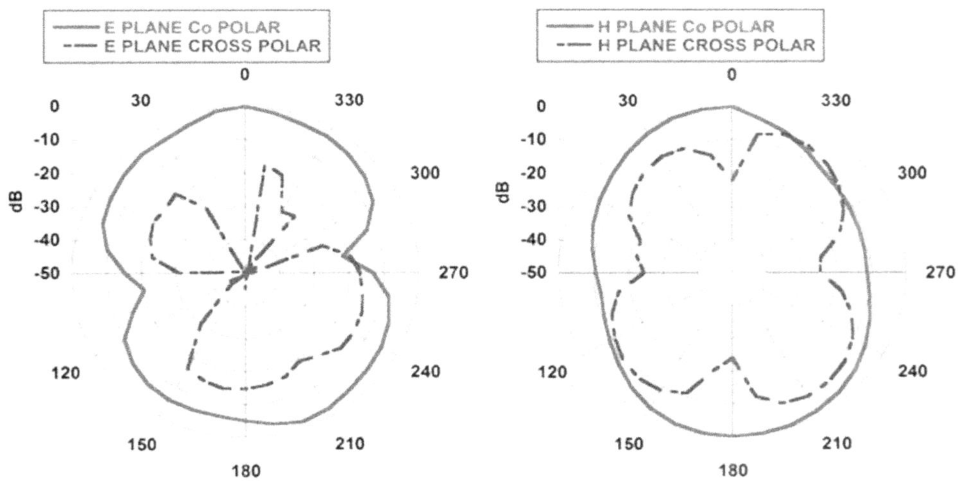

Discussion

In this study, a printed hexagonal wide slot antenna excited by a 50Ω microstrip line with extended fork like tuning stub has been proposed and experimentally investigated. Simulation and experimental results reveal that a 1.5:1 VSWR bandwidth of 900MHz and 2:1 VSWR bandwidth of 1340MHz has been achieved at operating

Figure 9. Peak antenna gain versus frequency for proposed antenna (Antenna2) (Chattopadhyay et al, 2012)

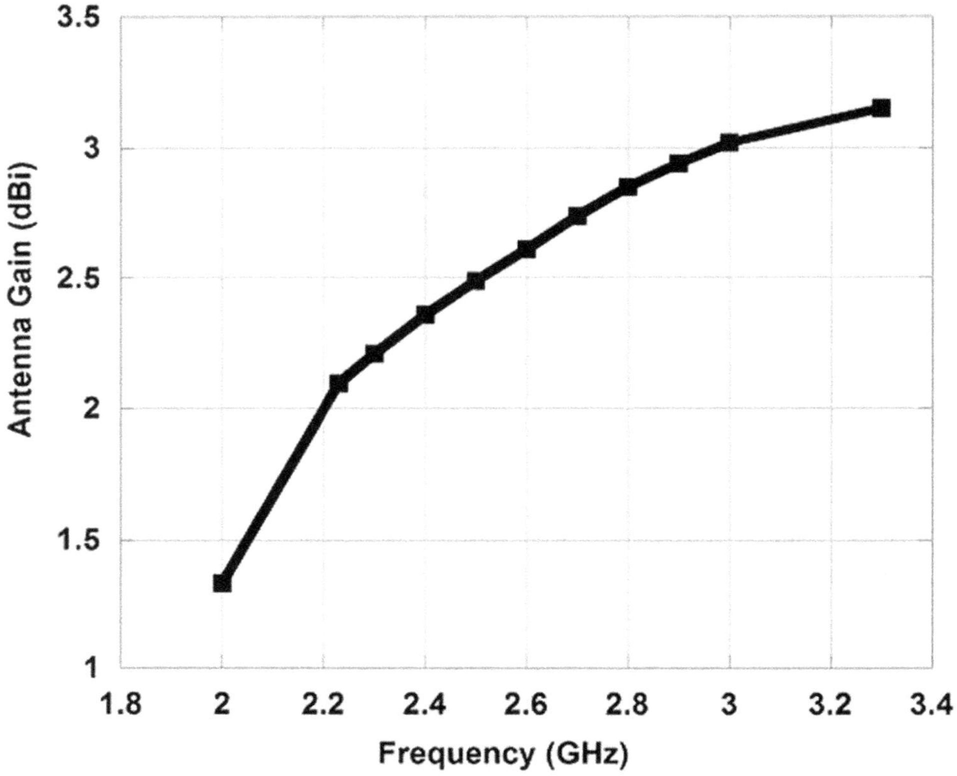

frequency around 2500MHz. The obtained bandwidth is three times larger than the reference hexagonal slot antenna with straight section tuning stub. The VSWR level 1.5 is resulting in more high power radiation in the range 2095MHz to 2995MHz. However, wider impedance bandwidth of about 1340MHz (2010MHz-3350MHz) is available if 2:1 VSWR bandwidth is accepted.

Studies on Bandwidth Enhancement of a Microstrip Line Fed Hexagonal Wide Slot Antenna Using Hexagonal Tuning Stub (Chattopadhyay et al., 2013)

In the previous section, studies on hexagonal wide slot configuration having microstrip feed line loaded by forklike tuning stub were carried out where improvement in bandwidth of about three times had been noticed. Further enhancement of bandwidth has been carried out in this section keeping in view its practical applications conforming to various telecommunication standards. The following study is in continuation of the previous work where the printed hexagonal wide slot antenna excited by a 50Ω microstrip line loaded with hexagonal tuning stub has been proposed as an alternative approach. The performance of the proposed antenna has been compared with respect to the performance of the same reference antenna demonstrated in previous section. The substrate parameters i.e the substrate material relative permittivity, thickness and loss tangent are the same for both the reference and proposed configuration. Improvement in bandwidth is described by 2:1 VSWR level as reference in case of both simulation and experimental studies. An optimum impedance bandwidth can be obtained by suitable coupling between feed structure and the slot. In case of proposed antenna structure a variation of hexagonal tuning stub size has been done to achieve optimum coupling that may result in enhanced bandwidth. The simulation study has been carried out by MoM based 3D simulation software IE3D. Detail of the proposed antenna study is presented and experimentally validated also.

Antenna Configuration

The geometry of the proposed hexagonal wide slot antenna has been exhibited in Figure. 10(a). The proposed antenna has been printed on an FR4 substrate having

Figure 10a. Configuration of microstrip line fed wide slot antenna with Hexagonal tuning stub(proposed antenna) (Chattopadhay et al, 2013)

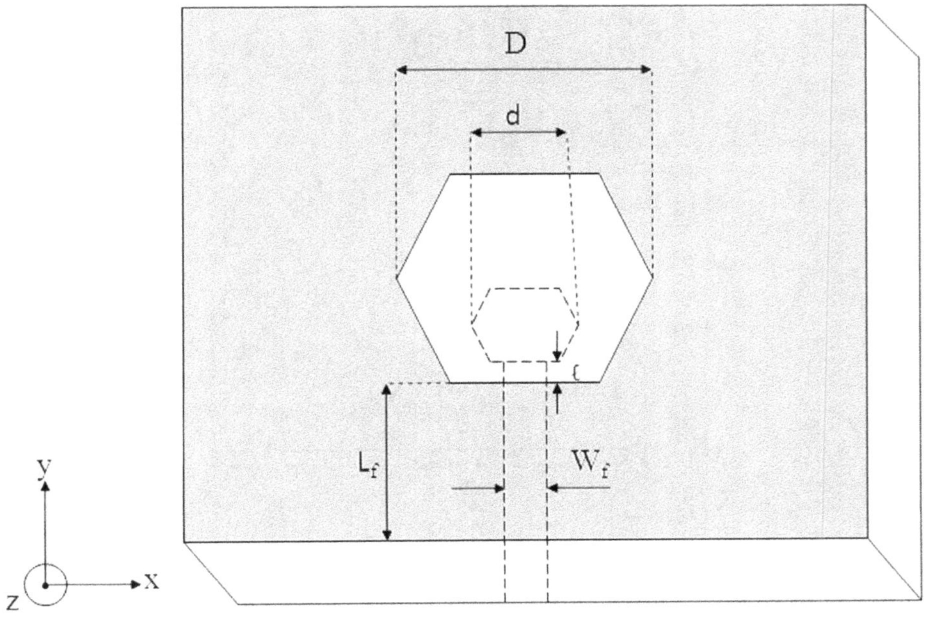

thickness of 1.6mm and relative permittivity 4.4. The diameter(D) of the hexagonal wide slot is 53.7mm. The wide slot has been excited by a 50Ω microstrip line loaded with hexagonal tuning stub, which is printed on opposite side of FR4 substrate. The feed line is having a length(L_f) 31.75mm and width(W_f) 3mm.

The tuning stub has been placed symmetrically with reference to the centre line along y-axis of the wide slot. The tuning stub contains a straight section length(ℓ) and one hexagonal section with a diameter(d). For a simple design the width of the straight section of the stub is kept the same as the width of the 50Ω microstrip feed line. By selecting suitable dimensions of hexagonal section(d) and straight section(ℓ) of the tuning stub satisfactory impedance matching over a larger bandwidth can be obtained. Antenna1, Antenna2, Antenna3, Antenna4 and Antenna5 have been configured for different dimensions of hexagonal section and straight section of tuning stub for fixed slot diameter(D), feed length(L_f) and feed width(W_f). The reference antenna has been exhibited in Figure 10(b). The reference hexagonal slot antenna parameters have been considered same as like the proposed antenna, except the tuning stub configuration. For the reference antenna the tuning stub is a straight extension of feed line containing a length(ℓ) of 20mm. The photography of the top and the bottom surfaces of the prototype has been exhibited in Figure 11(a) and 11(b).

Figure 10b. Configuration of microstrip line fed wide slot antenna with a simple straight tuning stub(reference antenna) (Chattopadhay et al, 2013)

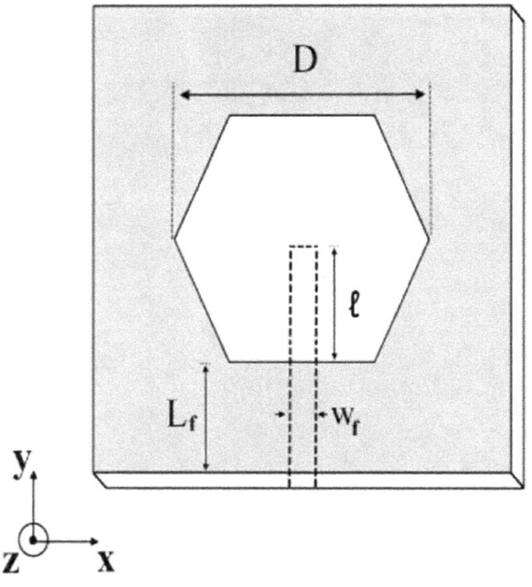

Figure 11a. Slot-side view of the prototype (Chattopadhyay et al, 2013)

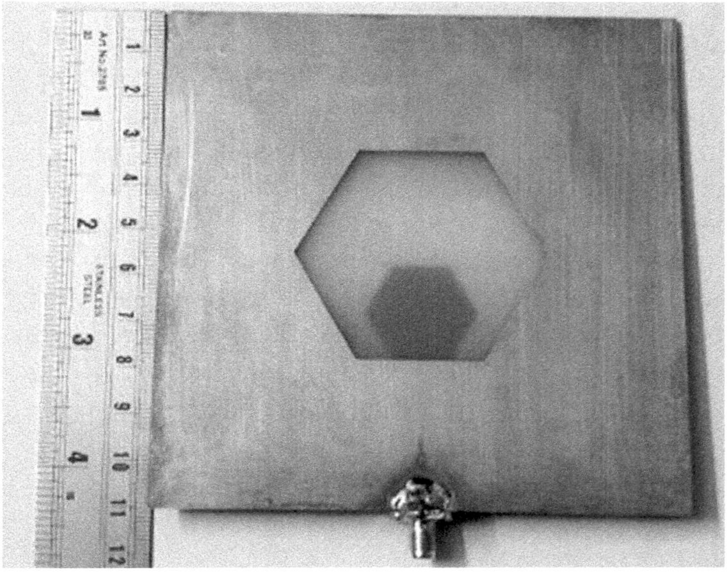

Figure 11b. Feed-side view of the prototype (Chattopadhyay et al, 2013)

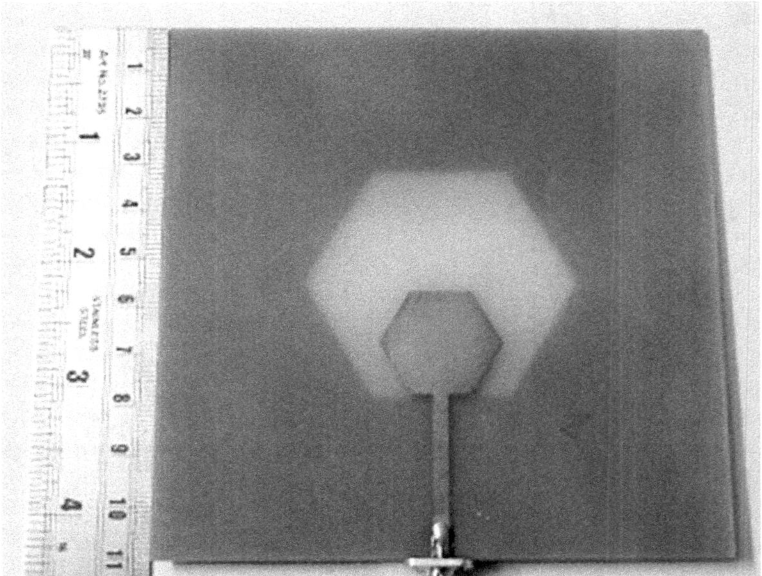

Discussion on Obtained Results

In this area the impedance, radiation and gain characteristics of the proposed antenna are discussed. In case of simulation the VSWR responses of the different configurations i.e Antenna1, Antenna2, Antenna3, Antenna4 and Antenna5 for different dimensions of the hexagonal section of tuning stub(d) and straight section of stub length(ℓ) have been studied and compared with respect to the reference antenna VSWR response, as exhibited in Figure 12.

The simulation results for different configurations in comparison with reference antenna's measured data are shown in Table 3. From the obtained simulation results it is observed that Antenna1 to Antenna5, all configurations exhibit larger bandwidth

Figure 12. Simulated VSWR responses of different configurations (Chattopadhyay et al, 2013)

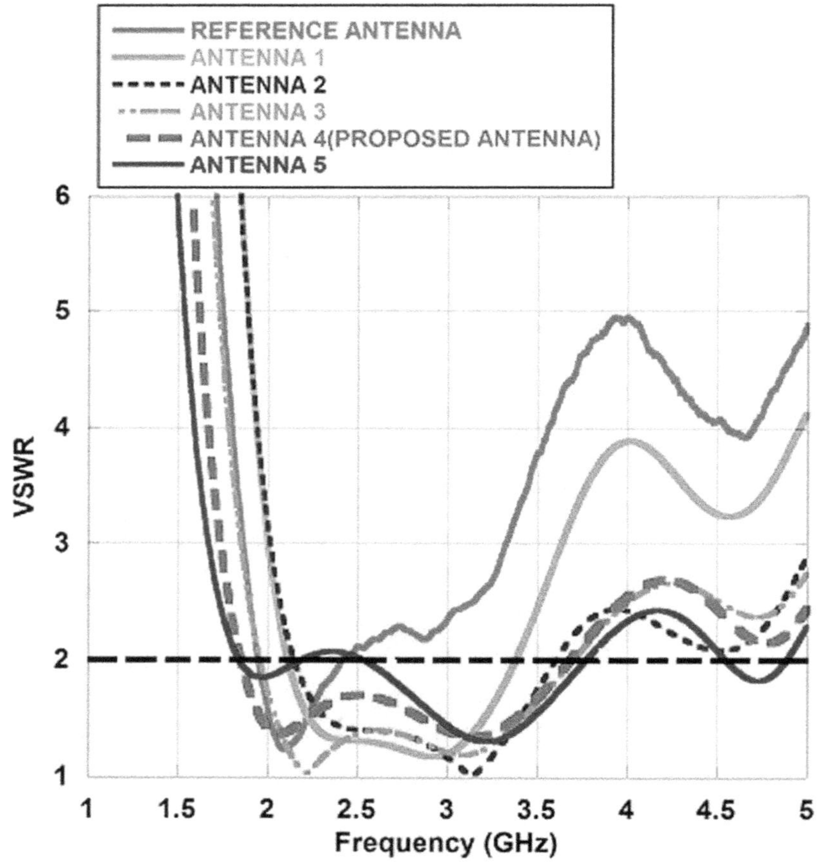

**For a more accurate representation see the electronic version.*

Table 3. Performance study of various configurations including proposed antenna(Antenna4) and reference antenna in view of 2:1 VSWR bandwidth (Chattopadhyay et al, 2013)

Antenna Types	Diameter of Hexagonal stub(d)mm	Straight section length of stub(ℓ) mm	VSWR Bandwidth (MHz, %)	Centre frequency f_c (MHz)
Reference Antenna	----------	20	470, 21.51	2185
Antenna1	10	4.9	1267, 44.07	2875
Antenna2	14	1.9	1457, 50.63	2877.5
Antenna3	20	0.5	1745, 61.63	2831.5
Antenna4 (Proposed Antenna)	24	0.5	1844, 66.76	2762
Antenna5	28	0.5	1242, 39.34	3157

SlotDiameter(D)=53.7mm, Feedlength(L_f)=31.75mm, Feedwidth(W_f)=3mm

than reference antenna. With the increase in diameter(d) of the tuning stub, the length of the straight section (ℓ) will have to be reduced to get suitable coupling with the radiating slot resulting in wider VSWR bandwidth. Antenna4 has been considered as proposed configuration as it exhibits largest bandwidth (validated by both simulation and experimental study shown later). The results have been compared with respect to 2:1 VSWR level for most acceptable antenna matching performances. Keeping the hexagonal section of stub diameter constant at d=24mm, if the straight section length(ℓ) is further reduced or increased, reduction in VSWR bandwidth is observed.

For explaining the phenomenon of bandwidth enhancement the simulated impedance plot on smith chart and magnetic current distribution of the proposed antenna have been presented in Figure 13(d) and Figure 14(a-d) respectively. With reference to Figure 13(a-d) it is observed that with the increase in diameter(d) of hexagonal stub, the size of the coupling loop is increasing. For the proposed configuration, diameter of the hexagonal section (d) of the tuning stub is 24mm, for which the tightest coupling loop is obtained within VSWR=2 circle around the centre of the smith chart within VSWR=2 circle. These results are ensuring an enhanced impedance bandwidth.

The bandwidth is enhanced because the other resonant modes in the vicinity of fundamental mode are excited due to presence of hexagonal stub of suitable dimensions. The magnetic current distributions shown in Figure 14(a-d) exhibit

Figure 13a. Input Impedance plot of reference antenna (Chattopadhyay et al, 2013)

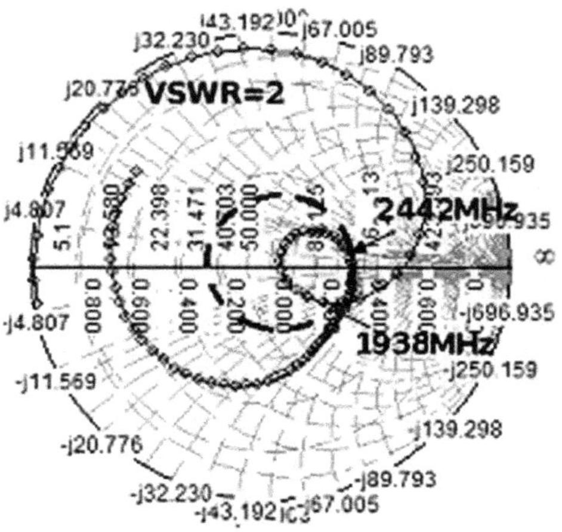

Figure 13b. Input Impedance plot of Antenna2 (Chattopadhyay et al, 2013)

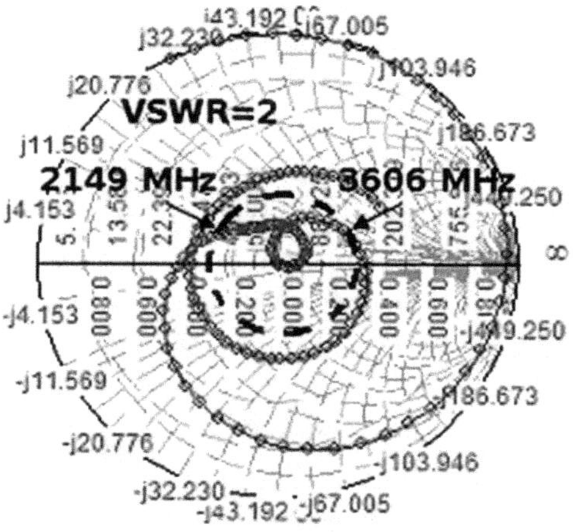

the other coupled resonant modes for the different path lengths of the geometry of hexagonal tuning stub and the hexagonal slot. Optimum tuning between different excited resonant modes is also resulting in enhanced bandwidth. The optimal diameter of the hexagonal tuning stub is 24mm to get largest bandwidth, which is found to be smaller than half of the slot diameter(D). However, the straight section length (ℓ) of the tuning stub is also to be adjusted corresponding to specified hexagonal stub diameter to get suitable matching by optimum coupling with the radiating slot.

Figure 13c. Input Impedance plot of Antenna3 (Chattopadhyay et al, 2013)

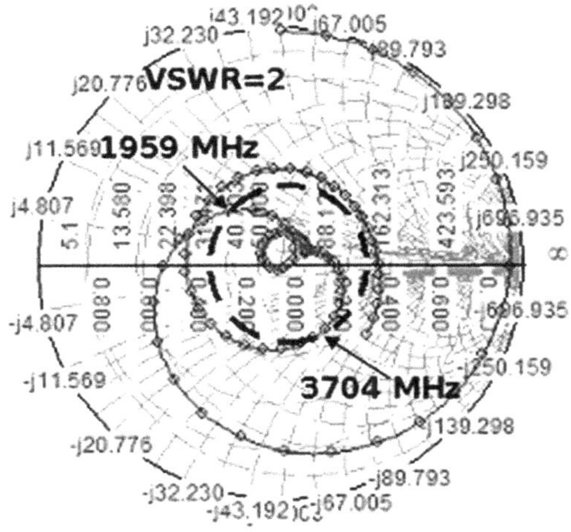

Figure 13d. Input Impedance plot of Antenna4(proposed antenna) (Chattopadhyay et al, 2013)

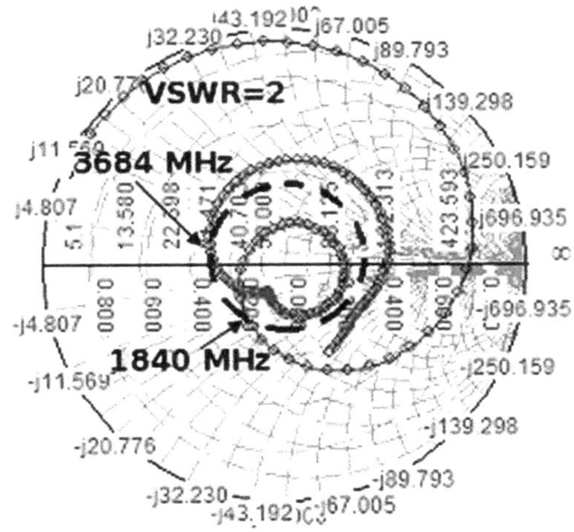

For experimental study the prototyping of both the proposed and reference antenna have been done. The VSWR responses of the fabricated antennas have been studied by Agilent Technologies PNA-L Network analyzer(Model No. N5230A, 10MHz-20GHz). The simulated and measured VSWR responses of the proposed antenna with respect to reference antenna for 2:1 VSWR bandwidth is exhibited in Figure 15. From the measured VSWR response of the proposed antenna, the available

Figure 14a. Magnetic current distribution of proposed antenna at 2000MHz (Chattopadhyay et al, 2013)

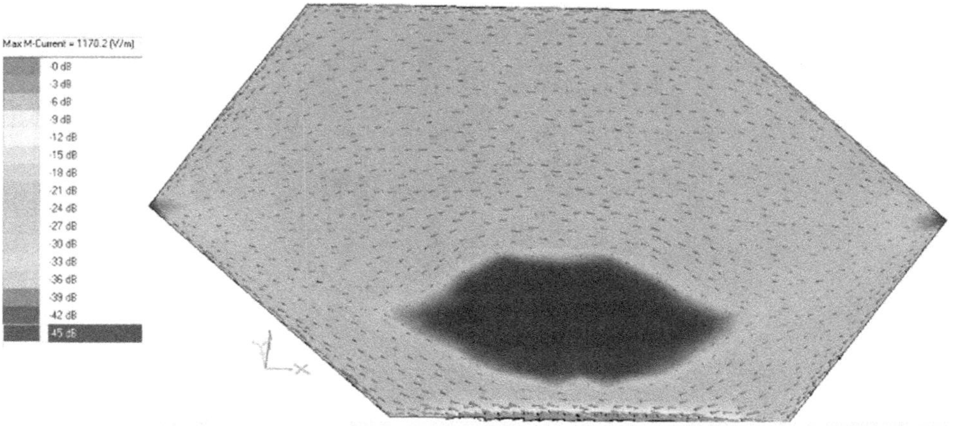

**For a more accurate representation see the electronic version.*

Figure 14b. Magnetic current distribution of proposed antenna at 2500MHz (Chattopadhyay et al, 2013)

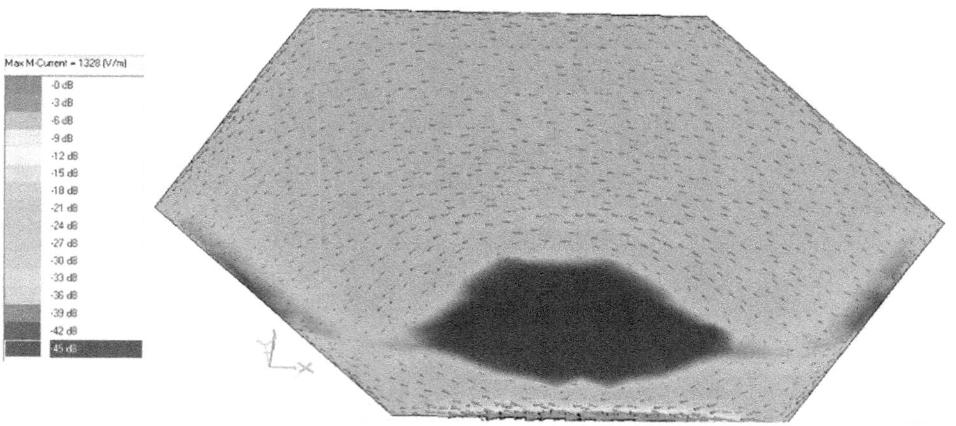

**For a more accurate representation see the electronic version.*

bandwidth is found to be 1751MHz(1.921GHz-3.672GHz). The obtained result is slightly smaller than the simulation result. This is probably due to tolerance in loss tangent value used in simulation and practically available substrate material along with the fabrication tolerances. The experimental study reveals that, increase in hexagonal stub diameter(d) beyond 24mm, results in reduction of bandwidth. The VSWR response with 26mm diameter of hexagonal section of stub, exhibits dual band property rather than a single wideband VSWR response.

Figure 14c. Magnetic current distribution of proposed antenna at 2790MHz (Chattopadhyay et al, 2013)

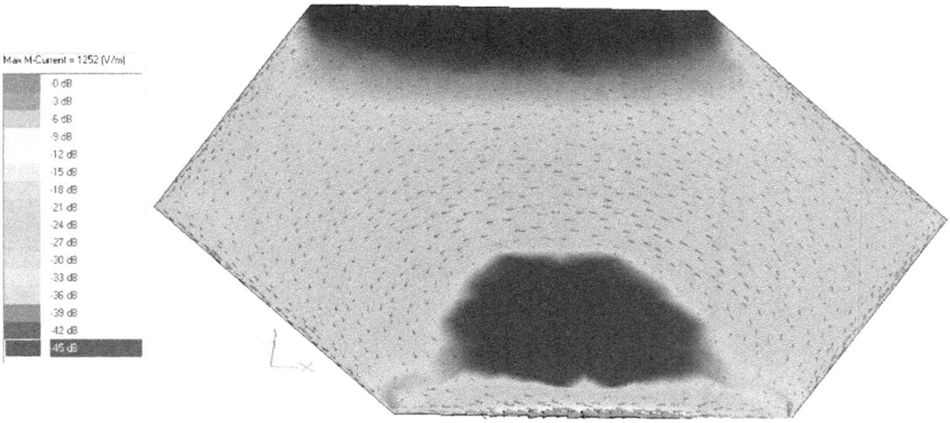

**For a more accurate representation see the electronic version.*

Figure 14d. Magnetic current distribution of proposed antenna at 3200MHz (Chattopadhyay et al, 2013)

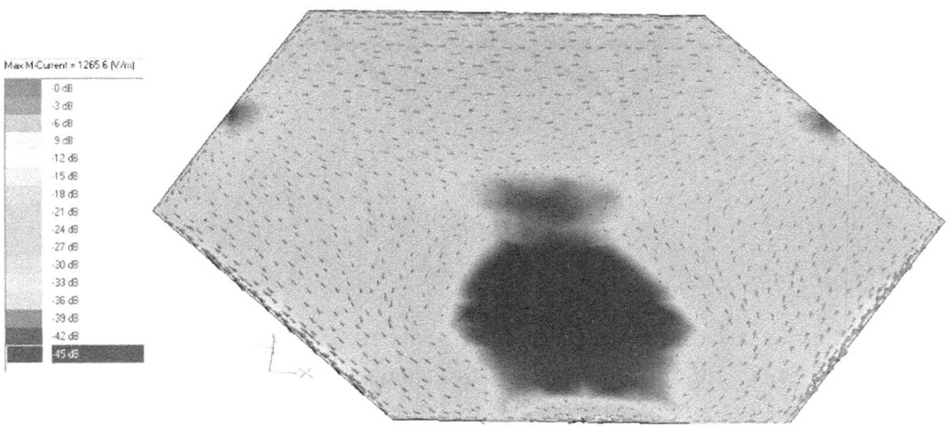

**For a more accurate representation see the electronic version.*

The radiation patterns of the proposed antenna for hexagonal stub section d=24mm and straight section length ℓ=0.5mm have been measured experimentally and plotted in Figure 16(a-c) by E-plane(y-z plane) and H-plane(x-z plane) for the frequencies 2135MHz(lowest VSWR value near lower band edge), 2796MHz(Centre frequency) and 3000MHz.

Figure 15. Study on simulated and measured VSWR response of the proposed antenna (Antenna4) (Chattopadhyay et al, 2013)

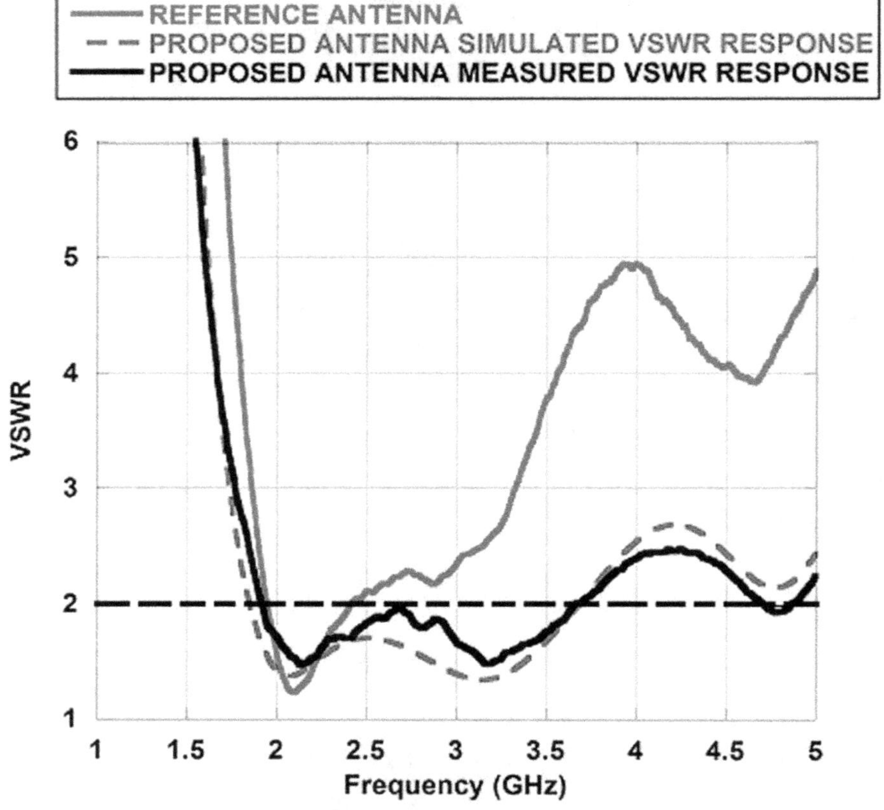

From the measured radiation pattern it is observed that the wide slot antenna is a bi-directional radiator and the radiation patterns are almost similar on both sides. From these results it can be seen that for operating frequencies across the impedance bandwidth within 1751MHz, the radiation patterns have same polarization planes and similar broadside radiation characteristics. It has been found that the cross-polarization levels are higher in H-plane(x-z plane) in comparison with the E-plane(y-z plane). With the increasing operating frequency, the cross-polarization level is also found to be increasing. The nature of cross-polarization also exhibits the excitation of higher order modes in respective directions. The conceived design with different values of hexagonal section of stub diameter(d) and straight section length(ℓ) show almost the same radiation characteristics.

Figure 16a. E-plane and H-plane radiation patterns at 2135MHz (Chattopadhyay et al, 2013)

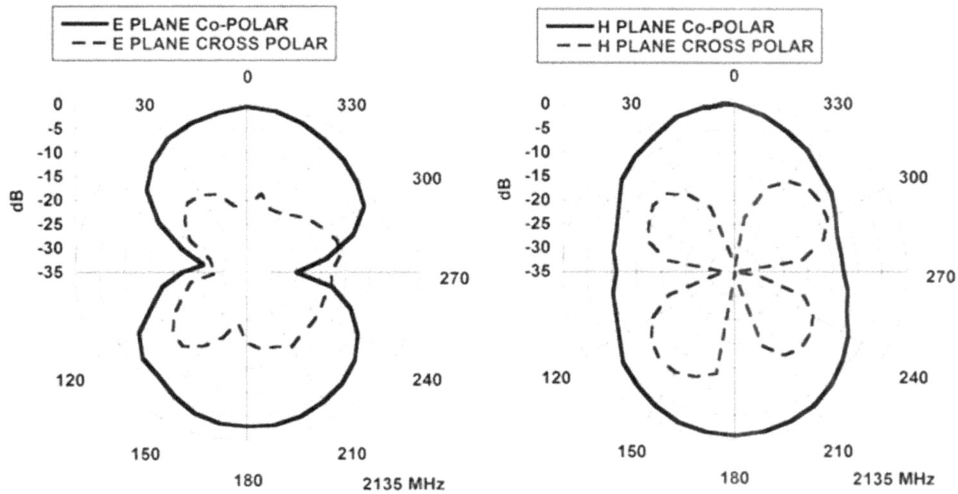

Figure 16b. E-plane and H-plane radiation patterns at 2796MHz (Chattopadhyay et al, 2013)

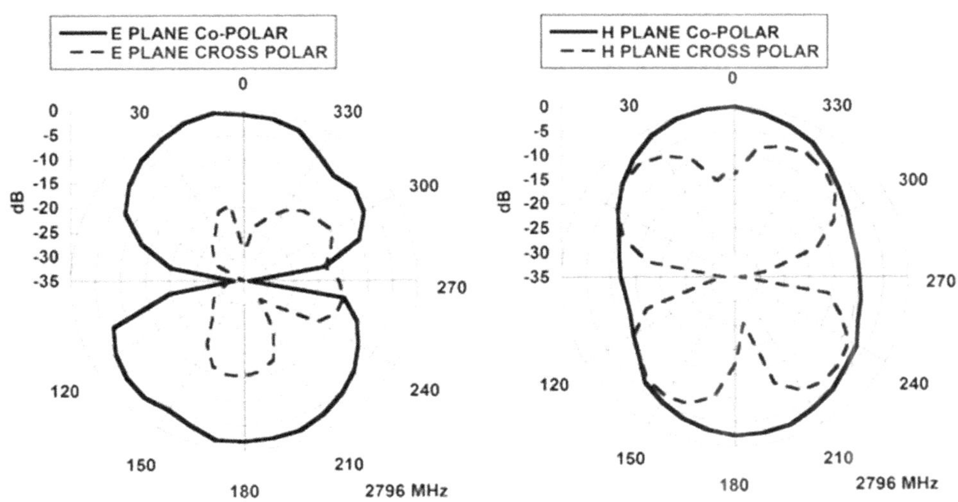

The variation in peak gain of the proposed antenna within 2:1 VSWR bandwidth is shown in Figure 17. The peak gain variation has been observed to be about 2dBi. The peak antenna gain available is 3.59dBi. The gain around the centre frequency is about 3dBi. From the study it is observed that the reference hexagonal slot antenna

Figure 16c. E-plane and H-plane radiation patterns at 3000MHz (Chattopadhyay et al, 2013)

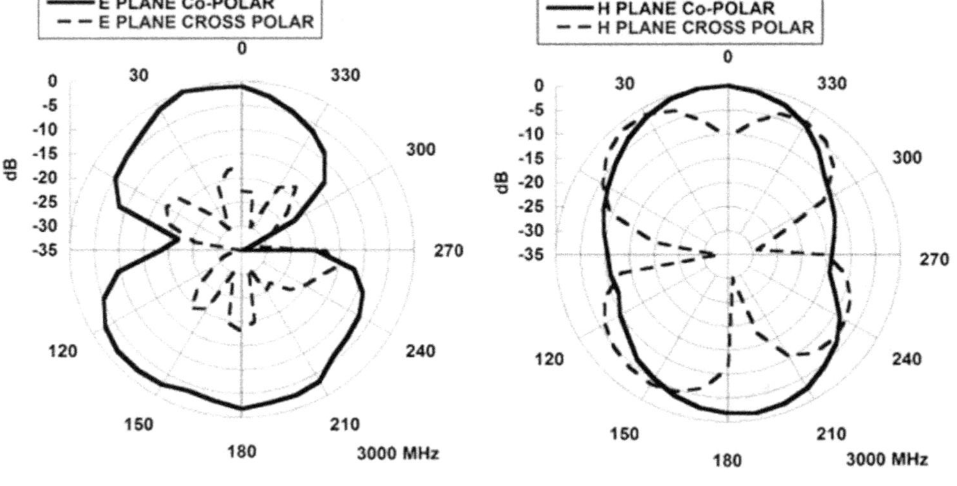

Figure 17. Peak gain response for the proposed antenna (Antenna4) (Chattopadhyay et al, 2013)

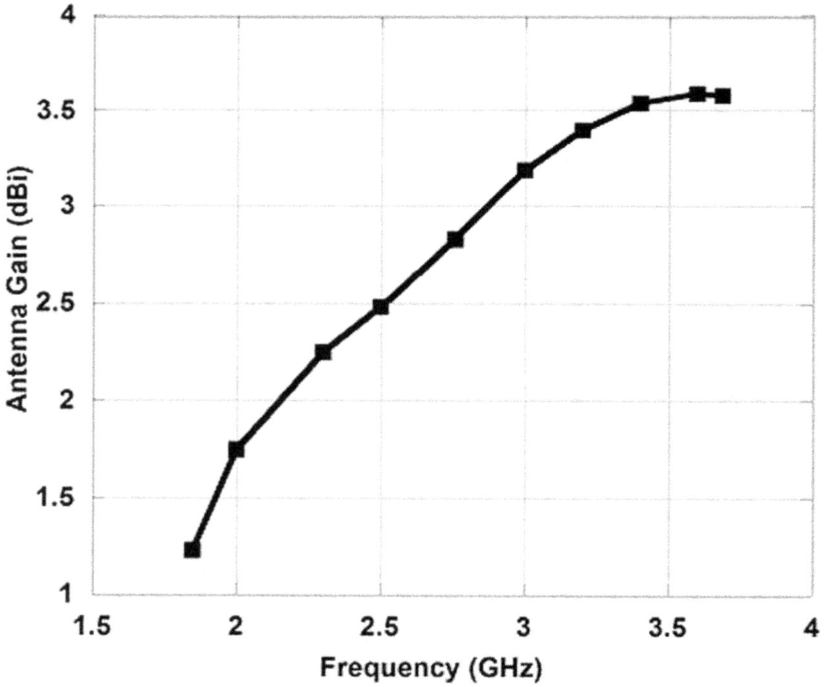

has a gain of 1.8dBi within its small bandwidth range. Slight improvement in gain is noticed around the center frequency of the wideband proposed hexagonal slot antenna. The lower gain is a common property of the wide slot antennas which may be suitable for receiver applications. This is probably due to non-uniform aperture field distribution in the large slot area.

Discussion

In this study, a printed hexagonal wide slot antenna fed by a 50Ω microstrip feed line loaded with hexagonal tuning stub towards bandwidth enhancement has been proposed and experimentally investigated. A hexagonal slot excited by simple 50Ω microstrip feed line with straight section tuning stub having a bandwidth of 470MHz is considered as reference antenna. The experimental results show that 2:1 VSWR bandwidth can be significantly improved by using the hexagonal tuning stub with a straight section. For the proposed antenna(Antenna4) with stub parameters d=24mm and ℓ=0.5mm, 2:1 VSWR bandwidth of 1751MHz(1921MHz-3672MHz) has been obtained which is about 62.63% with respect to centre frequency around 2796MHz. The available bandwidth is around four times greater than that of the reference antenna. The antenna can be useful for the UMTS(1.920GHz- 2.170GHz), 2.4GHz WLAN and 2.5GHz/3.5GHz WiMAX multi-band operations in various wireless communication services as receiver element.

Studies on Bandwidth Enhancement of a Printed Hexagonal Wide Slot Antenna Through Rotation of Slot (Chattopadhyay et al, 2013)

In this section, a simple variation of hexagonal wide slot antenna through rotation of slot has been presented towards more enhanced impedance bandwidth. The purpose is to design an antenna suitable for use in all the WLAN (2.4GHz/5.2GHz/5.8GHz) and WiMAX (2.5GHz/3.5GHz/5.5GHz) band applications. The proposed design of hexagonal shaped wide slot antenna excited by a microstrip line loaded by hexagonal tuning stub having a straight section, described in previous section, has been modified by rotation of wide slot excited by suitable geometry of hexagonal stub.

By choosing proper angle of rotation of the slot along with suitable geometry of hexagonal shaped stub section, more resonant modes will be excited. Within the operating bandwidth the availability of large number of resonant modes having similar radiation patterns with same polarization planes are resulting in significant bandwidth enhancement.

Antenna Configuration

The geometry and the dimensions of the proposed microstrip line fed hexagonal wide slot antenna are shown in Figure 18. The proposed antenna has been printed on an FR4 substrate with a thickness of 1.6mm, and relative permittivity 4.4. The hexagonal shaped wide slot has dimensions D1 (46.6mm) and D2 (53.7mm). The wide slot is fed by a 50Ω microstrip line loaded by hexagonal tuning stub, which is printed on opposite side of FR4 substrate. The printed hexagonal wide slot is rotated by an angle of 30° as compared to the reference antenna shown in Figure 19(a). To get matching between the slot and the feed, suitable hexagonal stub with a straight section is designed. The feed line is having length (Lf) 28.15mm and width (Wf) 3mm.

The tuning stub has been placed symmetrically with respect to the centre y-axis of the wide slot. The tuning stub has a straight section length ℓ and one hexagonal section having dimensions d_1 and d_2. For simplicity the width of the straight section of the stub has been maintained same as the width of the 50Ω microstrip line. By selecting suitable dimensions of d_1, d_2 and ℓ of the tuning stub along with the rotation angle of the hexagonal wide slot, acceptable impedance matching over a

Figure 18. Configuration of proposed microstrip line fed hexagonal slot antenna with thirty degree rotation of slot (Chattopadhyay et al, 2013)

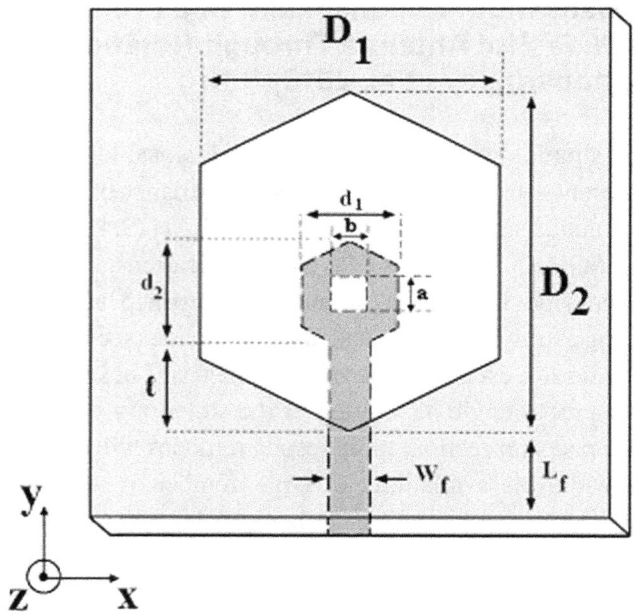

larger bandwidth can be obtained. To achieve acceptable impedance matching over a wide bandwidth easily, a square slot having dimensions a×b mm is etched out on the hexagonal section of the tuning stub. Antenna1, Antenna2, Antenna3 and Antenna4 has been created for different dimensions of d_1, d_2 and ℓ for fixed slot dimensions D_1, D_2, slot rotation angle 30°, feed length (L_f) and feed width (W_f). The reference antenna is shown in Figure 19(a) having zero degree rotation of the wide slot along with a straight section tuning stub.

This reference antenna is identical to the references used in previous investigations on wide slot antenna. The parameters of the reference antenna are the same as that of the proposed antenna, except the slot rotation angle and tuning stub shape, that is a straight extension of the feed line having length of 20mm. Figure 19(b) exhibits the 30° rotated hexagonal slot with the same feed line where the straight section length of tuning stub (ℓ) is slightly adjusted to get proper impedance matching with same lower band edge frequency of the reference antenna. For the 30° rotated slot, the length of the stub(ℓ) will be 24mm. The photographic view of the top and bottom surface of the proposed fabricated configuration is shown in Figure 20(a) and 20(b).

Figure 19a. Reference wide slot antenna with zero degree rotation of slot (Chattopadhyay et al, 2013)

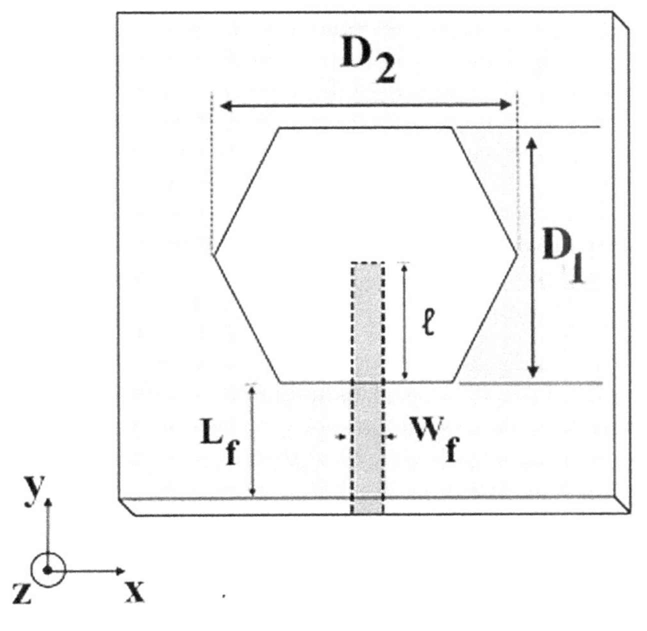

Figure 19b. Hexagonal slot antenna with thirty degree rotation of slot (Chattopadhyay et al, 2013)

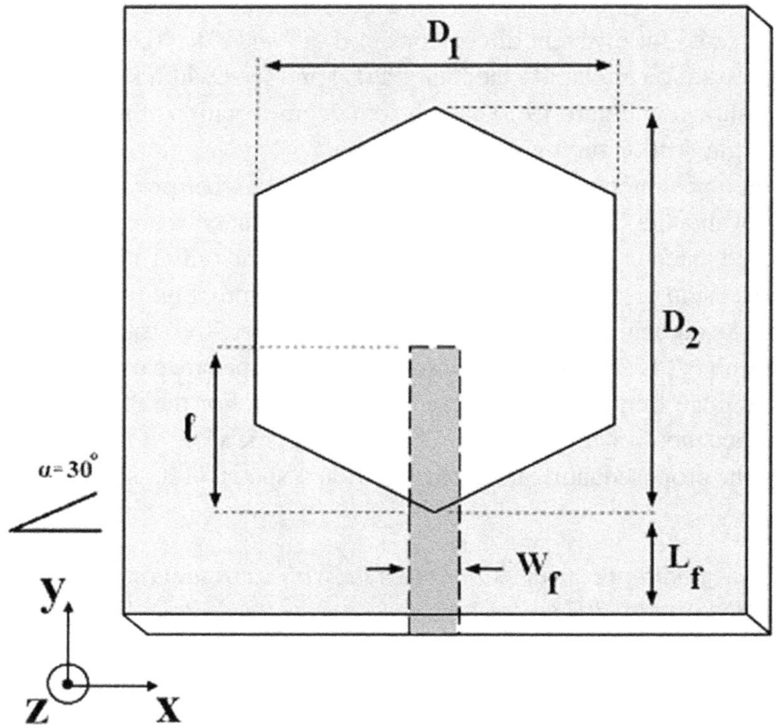

Discussion on Obtained Results

In this study, the impedance, radiation and gain characteristics of the proposed antenna has been exhibited. For the reference antenna, with straight section tuning stub the obtained bandwidth is 470MHz(1.95GHz-2.42GHz). With the rotation of hexagonal slot having rotation angle α=30°, bandwidth enhances by an amount of about 1070MHz(1.95GHz-3.02GHz), maintaining the same lower edge frequency of 1950MHz with the straight section tuning stub length(ℓ) adjusted to 24mm. This is the largest available impedance bandwidth with straight section tuning stub. Further study is carried out on the same rotated configuration towards more enhancement in bandwidth, with the hexagonal tuning stub having a straight section.

The simulation results on the study of VSWR responses of different configurations i.e Antenna1, Antenna2, Antenna3 and Antenna4 by varying the parameters d_1, d_2 and ℓ of the hexagonal stub, are compared with respect to the measured VSWR response of the reference antenna and the antenna with 30° rotation of hexagonal

Figure 20a. Slot-side photograph of the prototype (Chattopadhyay et al, 2013)

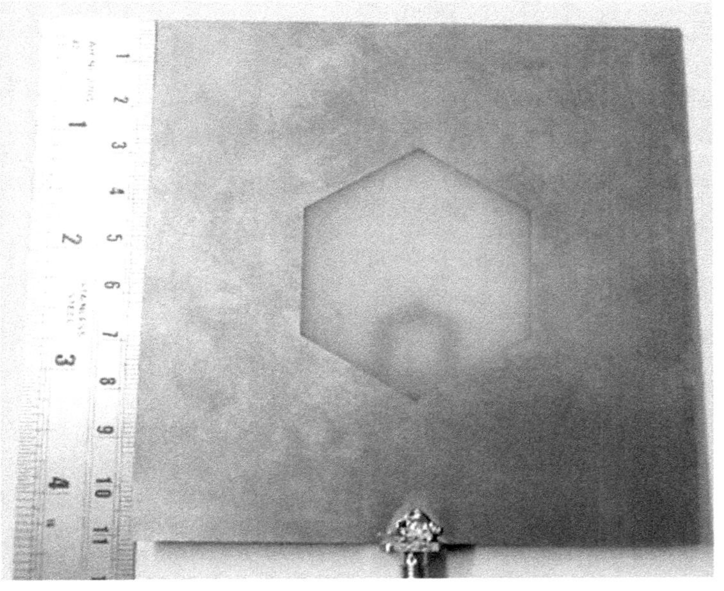

Figure 20b. Feed-side photograph of the prototype (Chattopadhyay et al, 2013)

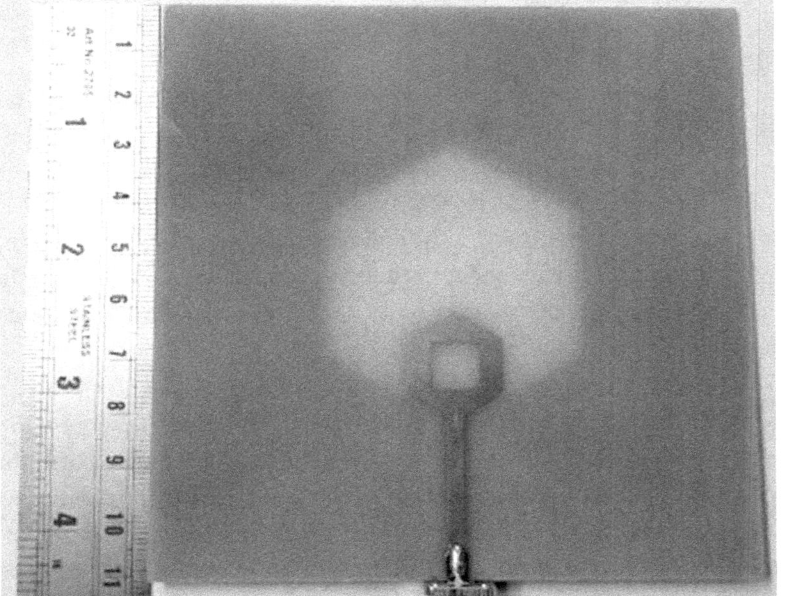

Figure 21. Simulated VSWR response of different configurations with measured result of reference antenna (Chattopadhyay et al, 2013)

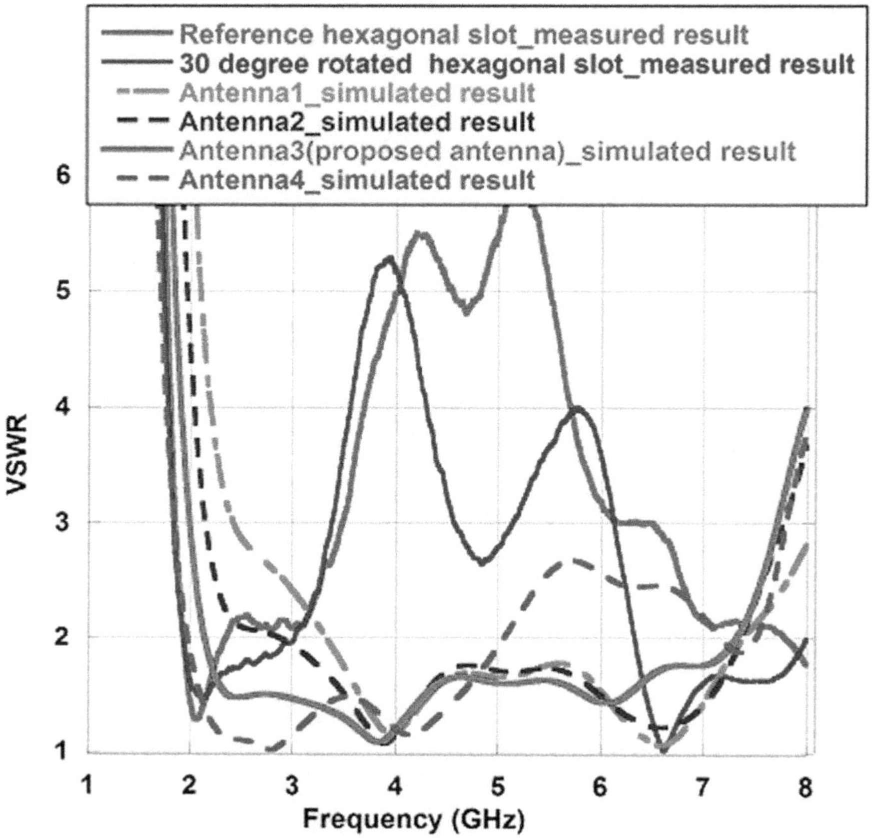

**For a more accurate representation see the electronic version.*

slot having straight section tuning stub. The comparison is shown in Figure 21. The simulation results of different configurations in comparison with the measured data of reference antenna and antenna with the slot rotation angle 30° having straight section tuning stub are shown in Table 4.

From the exhibited results in Table 4 it is observed that for Antenna3, the largest bandwidth of 5175MHz is obtained with respect to the centre frequency of 4728MHz. So the Antenna3 is considered as proposed antenna. Another study is carried out for different rotation angles of hexagonal slot along with the similar rotation of hexagonal section of stub. By varying the rotation angle of slot (α) and hexagonal stub section for different straight section length of stub (ℓ), the design parameters and the corresponding simulated data are listed in Table 5. From the obtained results it can be seen that the proposed antenna with α=30° and ℓ=1.5mm

Table 4. Performance study of different configurations including proposed antenna(antenna3) and reference antenna defined by 2:1 VSWR bandwidth (Chattopadhyay et al, 2013)

Antenna Types	Dimensions of Hexagonal stub (d_1,d_2)mm	Dimensions of Square slot on Hexagonal stub (a,b)mm	Straight Section length of stub (ℓ)mm	VSWR Bandwidth (MHz, %)	Centre Frequency fc (MHz)
Reference Hexagonal Slot	---------------	-----------------	20	470, 21.51	2185
Rotated Hexagonal Slot (30 degree)	---------------	------------------	24	1070, 43.05	2485
Antenna1	13.88,15.17	5, 5	1.8	4081, 76.26	5351
Antenna2	15.63, 17.2	5, 5	1.8	4511, 88.31	5108
Antenna3 (Proposed Antenna)	17.4, 19.15	9, 9	1.5	5175, 109.45	4728
Antenna4	19.05, 21.1	9, 9	2.8	3069, 87.26	3517

Proposed antenna D_1= 46.6mm, D_2=53.7mm

Table 5. performance study of different configurations for different rotation angles defined by 2:1 VSWR bandwidth. (Chattopadhyay et al, 2013)

Rotation Angle (α in degree)	Straight section length of Tuning stub (ℓ in mm)	f_H (MHz)	f_L (MHz)	Centre Frequency f c (MHz)	VSWR Bandwidth (MHz, %)
0 ≈60	0.5	3586	1950	2768	1636, 59.1
20≈80	1.5	5898	1990	3944	3908, 99.08
30≈90	1.5	7315	2140	4728	5175, 109.45
40	1.5	5729	1990	3859	3739, 96.89
45	1.5	5890	1942	3916	3948, 100.81
50	1.0	6550	1942	4246	4608, 108.52
70	0.8	6542	1942	4242	4600, 108.44

has largest impedance bandwidth of 5175MHz which is about eleven times larger than that of the reference antenna(470MHz) and about five times that of the 30° rotated hexagonal slot configuration(1070MHz) with straight section tuning stub.

For explaining the phenomenon of enhancement of bandwidth, experimental results of impedance plots on smith chart of the reference antenna and proposed antenna are presented in Figure 22. With reference to the said figures, the proposed

Figure 22. Measured Impedance plot on smith chart of (a) reference antenna and (b) proposed antenna (Chattopadhyay et al, 2013)

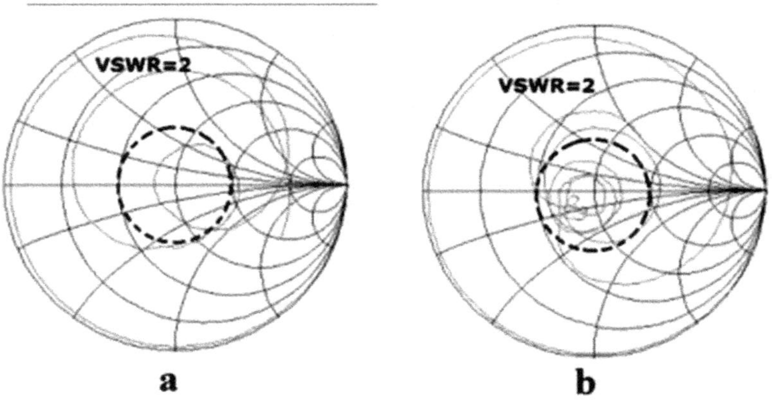

antenna not only exhibits different resonant loops close to the centre of the smith chart within VSWR=2 circle but also it has the tightest coupling loop to provide a great enhancement in impedance bandwidth. The bandwidth is enhanced because the other resonant modes in the vicinity of the fundamental mode are excited due to presence of hexagonal stub having suitable dimension.

The magnetic current distributions shown in Figure 23(a-g) exhibit the excitation of different coupled resonant modes having different path lengths excited by the geometry of hexagonal tuning stub and hexagonal shaped slot.

Figure 23a. Magnetic current distribution of the proposed antenna at 2455MHz (Chattopadhyay et al, 2013)

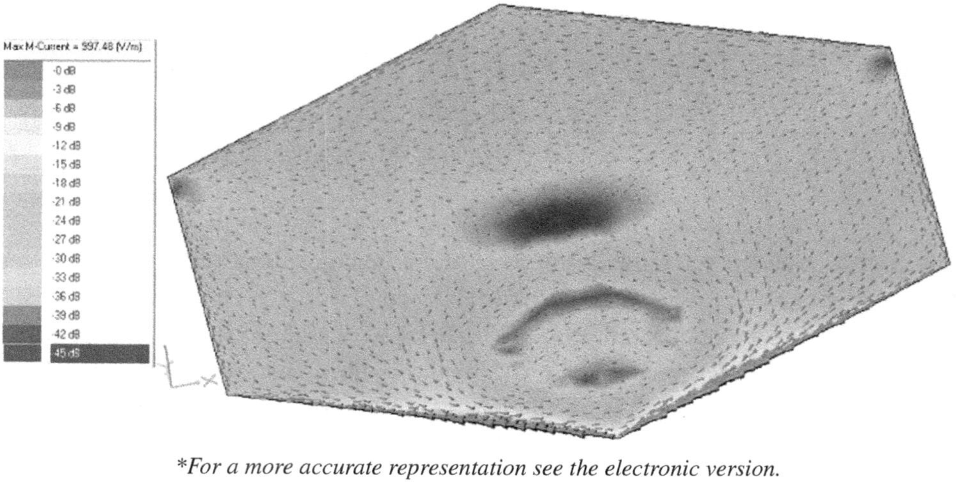

**For a more accurate representation see the electronic version.*

Figure 23b. Magnetic current distribution of the proposed antenna at 3155MHz (Chattopadhyay et al, 2013)

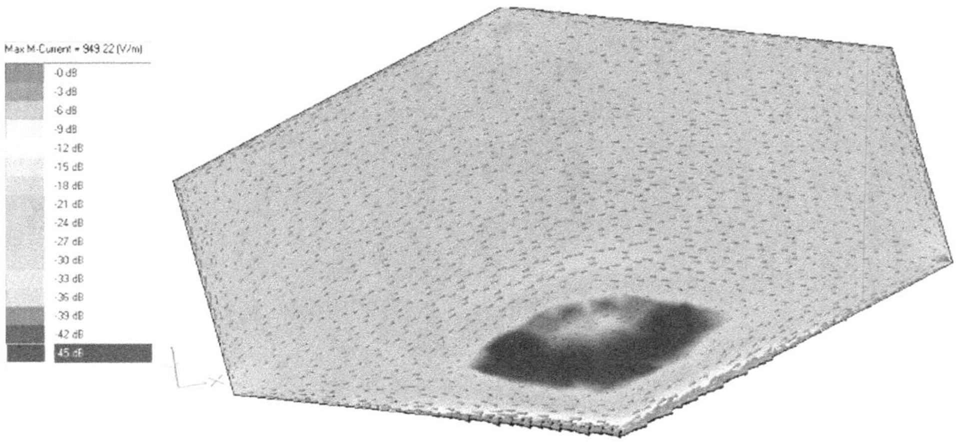

**For a more accurate representation see the electronic version.*

Figure 23c. Magnetic current distribution of the proposed antenna at 3750MHz (Chattopadhyay et al, 2013)

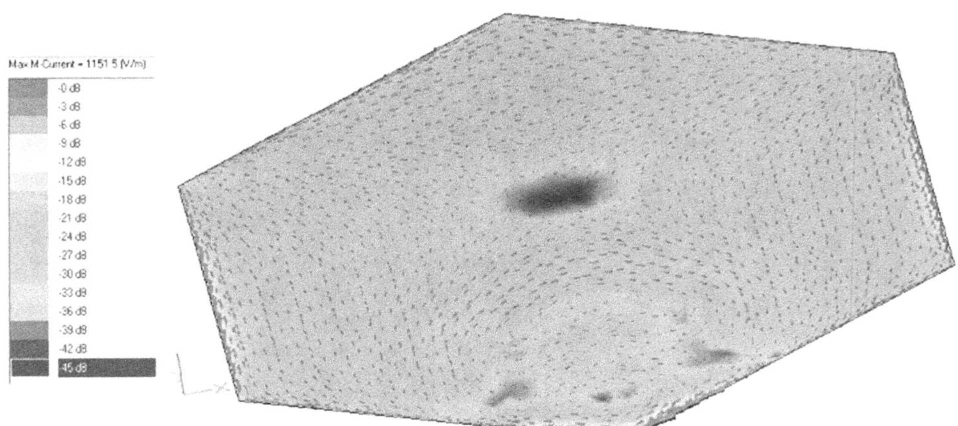

**For a more accurate representation see the electronic version.*

The optimal dimension of hexagonal tuning stub is d_1=17.4mm and d_2=19.15mm to get largest bandwidth which is found to be smaller than half of the slot dimensions D_1 and D_2. However, the length ℓ of straight section of tuning stub is also to be adjusted corresponding to specified hexagonal stub dimensions to get suitable matching by optimum coupling with the slot.

Figure 23d. Magnetic current distribution of the proposed antenna at 4590MHz (Chattopadhyay et al, 2013)

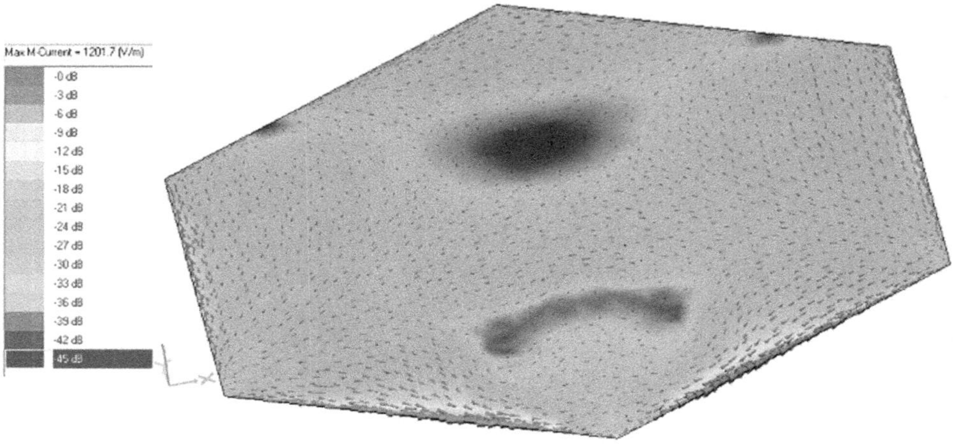

**For a more accurate representation see the electronic version.*

Figure 23e. Magnetic current distribution of the proposed antenna at 5500MHz (Chattopadhyay et al, 2013)

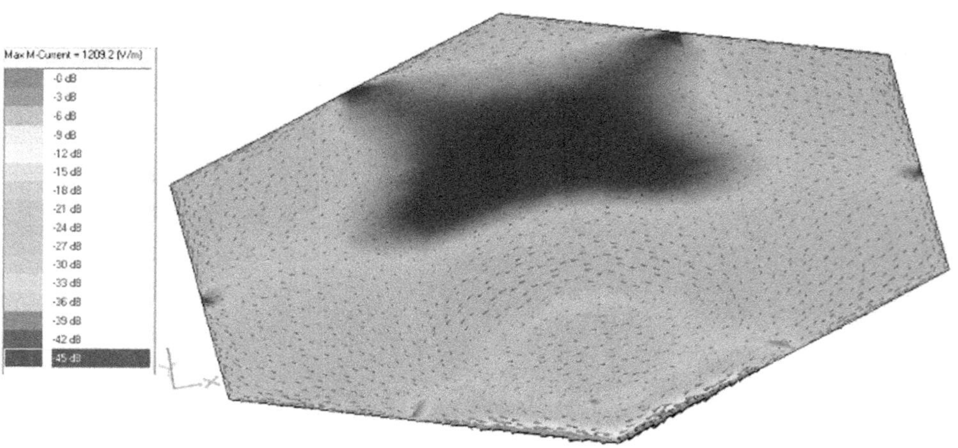

**For a more accurate representation see the electronic version.*

The experimental study on the proposed fabricated antenna configuration has been carried out by Agilent Technologies PNA-L Network Analyzer(Model No. N5230A,10MHz–20GHz). The simulation results and experimentally measured VSWR responses of the proposed antenna with respect to the reference antenna for 2:1 VSWR bandwidth is shown in Figure 24. The measured VSWR response of the thirty degree rotated slot with straight section tuning stub is also shown

Figure 23f. Magnetic current distribution of the proposed antenna at 6200MHz (Chattopadhyay et al, 2013)

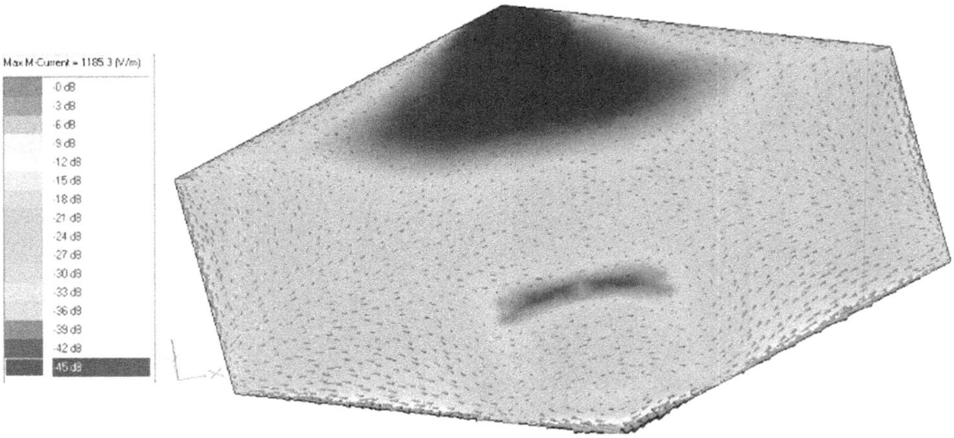

**For a more accurate representation see the electronic version.*

Figure 23g. Magnetic current distribution of the proposed antenna at 6750MHz (Chattopadhyay et al, 2013)

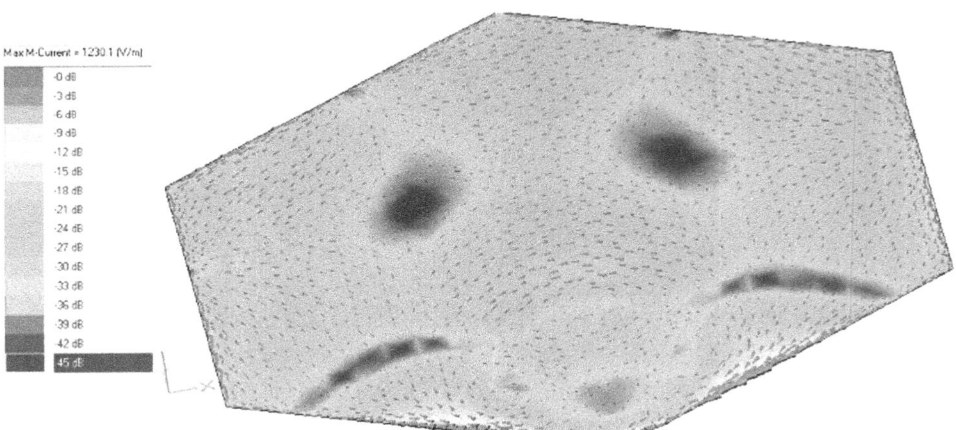

**For a more accurate representation see the electronic version.*

for comparison. From the experimental result of VSWR study on the proposed antenna, the available bandwidth is 5165 MHz (2.117GHz-7.282GHz) around centre frequency of 4699.5MHz. The available bandwidth is about 110% with respect to centre frequency.

The antennas with different rotation angle of slot having suitable hexagonal section and straight section of stub show the same radiation characteristics. Thus

Figure 24. Studies on simulated and measured VSWR response of proposed antenna along with the reference hexagonal slot antenna (Chattopadhyay et al, 2013)

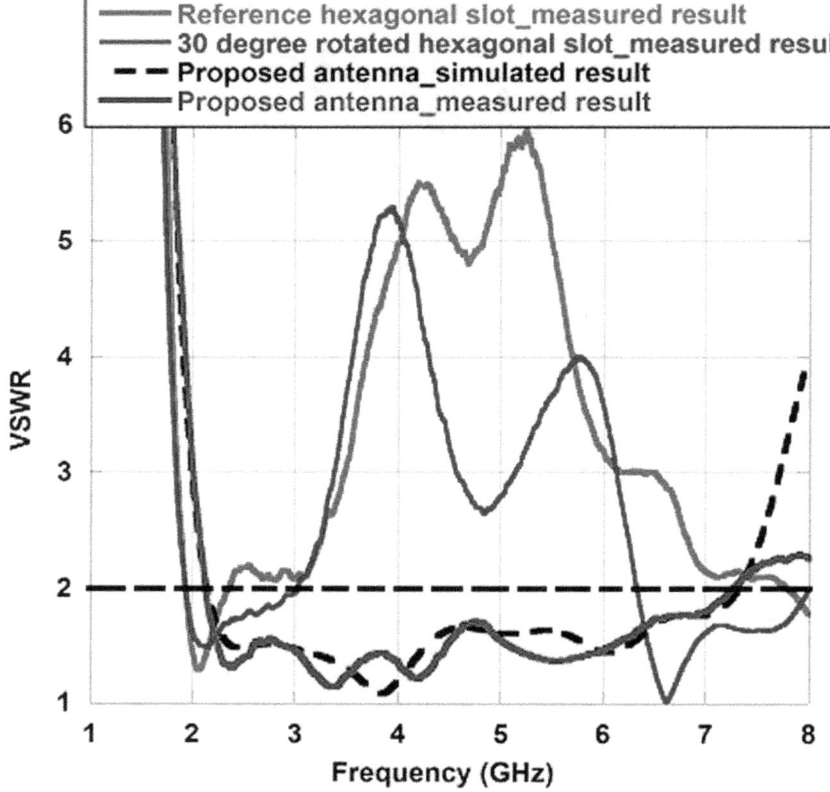

**For a more accurate representation see the electronic version.*

the radiation patterns only for the proposed antenna are plotted by E-plane(y-z plane) and H-plane(x-z plane). Figure 25(a-e) show the typical measured radiation patterns at the operating frequencies 2500MHz, 3500MHz, 5200MHz, 5500MHz and 5800MHz.

From the Figure 25(a-e), it is observed that the printed wide slot antenna without a reflecting plate is a bidirectional radiator. So the radiation patterns on both sides of this antenna are similar. From these results it can be seen that for the operating frequencies across the impedance bandwidth within 2117MHz to 7282MHz, the antenna maintains same polarization and exhibits similar broadside radiation characteristics. This has also been observed that cross polarization levels are higher in H-plane (x-z plane) as compared to E-plane (y-z plane) because of the large height of the slot. Higher order modes have been found to be excited in cross polarization direction.

Figure 25a. E-plane and H-plane radiation patterns at 2500MHz (2.5 GHz) (Chattopadhyay et al, 2013)

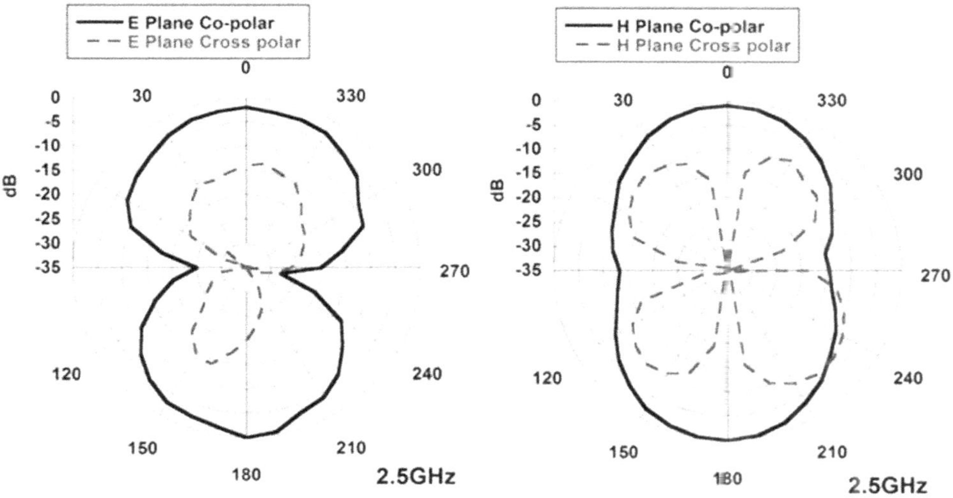

Figure 25b. E-plane and H-plane radiation patterns at 3500MHz (3.5 GHz) (Chattopadhyay et al, 2013)

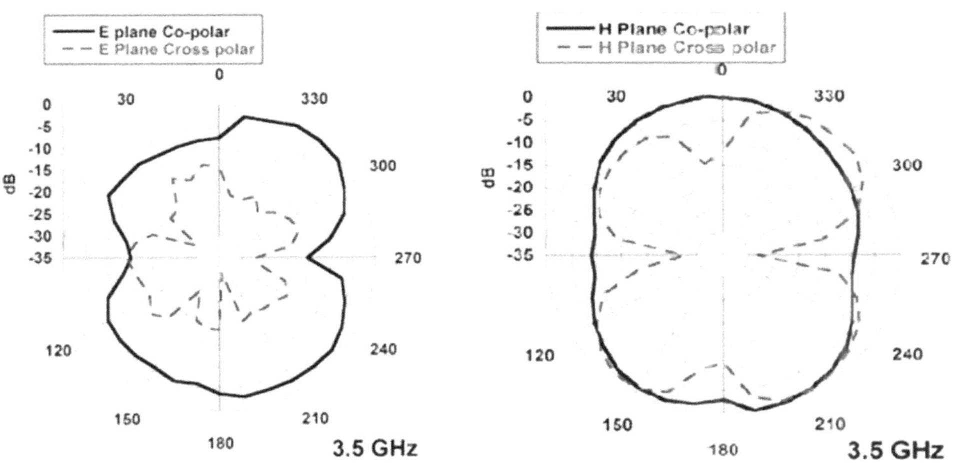

The variation in peak gain of the proposed antenna within 2:1 VSWR bandwidth is shown in Figure 26. The variation of peak gain within the wide bandwidth is observed to be about 3.3 dBi. Within the wide bandwidth range the radiation pattern remains broadside and stable. The maximum value of the available gain is 5.6dBi.

Figure 25c. E-plane and H-plane radiation patterns at 5200MHz (5.2 GHz) (Chattopadhyay et al, 2013)

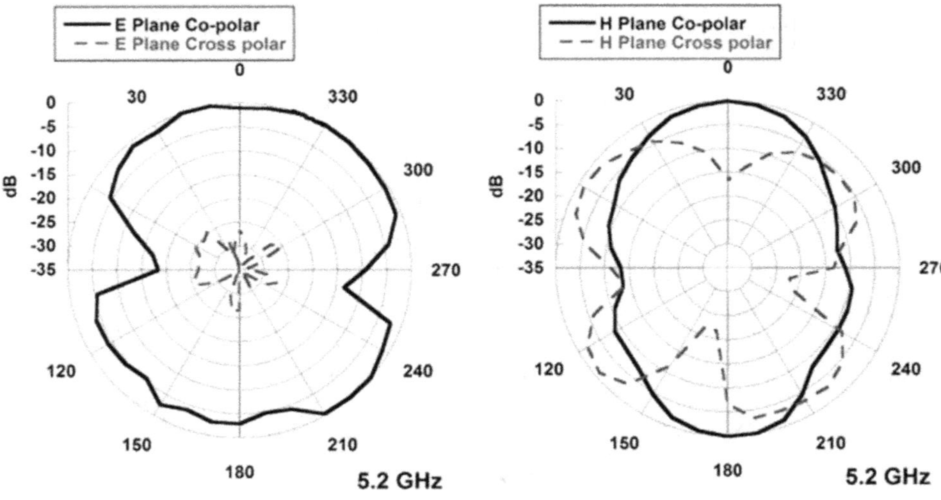

Figure 25d. E-plane and H-plane radiation patterns at 5500MHz (5.5GHz) (Chattopadhyay et al, 2013)

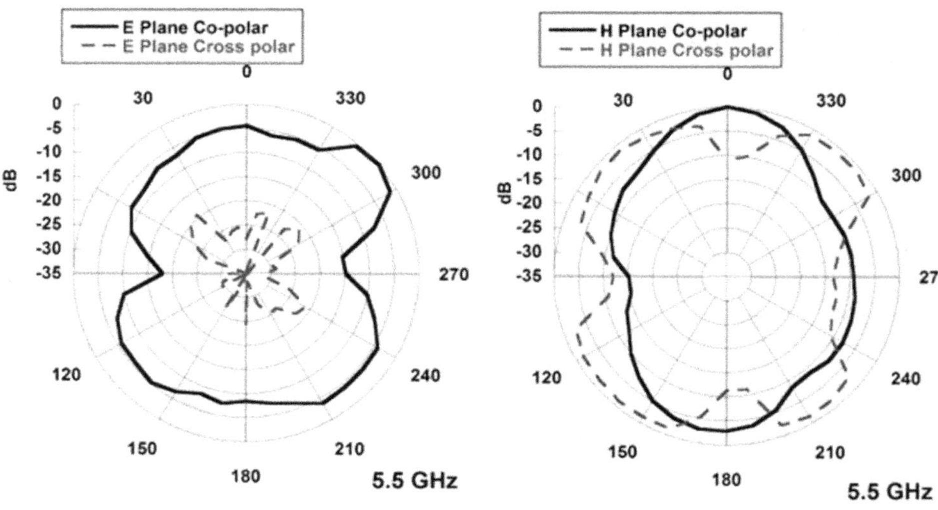

Figure 25e. E-plane and H-plane radiation patterns at 5800MHz (5.8 GHz) (Chattopadhyay et al, 2013)

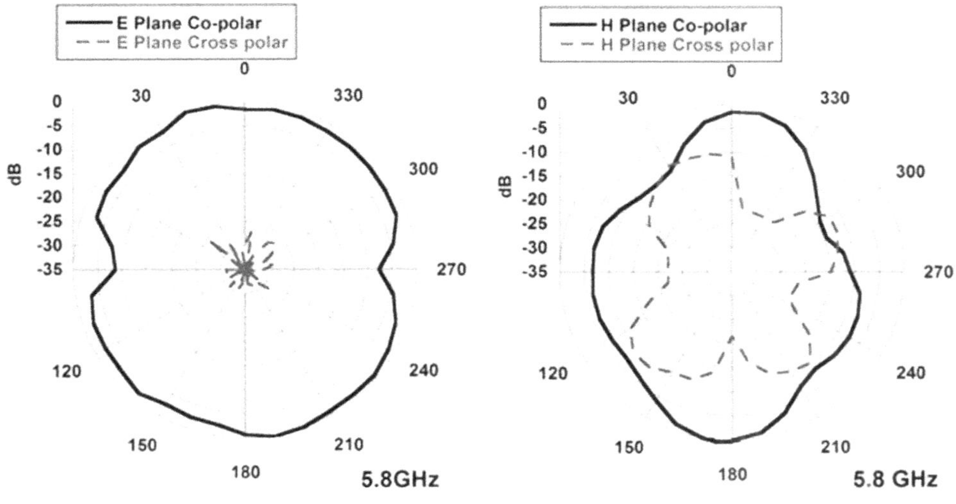

Figure 26. Peak gain response of the proposed antenna (Chattopadhyay et al, 2013)

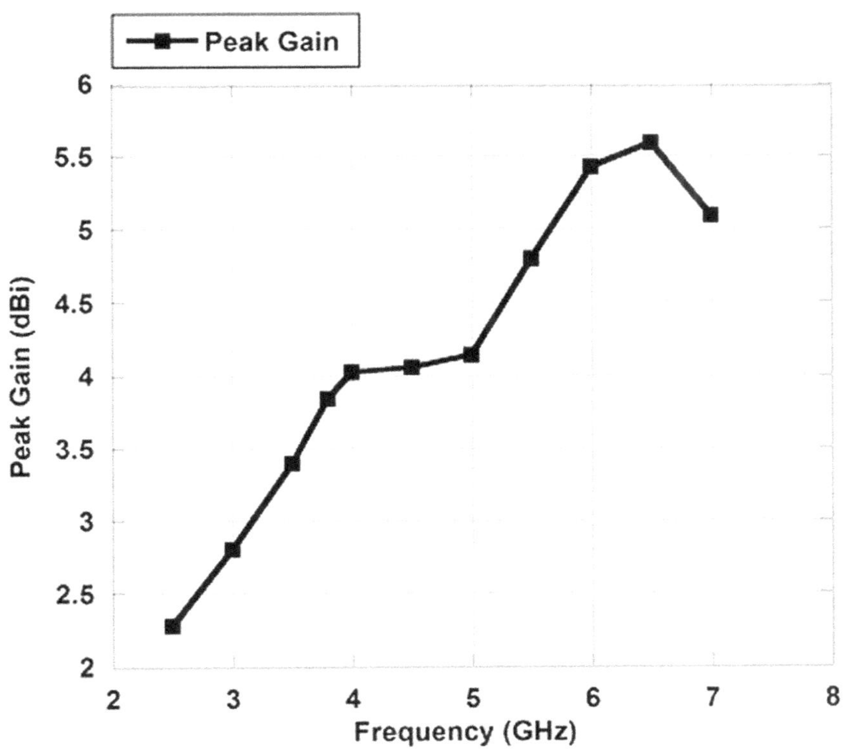

Discussion

In this investigation, a printed hexagonal wide slot antenna having 30° rotation angle of slot and excited by 50Ω microstrip line, loaded with hexagonal tuning stub for bandwidth enhancement has been presented. Significant improvement in bandwidth to the tune of 110% with respect to centre frequency has been achieved. The impedance bandwidth determined by VSWR=2 is found to have 5165MHz(2.117GHz-7.282GHz) for the proposed antenna around the centre frequency of 4699.5MHz. The obtained bandwidth is about eleven times wider than that of the corresponding reference antenna. Within this wide impedance bandwidth, the gain variation is about 3.3dBi. The broadside radiation pattern is available for the entire bandwidth range. The proposed antenna configuration is suitable for all the WLAN(2.4GHz/5.2GHz/5.8GHz) and WiMAX (2.5GHz/3.5GHz/5.5GHz) application bands.

REFERENCES

Agrawall, N. P., Kumar, G., & Ray, K. P. (1998). Wide Band Planar Monopole Antennas. *IEEE Transactions on Antennas and Propagation*, *46*(2), 294–295. doi:10.1109/8.660976

Chattopadhyay, K., RoyChoudhury, S., Das, S., & Bhadra Chaudhuri, S.R. (2013). Wideband microstrip-line-fed hexagonal slot antenna for WiMAX and Wireless Local Area Network applications. *The Journal of Engineering, IET, UK.*

Chattopadhyay, K., Das, S., & Bhadra Chaudhuri, S. R. (2012). Bandwidth Enhancement of a Microstrip line Fed Hexagonal Wide-slot Antenna Using Fork-like Tuning Stub. *International Journal of Soft Computing and Engineering, 2*(2).

Chattopadhyay, K., Mitra, T., Das, D., Das, S., & Bhadra Chaudhuri, S. R. (2013). Bandwidth Enhancement of Hexagonal Wide-slot Antenna Using a Tuning Stub. *Journal of the Institution of Electronics and Telecommunication Engineers*, *59*(6), 732–738. doi:10.4103/0377-2063.126976

Chen, W. F., Tsai, C. C., & Huang, C. Y. (2008). Compact Wide- slot antenna for ultra-wideband Communications. *Electronics Letters*, *44*(15), 892. doi:10.1049/el:20081563

Dastranj, A., & Abiri, H. (2010). Bandwidth Enhancement of Printed E-Shaped Slot Antennas Fed by CPW and Microstrip Line. *IEEE Transactions on Antennas and Propagation*, *58*(4), 1402–1407. doi:10.1109/TAP.2010.2041164

Dastranj, A., Imani, A., & Moghaddasi, M. N. (2008). Printed Wide Slot Antenna for Wideband Applications. *IEEE Transactions on Antennas and Propagation*, *56*(10), 3097–3102. doi:10.1109/TAP.2008.929459

Jan, J. Y., & Su, J. W. (2005). Bandwidth Enhancement of a Printed wide slot Antenna With a Rotated Slot. *IEEE Transactions on Antennas and Propagation*, *53*(6), 2111–2114. doi:10.1109/TAP.2005.848518

Jan, J. Y., & Wang, L. C. (2009). Printed Wideband Rhombus Slot Antenna With a Pair of Parasitic Strips for Multiband Applications. *IEEE Transactions on Antennas and Propagation*, *57*(4), 1267–1270. doi:10.1109/TAP.2009.2015859

Jang, Y. W. (2001). Experimental study of Large Bandwidth Three-Offset Microstripline-Fed Slot Antenna. *IEEE Microwave and Wireless Components Letters*, *11*(10), 425–427. doi:10.1109/7260.959318

Kahrizi, M., Sarkar, T. K., & Maricevic, Z. A. (1993). Analysis of a wide Radiating slot in the Ground plane of a Microstrip line. *IEEE Transactions on Microwave Theory and Techniques*, *41*(1), 29–37. doi:10.1109/22.210226

Kim, H., & Yoon, Y. J. (2005). Microstrip-Fed Slot Antennas With Suppressed Harmonics. *IEEE Transactions on Antennas and Propagation*, *53*(9), 2809–2817. doi:10.1109/TAP.2005.854557

Li, P., Liang, J., & Chen, X. (2006). Study of printed elliptical/circular slot antennas for ultrawideband applications. *IEEE Transactions on Antennas and Propagation*, *54*(6), 1670–1675. doi:10.1109/TAP.2006.875499

Liu, Y.F., Lau, K.L., Xue, Q., & Chan, C.H. (2004). Experimental studies of Printed Wide-Slot Antenna for Wide-Band Applications. *IEEE Antennas and wireless propagation Letters, 3*, 273–275.

Qu, S. W., Li, J.-L., Chen, J. X., & Xue, Q. (2007). Ultrawideband Strip-Loaded Circular Slot Antenna With Improved Radiation Patterns. *IEEE Transactions on Antennas and Propagation*, *55*(11), 3348–3353. doi:10.1109/TAP.2007.908847

Sung, Y. (2011). A Printed Wide-Slot Antenna With a Modified L-Shaped Microstrip Line for Wideband Applications. *IEEE Transactions on Antennas and Propagation*, *59*(10), 3918–3922. doi:10.1109/TAP.2011.2163748

Sung, Y. (2012). Bandwidth Enhancement of a Microstrip Line-Fed Printed Wide-Slot Antenna with a Parasitic Centre Patch. *IEEE Transactions on Antennas and Propagation*, *60*(4), 1712–1716. doi:10.1109/TAP.2012.2186224

Sze, J. Y., & Wong, K. L. (2001). Bandwidth Enhancement of a Microstrip line fed Printed wide slot Antenna. *IEEE Transactions on Antennas and Propagation*, *49*(7), 1020–1024. doi:10.1109/8.933480

Yoshimura, Y. (1972). A Microstripline slot Antenna. *IEEE Transactions on Microwave Theory and Techniques*, 760–762.

Chapter 5
Fractal-Inspired Ultra-Wideband Diversity Slot Antenna for Wireless Communication Systems

Anirban Karmakar
Tripura University, India

Anuradha Saha
Netaji Subhash Engineering College, India

ABSTRACT

A novel compact ultra-wideband (UWB) multiple-input multiple-output (MIMO) slot antenna with band notch characteristics is presented for portable wireless UWB applications. The antenna comprises of co-planar waveguide feed (CPW) and two radiating monopoles oriented in orthogonal orientation for providing orthogonal radiation patterns. A Minkowski fractal parasitic stub along with a Minkowski fractal grounded stub has been placed at 45° between the monopoles to reduce the coupling between them, which in turn establishes high isolation between the radiators. An excellent band notch characteristic is obtained at 5.5 GHz by etching a modified E-shaped compact slot on the radiators. Results show that the designed antenna meets -10 dB impedance bandwidth and -17 dB isolation throughout the entire operating band (3.1 -12 GHz). Novelty of this design lies in improving isolation using fractal which occupies less space in compared to other isolation mechanisms in MIMO structures. The simulated and measured results depict that the proposed antenna is convenient for MIMO diversity systems.

DOI: 10.4018/978-1-5225-8531-2.ch005

Copyright © 2019, IGI Global. Copying or distributing in print or electronic forms without written permission of IGI Global is prohibited.

INTRODUCTION

In digital communication systems, MIMO (Multiple Input Multiple Output) has created a revolutionary importance. All types of wireless technologies faced some difficulties such as multipath, signal fading, limited spectrum, interference etc. MIMO has solved most of the difficulties by providing higher data throughput to multipath, increase in range and reliability etc., without consuming additional spectrum. IEEE standard of MIMO technology is 802.11n. MIMO antenna introduces antenna diversity like pattern, polarization and spatial diversity to increase the transmitted signal power and to improve Signal to Noise Ratio (SNR). To increase the data rate in MIMO, spatial multiplexing is used. In addition beam forming is used to enhance the data rate. MIMO has a wide range of applications in mobile, Digital Television (DTV), Metropolitan Area Network (MAN), Wireless Local Area Network (WLAN) etc. Basically the system which have multiple transmit and multiple receive antennas is known as MIMO system. The wireless technology has remarkably improved the communication range and maximizes the capacity by using MIMO system. MIMO system analyses multi-path propagation by using different transmission paths to the receiver. These paths can be used to provide redundancy of transmitted data, which improves the reliability of transmission or by increasing the number of simultaneous transmitted data streams and increasing the data rate of the system. The capacity of these systems depends upon increasing the number of transmit antennas and also depends upon the number of receiving antennas, which must be greater than or equal to the number of transmitting antennas. So an enhancement in capacity means capability of faster communication and as a consequence this capacity improvement over conventional one-antenna systems has fuelled intense research interest in MIMO techniques, which results the development of MIMO systems.

Mathematically, the MIMO communication is performed through a matrix, so it may be possible to transmit multiple parallel signal streams or data streams simultaneously in the same frequency band and so that increase in spectral efficiency

Figure 1. Block Diagram of a MIMO System

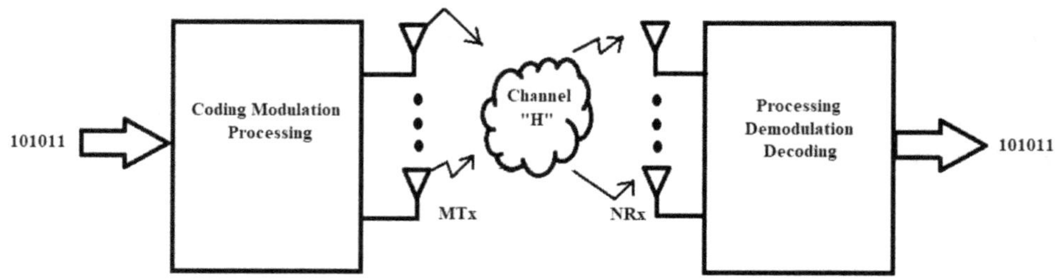

occurs. This technique is known as Spatial Multiplexing, which is shown in Figure 2. The data stream is encoded with vector encoder and then transmitted by M transmitters. The coded signal experienced distortion while passing through MIMO radio channel. The receiver has N number of antennas. Each antenna element receives the coded signals from all M transmit antennas, and in tern the received signals exhibit inter-channel interference. The received signals are then down converted to the base band and sampled per symbol interval. The MIMO processing unit recovers the transmitted data streams from the sampled base-band signals.

There are M number of transmitting antenna elements and N number of receiving antenna elements. So there are $M * N$ sub-channels between transmitter and receiver. Each sub-channel represents a selective fading and as a result, it is modelled as a FIR filter with complex coefficients. In the previous case, the received signal by the jth receive antenna can be shown as:

$$y_j = \sum_{i=1}^{M} h_{ij} x_i + n_j \tag{1}$$

where, x_i =transmitted signal from i^{th} antenna, y_j= received signal at j^{th} antenna.

Variable n_j indicates samples of symmetric complex Gaussian noise with variance σ_n^2 at j^{th} receiver. The fading channel is explained as a sum of complex paths h_{ij} between the receiving and transmitting antennas. The complex gain coefficient h_{ij} satisfies with Gaussian distribution. The matrix form of equation (1) will be,

$$y = Hx + n \tag{2}$$

where, y represents the column vector of the received signals, H is channel matrix, x is the column vector of the transmitted signal, n is a column vector of the additive white Gaussian noise. The channel capacity can be expressed as (Sharawi, 2017),

Figure 2. Block diagram of MIMO system using Spatial Multiplexing

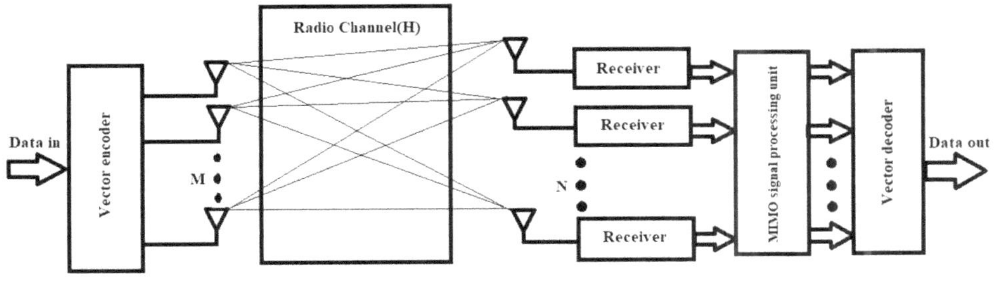

$$C = \log_2 \left[\det \left(I + \frac{\rho}{M} HH^* \right) \right] \tag{3}$$

where, $\rho = \frac{P}{\sigma_n^2}$ and P is the cumulative power transmitted by all antennas and σ_n^2 represents the noise power at each receive antenna. H is the matrix presenting quasi-static channel response and the superscript * denotes transpose conjugate of channel matrix H. When uncorrelated paths are assumed, all eigen values of the product HH* are non-zero and approximately equal. Hence the capacity can be expressed as:

$$C = \sum_{i=1}^{M} \log_2 \left(1 + \frac{\lambda_i}{M} \rho \right) \approx A_{\min} \log_2 \left(1 + \frac{N}{A_{\min}} \rho \right) \tag{4}$$

where, $A_{\min} = \min(M,N)$ .The availability capacity in MIMO channel increases linearly with $A_{\min}$, without increasing transmit power.

In MIMO systems, the signals transmitted by multiple antenna elements are uncorrelated. In reality, the current induced on one antenna produces a voltage at the terminals of nearest antenna element, which is known as mutual coupling. So mutual coupling always exists between nearby antenna elements. However, for MIMO applications, the mutual coupling should be reduced as low as possible. The port-to-port isolation in MIMO is defined as the transmission of power between two of the input ports of the multiport antenna under test (Sharawi, 2017). It is denoted by $|S_{21}|$ parameter. Isolation is a positive quantity which can be explained as (Sharawi, 2017):

$$Isolation = -10 \log_2 |S_{21}|^2 \tag{5}$$

To enhance the energy radiated by an antenna, it should be confirmed that a very little amount of transmitted energy is lost into the ports of other antennas, which are terminated by the matched terminations. As antenna isolation is directly related to the antenna efficiency (η), hence the MIMO systems require the $|S_{21}|$ to be minimized to as low as possible to enhance the isolation. The performance of MIMO systems is also reflected by the Mean effective gain (MEG) of the antenna. The MEG is a measure of the antenna gain which can be defined “as the ratio of the mean received power of the antenna and the total mean incident power". It can be expressed by the following equation as (Sharawi, 2017),

$$MEG = \int_0^{2\pi}\int_0^{\pi}\left(\frac{XPR}{XPR+1}G_\theta\left(\theta,\phi\right)P_\theta\left(\theta,\phi\right)+\frac{1}{1+XPR}G_\phi\left(\theta,\phi\right)P_\phi\left(\theta,\phi\right)\right)\sin\theta d\theta d\phi \quad (6)$$

where, P_θ and P_φ are the angular diversity functions of the incident power with respect to θ and φ directions respectively, G_θ and G_φ are the gains with respect to θ and φ directions respectively. XPR denotes the cross-polarization power gain which is expressed as (Sharawi, 2017):

$$XPR = \frac{\int_0^{2\pi}\int_0^{\pi} P_\theta\left(\theta,\phi\right)\sin\theta d\theta d\phi}{\int_0^{2\pi}\int_0^{\pi} P_\phi\left(\theta,\phi\right)\sin\theta d\theta d\phi} \quad (7)$$

The MEG is a normalized measure of the received power, where the powers (either P_θ or P_φ) are normalized as (Sharawi, 2017):

$$\int_0^{2\pi}\int_0^{\pi} P_\theta\left(\theta,\phi\right)\sin\theta d\theta d\phi = 1 \quad (8)$$

And gains are normalized as (1):

$$\int_0^{2\pi}\int_0^{\pi}\left(G_\theta\left(\theta,\phi\right)+G_\phi\left(\theta,\phi\right)\right)\sin\theta d\theta d\phi = 4\pi \quad (9)$$

The envelope correlation coefficient (ECC) can be computed from 3D radiation patterns. Antennas with an envelope correlation coefficient less than 0.5 are known to provide significant diversity performance. The correlation coefficient can also be calculated from S-parameters using the following equation (Sharawi, 2017):

$$\rho_e = \left|\frac{S_{11}^* S_{12} + S_{21}^* S_{22}}{\sqrt{1-\left|S_{11}\right|^2-\left|S_{21}\right|^2}\sqrt{1-\left|S_{22}\right|^2-\left|S_{12}\right|^2}}\right|^2 \quad (10)$$

Diversity gain (DG) is used to quantify the performance level of any diversity techniques. Mathematically, it can be shown as (Sharawi, 2017),

$$DG = \frac{\left(SNR\right)_c}{\left(SNR\right)_r} \tag{11}$$

where, indices *c* and *r* are used for the combined and the reference. The Diversity Gain is related to the correlation coefficient (ρ_c). The higher the diversity gain the lower the correlation coefficient. The relation between DG and correlation coefficient (ρ_c) can be expressed by following equation (Sharawi, 2017),

$$DG = 10\sqrt{1 - \left|\rho\right|^2} \tag{12}$$

In place of simple reflection coefficient in MIMO systems, the array's total active reflection coefficient (TARC) can also be used so that it determines both coupling and random signal combination. In case of a port excitation, summation of the available power at all excitation ports is assumed as incident power, radiated power as transferred power, and the difference between these two as reflected power. The square root of the ratio of reflected power and incident power is defined as the TARC (Sharawi, 2017), which is mathematically expressed by:

$$\Gamma_a^t = \frac{available\ power - radiated\ power}{available\ power} \tag{13}$$

In recent years, research in the area of Ultrawideband (UWB) systems ("Federal Communications Commission", 2002) has generated a huge amount of interest among microwave engineers. In february 2002, the allocation of the frequency band between 3.1 GHz to 10.6 GHz and the Equivalent Isotropic Radiated Power (EIRP) less than -41.3 dBm/MHz for the unlicensed indoor UWB wireless communication system by the Federal Communication Commission (FCC), USA ("Federal Communications Commission", 2002) sparked a everlasting interest in industries, universities as well as government sectors. According to aforementioned regulation, UWB technology which is based on transmitting very narrow pulses with a few nanoseconds duration or less has received great attention in various fields for the short-distance (< 10 m) wireless communication systems. Its needless to mention that UWB technology has many benefits like High data rate, large channel capacity, Immunity to multipath interference, low complexity, minimal cost, low power consumption along with Coexistence with other wireless communication systems ("Federal Communications Commission", 2002).

UWB communication system need to use the large bandwidth which is capable of working in harsh communication channels with low SNR. Thus, these conditions offer a large channel capacity which causes high data rates in UWB. UWB communication system has another advantage of excellent immunity to multipath interference. Basically, in the narrow band communication system, the fading which is caused by reflected signals from various objects is unavoidable which can weaken the received signal up to -40 dB. So, it can be concluded that UWB communication system can resolve the fading problem and is good in multipath diversity like MIMO system.

Furthermore, UWB system is simple and is of lower cost due to the carrier-free nature of the signal transmission. However, the wideband nature of the UWB signal means it occupy frequencies commonly used as carrier frequency. Thus, the UWB signal propagates well in the transmitters and will be received well in the receivers without any requirement of RF mixing stage for the up/down-conversion. Quite low power density obtained through the FCC's radio regulation emission mask within the limit of -41.3 dBm/MHz (75 nanowatts/MHz) for UWB system is the other advantage. It causes the UWB system to consume lower power and to coexist with existing narrow-band systems with minimal or no interference which gained strong foothold in WPAN applications. Consequently, UWB communication system is a good candidate for the short distance wireless communication due to the above-mentioned advantages. Also, high data rate cable less transmissions between modern gadgets will provide new experience in home as well as in personal entertainment. Also, sensors of all types also offered an opportunity for UWB to flourish. The cost of installation and maintenance can be reduced significantly by using UWB sensor networks due to being devoid of wires. This merit is especially attractive in medical imaging systems because a UWB sensor network frees the patient from being shackled by wires and cables when extensive medical monitoring is required. UWB system also provides an excellent solution for indoor location with a much higher degree of accuracy than a conventional GPS system ("Federal Communications Commission", 2002).

In USA, the FCC approved a UWB spectral mask specified 7.5 GHz between 3.1 GHz and 10.6 GHz for communication devices and protected existing users operating within this spectrum by limiting the UWB signals EIRP level to -41.3 dBm/MHz which is known as Part 15 Limit. One of the major components in UWB system is the antenna, which is a challenging task to design due to its wide 110% impedance bandwidth. However, there are some narrowband communication systems that coexist with the UWB communication system, which severely interferes with the functioning of the UWB systems. The most notable among them is 5.15-5.825 GHz band assigned for IEEE802.11a and HIPERLAN/2 as shown in the Figure 3.

Conventional UWB technologies face the challenges like multipath fading and high bit error rate ("Federal Communications Commission", 2002). To resolve this limitation, MIMO technology has been evolved (Kaiser, Zheng, & Dimitrov, 2009). MIMO employs multiple antennas at the transmitting and receiving ends for improving the data rate and quality of communication without utilization of additional power (Kaiser, Zheng, & Dimitrov, 2009). Designing compact planar MIMO antenna with high isolation operating over the entire UWB bandwidth is a growing challenge now a day. One should also ensure that the isolation improvement mechanism should not influence the wideband impedance characteristics of the proposed antenna. Some of the commonly used techniques for suppressing mutual coupling are diversity techniques (Zhang, Lau, Sunesson, & He, 2012) which are already reported as decoupling structures such as tree like structures (Zhang, Ying, Xiong, & He, 2009), Parasitic meander-lines (Iqbal, Saraereh, Ahmad, & Bashir, 2018) and inserting stubs (Ghosh, Ghosal, Mitra, & Chaudhuri, 2016) etc. The UWB spectrum also consists of several pre-assigned (by FCC) narrow bands some of which are Wi-Max (3.3-3.7 GHz), IEEE 802.11 wireless local area network (WLAN) operating at 5.15 GHz to 5.85 GHz, etc as shown in Figure 3. Which

Figure 3. Existence of other wireless systems along with UWB ("Federal Communications Commission", 2002)

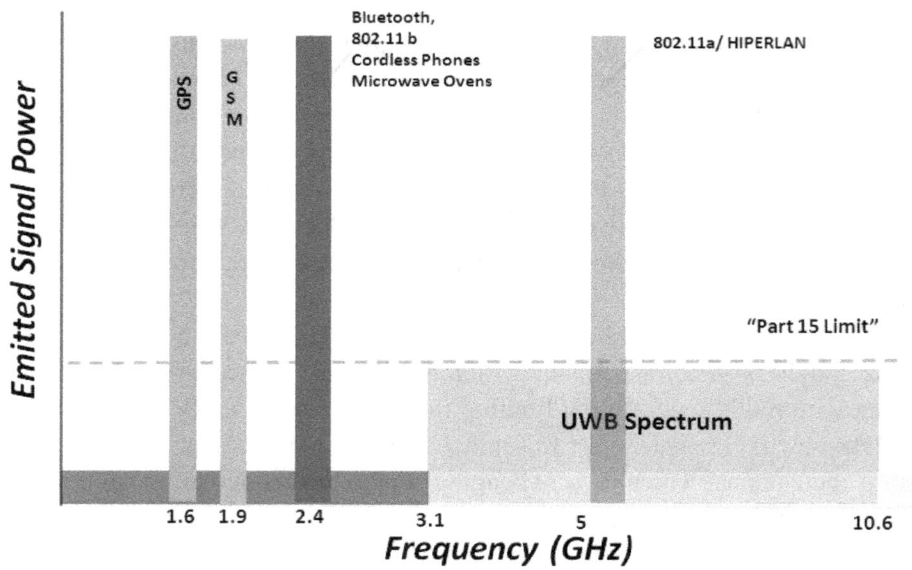

interfere with UWB applications. Hence, the UWB antennas are required to filter out the undesired bands. To acquire this, several band notching techniques are proposed which includes inserting T-shaped strips (Gallo et al., 2012), electromagnetic band gap (EBG) structures (Li, Feresidis, Mavridou, & Hall, 2015), split-ring resonators (SRR's) (Li, Zhou, Hong, & Wang, 2011), and many more. In this design, band notch characteristic at 5.5 GHz has been achieved in UWB MIMO by etching a modified E-shaped slot on the orthogonal radiators.

In this chapter, a compact band notched UWB diversity antenna is proposed which consists of CPW feed radiating planar monopoles placed at orthogonal orientation. A Minkowski fractal shaped parasitic element is used in this endeavour to reduce the coupling between the antenna elements. A grounded rectangular stub with a Minkowski fractal shaped top is placed at 45° between the CPW's to establish and ensure high isolation between the radiating elements. The impedance matching at the UWB spectrum is achieved by using conventional half circular slots on top of the radiating monopoles and by employing stepped impedance transformers at bottom. An excellent band notch characteristic has been obtained at 5.5 GHz by etching a modified E-shaped slot on the radiating monopoles. The notch band reduces the electromagnetic interference occurring at the WLAN band (5.15-5.85 GHz). The radiation patterns and ECC depicts that the antenna is suitable for MIMO/diversity systems.

ANTENNA DESIGN

Antenna Configuration

The geometry of the proposed UWB MIMO antenna (Debnath, Karmakar, Saha, & Huda, 2018) is shown in Figure 4. The antenna is fabricated on a low cost FR4 substrate, with a relative dielectric constant of 4.4, thickness of 0.8mm and a loss tangent of 0.02. The total dimension of the antenna is $46 \times 46 \times 0.8$ ($L \times W \times h$) mm^3. The proposed antenna comprises of two orthogonal planar monopoles, a partial ground plane, a parasitic Minkowski fractal element and a grounded rectangular stub with a Minkowski fractal top. Both the monopoles have been feed by a pair of co-planar waveguide (CPW) feed as shown in Figure 4.

The Minkowski fractal shaped parasitic element creates dissipative coupling to reduce mutual coupling and a rectangular stub extended from the ground plane symmetrically with respect to the ground plane at 45° is to extend the effective current route which in turn enhances the wideband isolation. Each radiator has a modified E-shaped slot etched on it which provides the band rejection characteristics along

Figure 4. Schematic view of the UWB MIMO antenna [Source: Reproduced courtesy of The Electromagnetics Academy]

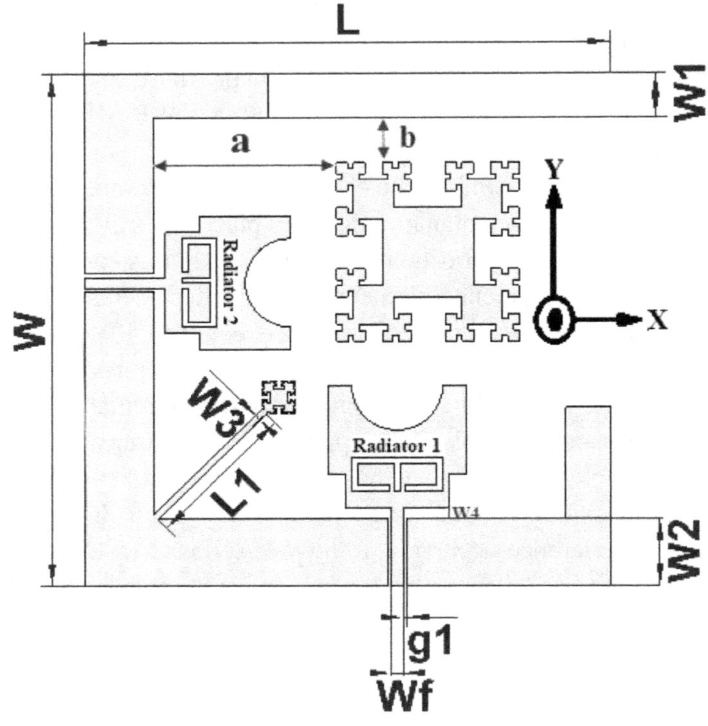

with compactness of notch structure as shown in fig. 5. To obtain the numerical analysis and suitable geometrical dimensions, HFSS electromagnetic software has been exploited.

FRACTAL STRUCTURES FOR ISOLATION ENHANCEMENT

Application of fractal geometry on antenna optimizes the shape of antenna so as to enhance the electrical path length and reduce the size of the overall radiator (Abed, 2018). In this MIMO configuration, a parasitic Minkowski fractal structure has been employed as shown in Figure 6, which creates a dissipative coupling to cancel mutual coupling between radiating elements and aids in enhancing the antenna performance. As the separation of the monopoles is small, the space wave and surface wave may result in coupling enhancement between them. Introduction of Minkowski fractal parasitic element captures these waves and converts them into surface current results

Figure 5. Monopole with modified-E shaped slot [Source: Reproduced courtesy of The Electromagnetics Academy]

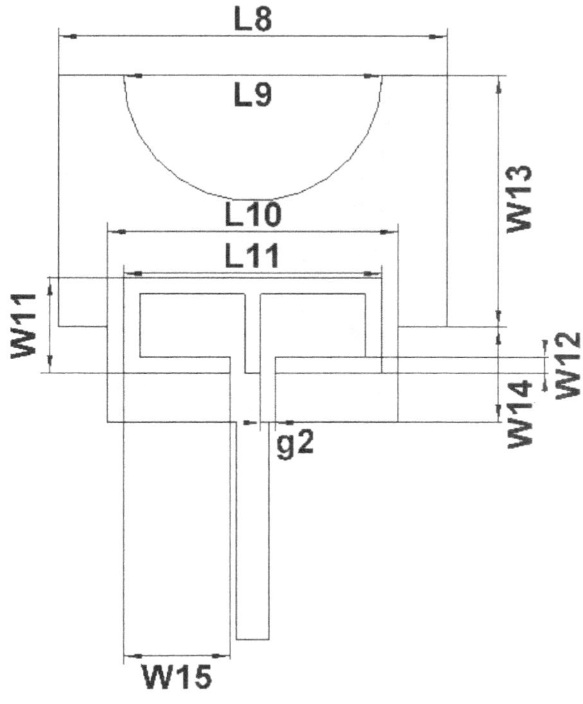

in reduced mutual coupling (Khan, Capobianco, Shafique, Ijaz, & Naqvi, 2015; Li, Du, Takahashi, Saito & Ito, 2012). When the 1st monopole is excited, due to small separation, current is coupled to the second monopole in absence of any isolation element. The presence of fractal parasitic element introduces two coupling paths between the monopoles. One is the coupling path from 1st monopole to parasitic element and the other coupling path is from fractal parasitic element to the second monopole.

As the parasitic element is a fractal, it captures most of the energy in its elongated fractal path which intern weakens coupled current component from 1st monopole to 2nd monopole. The geometrical construction of fractal Minkowski curve is shown in Figure 7 where each side of the basic structure is divided into three equal line segments and the middle line segment of each side is replaced by an inward projection with two vertical and two horizontal line segments of equal length. This process is repeated for successive iterations while maintaining the basic geometry of the antenna (Debnath, Karmakar, Saha, & Huda, 2018). Minkowski fractal algorithm (MFA) (Munish Kumar & VandanaNath, 2017) using IFS for creating the scaled

Figure 6. Parasitic Minkowski fractal Element [Source: Reproduced courtesy of The Electromagnetics Academy]

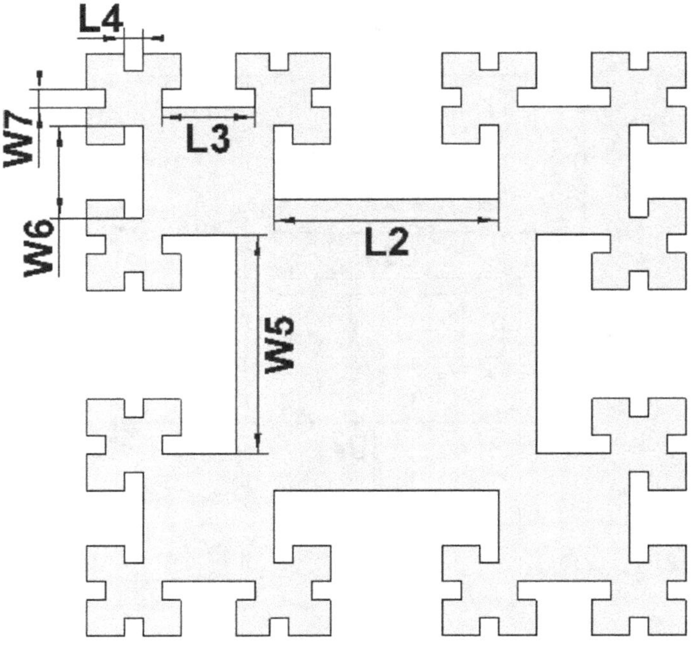

down copies of the original geometry has been described below. The MFA can be implemented by three parameters as shown in Figure 7. Here, '*a*' is the length of the generator, '*b*' is the indentation width, and '*h*' is the fractal or indentation depth along with first four iterations of Minkowski fractal (Munish Kumar & VandanaNath, 2017). The structure is obtained starting with a square patch (or initiator). Minkowski fractal geometry is deployed to increase the effective current path length from '*a*' to '*a*+2*h*', thereby increasing electrical path length for a given frequency. The perimeter of the Minkowski fractal after nth iteration can be found from Eq. (14), as mentioned below (Munish Kumar & VandanaNath, 2017):

$$L^{n}_{perimeter} = \left(1 + 2a_2\right) L^{n-1}_{perimeter} \tag{14}$$

Where, $a_2 = \dfrac{h}{a}$ (15)

Figure 7. Recursive generation of Minkowski fractal structure (a) Generator (b) Initiator (c) 1st Iteration (d) 2nd Iteration (e) 3rd Iteration [Source: Reproduced courtesy of The Electromagnetics Academy]

Mutual coupling between neighbouring radiating monopoles in the diversity antenna introduces unwanted disturbances in the radiation characteristics due to the introduction of surface waves. So, there is a requirement for additional isolation mechanism especially where strong surface current on the ground plane may be the main cause of coupling of electromagnetic energy between monopoles. To further intensify the isolation process, a grounded rectangular stub has been inserted and extended at 45° with respect to CPW feed to further enhance the isolation of the overall system. In this structure, a Minkowski fractal shaped top has been added to the stub which provides high isolation as shown in figure 8. When grounded fractal stub is introduced, more current through L-shaped ground concentrates on the grounded stub and captures in the fractal top due to their larger electrical length.

MODIFIED E-SHAPED SLOT FOR BAND NOTCH CHARACTERISTICS

The UWB spectrum is severely interfered by several pre-assigned narrow band systems such as Wi-Max (3.3 - 3.7 GHz), WLAN (5.15 - 5.85 GHz). The co-existence of this narrow band may cause effective electromagnetic interference to the UWB applications. To solve this issue, modified E-shaped slot has been designed and implemented over the planar monopoles as shown in Figure. 5. This modified E-shape has been specially considered as it will occupy lesser space than a Euclidian notch (Tripathi, Mohan, & Yadav, 2015). Each slot etched on the radiators acts as a quarter (guided) wavelength parallel open circuit transmission line that short the antenna at the relevant notch frequency of 5.5 GHz. The total length of this E-shape slot is about half wavelength of the centre frequency of the notch band which can be calculated by (Gao, He, Wei X et al., 2014):

Figure 8. Parasitic Minkowski fractal Element [Source: Reproduced courtesy of The Electromagnetics Academy]

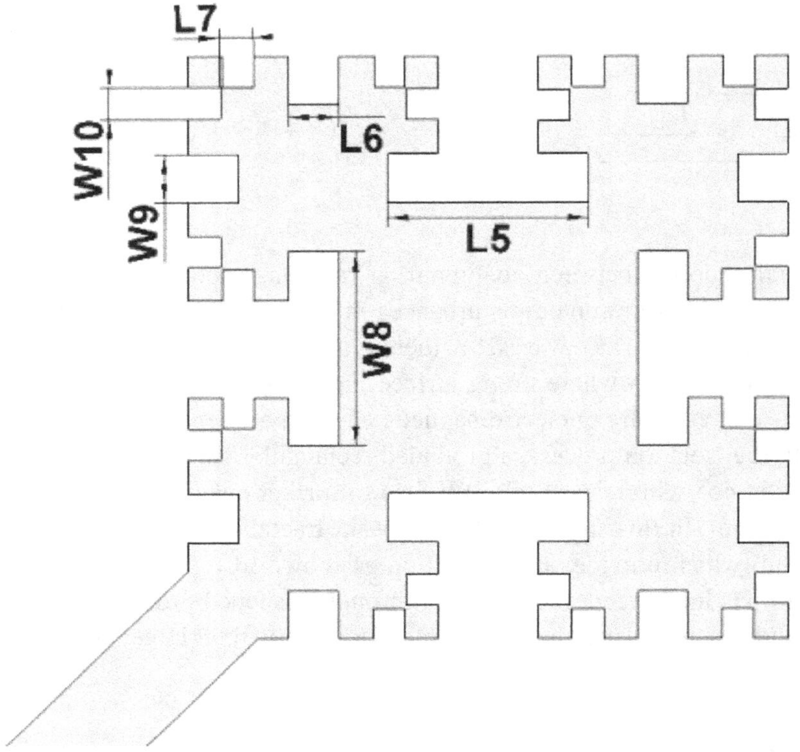

$$S = \frac{c}{2f_{notch}\sqrt{\varepsilon_{eff}}} \approx 3 \cdot W_{11} + 2 \cdot L_{11} - 2 \cdot g_2 \tag{16}$$

Here, the effective dielectric constant (ε_{eff}) can be approximated to half of the actual ε_{eff} of the FR4 material, due to the lack of ground plane. Here, c is the speed of the light, f_{notch} is the centre frequency of the notch band and S is the total length of the E-shaped slot. W_{11}, L_{11}, g_2 and L_{22} are the essential design parameters as shown in Figure 5. According to Equation 22, the calculated length of the slot is 20 mm, which is close to our designed value of 23 mm.

RESULTS AND DISCUSSION

Impedance Characteristics

The Simulated S_{11} characteristics of the antenna is shown in Figure. 9 which shows that the proposed antenna has an impedance bandwidth (S_{11}<-10 dB) ranging from 3.1 to 12 GHz except the WLAN (5.15 GHz–5.85 GHz) notch band. Band notching at this frequency has been achieved by etching a compact E-shaped slot on both the radiators.

It is seen from Figure 9 that adding fractal grounded stub and parasitic element for enhancing isolation have negligible effect on the wideband impedance matching of the UWB MIMO antenna. The proposed MIMO antenna has been successfully fabricated as shown in Figure 10 followed by measurement using a Rhode and Schwarz ZVA-40 vector network analyzer (VNA).

Figure 9. Simulated S_{11}/S_{22} Parameter with and without isolation structures [Source: Reproduced courtesy of The Electromagnetics Academy]

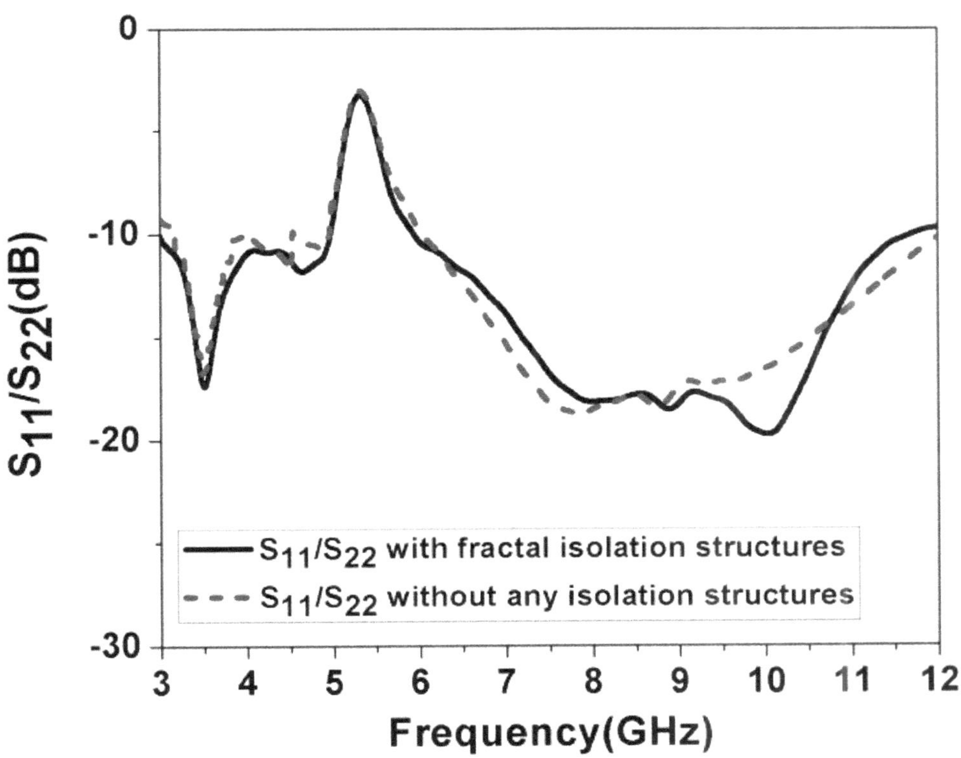

Figure 10. Fabricated prototype of the proposed antenna

While measuring the return loss characteristics, port-1 is excited and the port-2 is terminated with 50Ω load. Here the port-1 is designated as the port connected with radiator-1 as shown in Figure 4. The measured and simulated S-parameters are in good agreement as shown in Figure 11. The proposed UWB MIMO antenna covers the entire UWB bandwidth showing satisfactory notch characteristics at WLAN band.

In the previous section it is clear that by inserting Minkowski fractal isolation stub and parasitic fractal element, most of the current between two radiators has been trapped. This helps to minimize the mutual coupling among the adjacent antenna elements. It is evident from the parametric study as shown in Figure 12 that the value of S_{21} reduces to less than −17 dB for most of the operating band after the application of fractal stub along with parasitic fractal structure up to 3rd iteration. It is can be easily inferred from Figure 12 that as the number of iterations of both the fractal structures increases, the S_{12}/S_{21}characteristics decreases significantly with increasing isolation. This is for the intricacy of fractal structures which can confine more energy on them due to their larger electrical length than a Euclidian structure. We have limited this optimization up to 3rd iteration of fractal as higher iterations may complicate the fabrication process.

Figure 13 represents the parametric variation of S_{11} characteristics for the E-shaped slot. It is seen that with the increase of the length of the slot, the centre

Figure 11. Simulated and Measured S_{11}/S_{22} of the proposed antenna [Source: Reproduced courtesy of The Electromagnetics Academy]

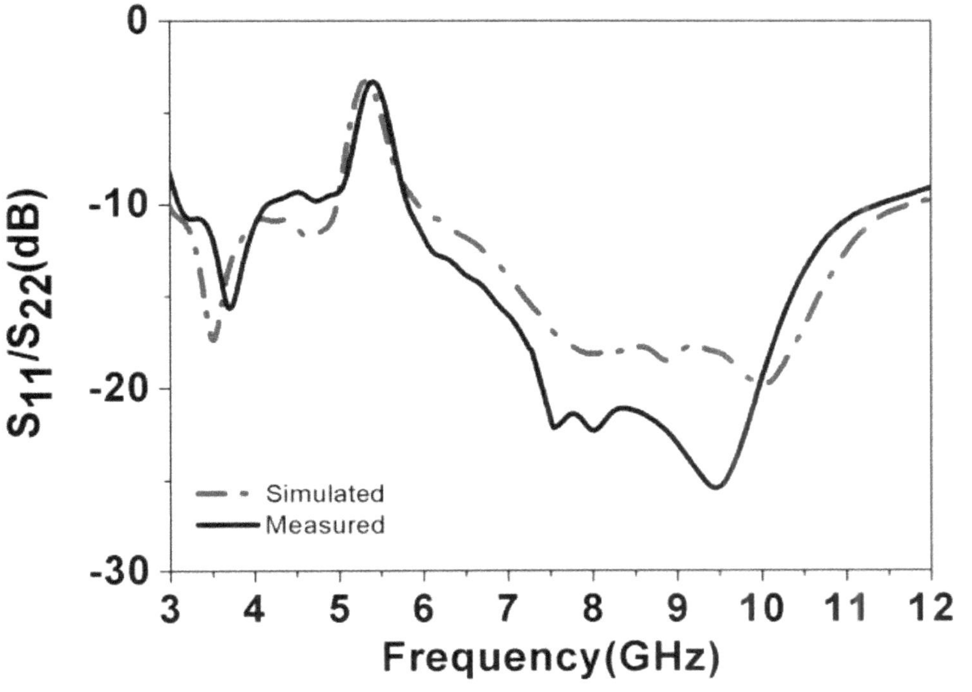

frequency of the notch band shifts towards lower end of the operating band which intern proves tuneable characteristics of the notch. Notch length of 23mm is chosen as it suites the suitable notch characteristics at 5.5 GHz. Final optimized dimension of the proposed antenna is described in Table 1.

SURFACE CURRENT DISTRIBUTION

To elaborately express the influence of the Minkowski fractal stub and Minkowski fractal parasitic elements, the simulated surface current distribution at three distinct frequencies are carried out and shown in Figure 14. Figure 14(a) shows the UWB MIMO antenna with and without fractal stub at a lower edge frequency of 4 GHz, Where port 1 is excited, port 2 is being terminated at 50 Ω impedance. It can be seen that without the fractal stub and parasitic element, strong current is coupled from one port to another port by direct current through common ground plane as well as induced current through radiating monopoles which may result in high

Figure 12. Simulated S_{12}/S_{21} characteristics based upon four iterations of Fractal Stub and Parasitic Fractal Element [Source: Reproduced courtesy of The Electromagnetics Academy]

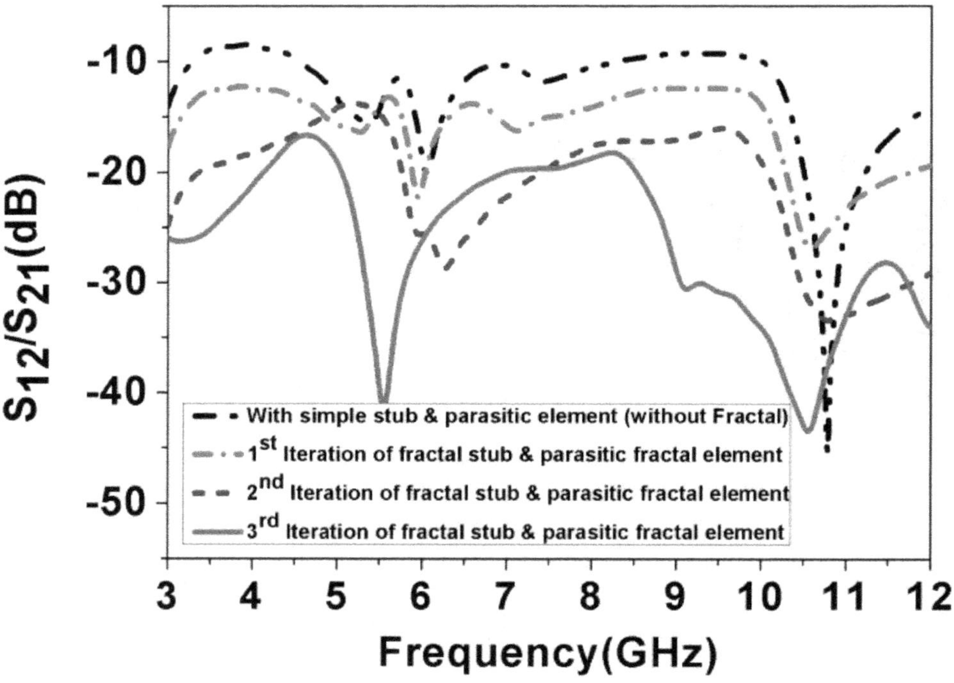

mutual coupling. When grounded fractal stub is introduced, more current through ground concentrates on the grounded stub and captures in the fractal for its larger electrical length. Side by side, induced current through radiating monopoles also captured by the parasitic fractal element and concentrates on the fractal which intern reduces the mutual coupling. The same effect can be seen at higher frequency of 7.5 GHz as shown in Figure 14 (c). Thus, the grounded fractal stub in combination with parasitic element causes high isolation (-17dB) between the two ports which is higher in comparison to other reported works (Zhang, Ying, Xiong, & He, 2009; Gallo et al., 2012; Li, Feresidis, Mavridou, & Hall, 2015; Debnath, Karmakar, Saha, & Huda, 2018; Khan, Capobianco, Shafique, Ijaz, & Naqvi, 2015; Ren, Hu, Yin, & Fan, 2014; Liu, Cheung, & Yuk, 2013; Zhang, Zhang, Tian, & Luo, 2015; Kang, Li, Wang, & Shi, 2015). At the notch frequency of 5.5 GHz as shown in Figure 14 (b), the current is concentrated around the E-shaped slots absorbing radiative energy making the antenna non radiative on that specific frequency, where the band-notching characteristics is obviously produced.

Figure 13. Simulated S_{11}/S_{22} of proposed antenna with different length of the E-shaped slot from antenna edge[Source: Reproduced courtesy of The Electromagnetics Academy]

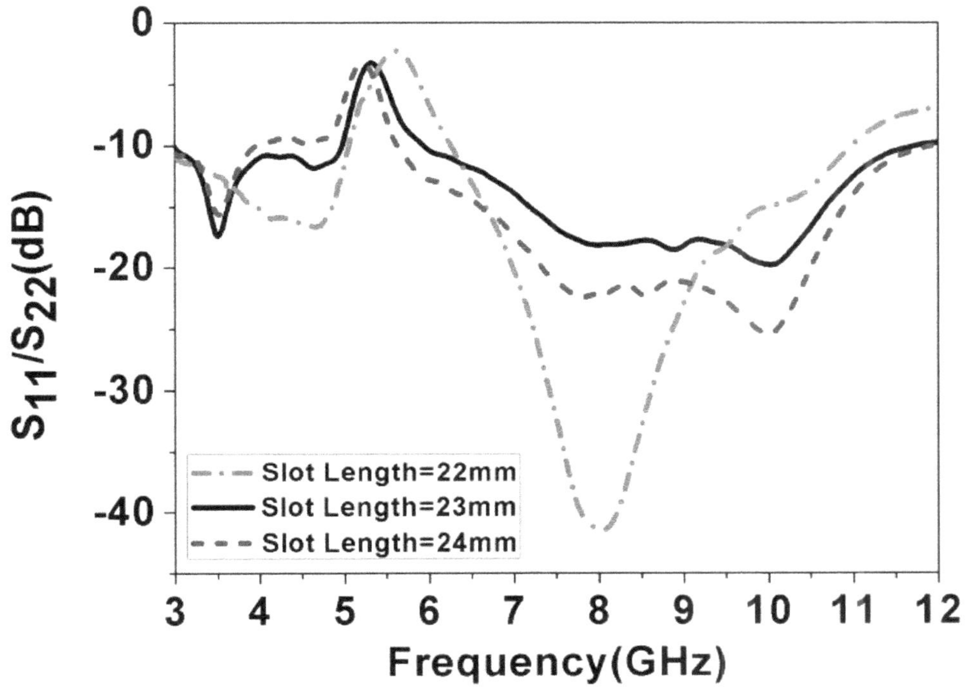

Table 1. Descriptive table monitoring the final dimension of the Proposed Antenna (Unit: mm)

Parameters	W	W_1	W_2	W_3	W_4	W_5	W_6	W_7
Unit(mm)	46	4	6	0.5	1	6	2.5	0.5
Parameters	W_8	W_9	W_{10}	W_{11}	W_{12}	W_{13}	W_{14}	W_F
Unit(mm)	1	0.25	0.16	3	0.5	8	3	1
Parameters	a	b	g_1	g_2	L	L_1	L_2	L_3
Unit(mm)	16	4	0.3	0.45	46	13.38	6	2.5
Parameters	L4	L_5	L_6	L_7	L_8	$L_9=L_{22}$	L_{10}	$L_{11}=W_{22}$
Unit(mm)	0.5	1	0.25	0.16	12	8	9	8

Figure 14. Simulated Surface Current Distribution with and without fractal stub and parasitic fractal element at (a) 4 GHz, (b) 5.5 GHz, (c) 7.5 GHz where port-1 is excited and port-2 is terminated by 50 Ω termination [Source: Reproduced courtesy of The Electromagnetics Academy]

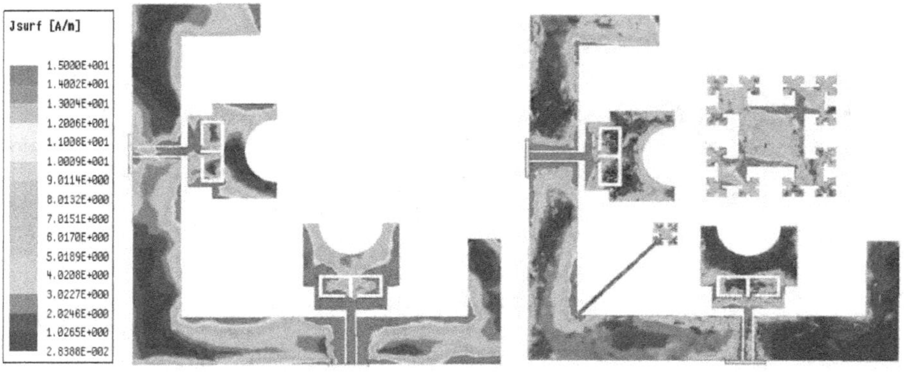

(a)

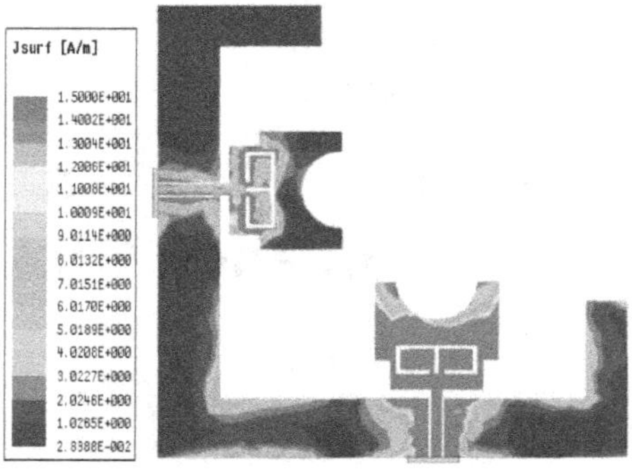

(b)

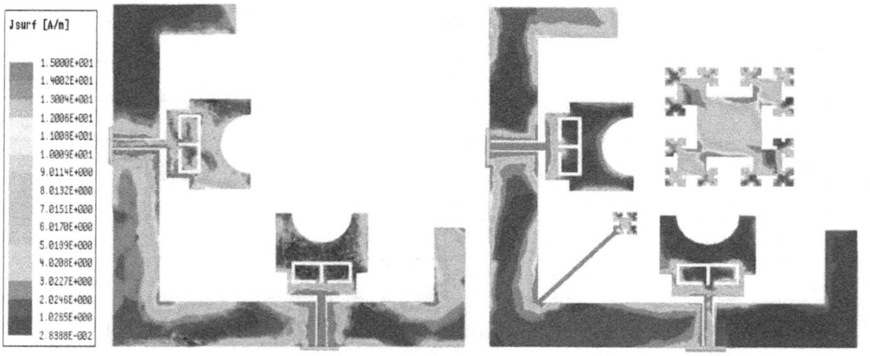

(c)

**For a more accurate representation see the electronic version.*

DIVERSITY CHARACTERISTICS

One of the primary requirements for diversity antenna elements to be effectively deployed in pattern diversity is that their radiation patterns should be uncorrelated. The mutual coupling between the adjacent radiating elements and the amount of correlation between each radiator can be studied in terms of envelope correlation coefficient (*ECC*) which can be calculated using S-parameters as follows (Sharawi, 2017):

$$ECC = \frac{\left|S_{11}^{*}S_{12} + S_{21}^{*}S_{22}\right|^{2}}{\left(1-\left|S_{11}\right|^{2}-\left|S_{21}\right|^{2}\right)\left(1-\left|S_{22}\right|^{2}-\left|S_{12}\right|^{2}\right)} \tag{17}$$

Under normal operating conditions, the value of *ECC* should be below 0.5 that depicts acceptable limit of signal distortion (Sharawi, 2017). It can be observed from Figure 15 that the simulated *ECC* is less than 0.02 throughout the operating bandwidth. The obtained value of *ECC* is well underneath the pragmatic threshold value of 0.5.

Figure 15. Simulated and measured ECC of the proposed antenna [Source: Reproduced courtesy of The Electromagnetics Academy]

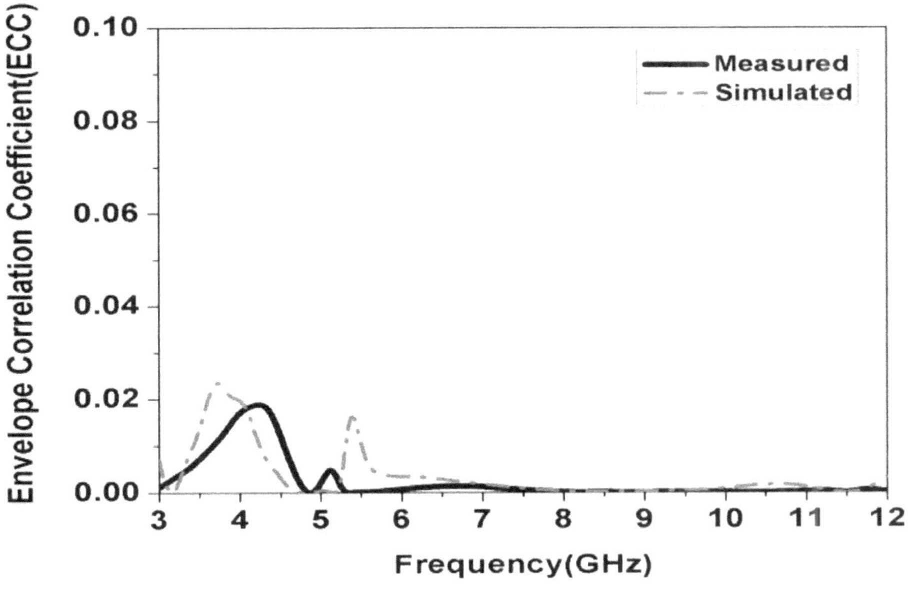

The simulated and measured peak gain and radiation efficiency of the proposed antenna (with port 1 excited) are shown in Figure 16 and Figure 17 throughout the operating band. It can be seen that except the notch band at 5.5 GHz, the measured peak gain ranges from 3 to 5 dBi across the entire UWB spectrum with a radiation efficiency of > 75%. A sharp decreasing trend is clearly observed at 5.5 GHz for both plots, which demonstrate the effective notch characteristics at WLAN band. It is also observed that except the notch band, the gain and radiation efficiencies are nearly stable across the entire UWB bandwidth.

The radiation characteristics in terms of normalized gain of the MIMO antenna has been simulated and measured with port-2 terminated with 50 ohm termination and port-1 excited, are shown in Figure 18. The proposed antenna shows a nearly omnidirectional radiation pattern over the desired operating band. Nearly omnidirectional patterns have been observed in the H-plane which is typical for monopole antennas.

Figure 16. Simulated and Measured peak gain of the antenna (Port 1 excited) [Source: Reproduced courtesy of The Electromagnetics Academy]

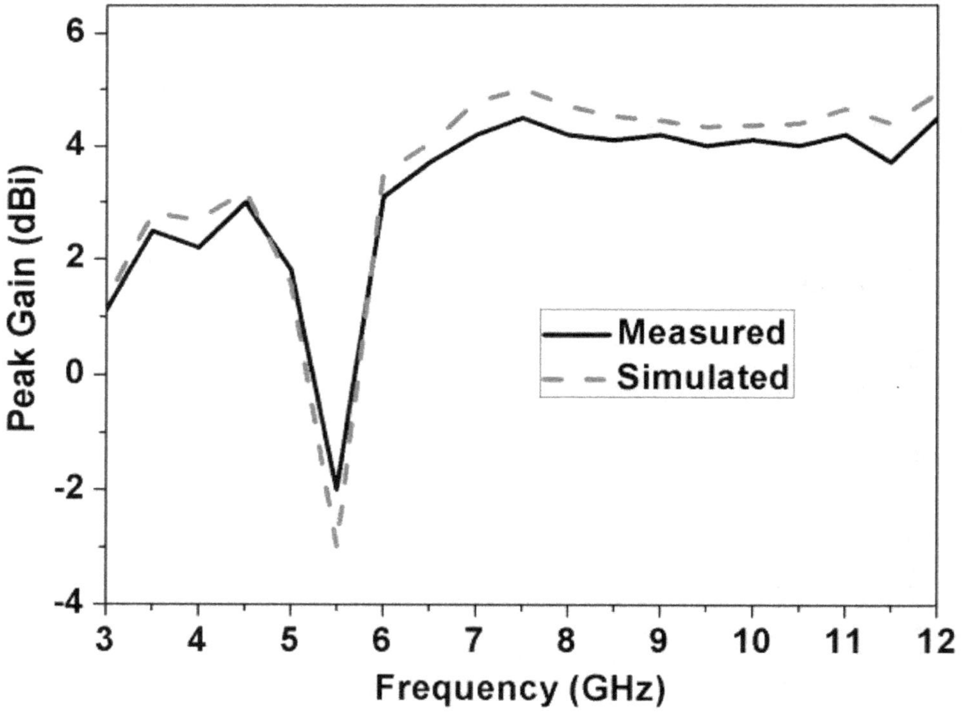

Figure 17. Simulated and measured radiation efficiency of the proposed antenna (Port 1 excited) [Source: Reproduced courtesy of The Electromagnetics Academy]

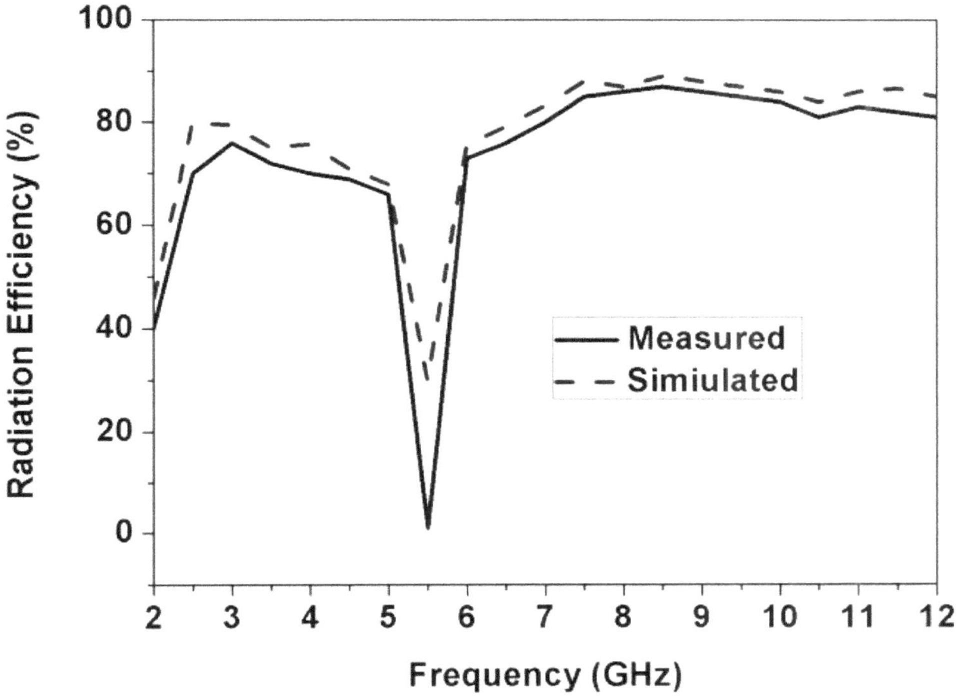

Radiation patterns in E-plane are not dumb-bell shaped and do not have significant null which may be different from these kind of typical single monopole antennas which caused for the L-shaped ground plane that makes the current distribution different from typical monopole antennas. At higher frequencies, radiation patterns are less omnidirectional due to higher order modes.

CONCLUSION

A compact printed UWB MIMO antenna with high isolation and band notch characteristics has been presented and analysed for portable UWB devices applications. The proposed antenna has an impedance bandwidth ($S_{11} \leq -10$ dB) ranging from 3.1 to 12 GHz excluding a rejection band centred at 5.5 GHz. The numerical and empirical results are in good agreement with each other. The results

Figure 18. Normalized radiation characteristics of the proposed antenna (with Port-1 excited and Port-2 terminated with 50 ohm termination) [Source: Reproduced courtesy of The Electromagnetics Academy]

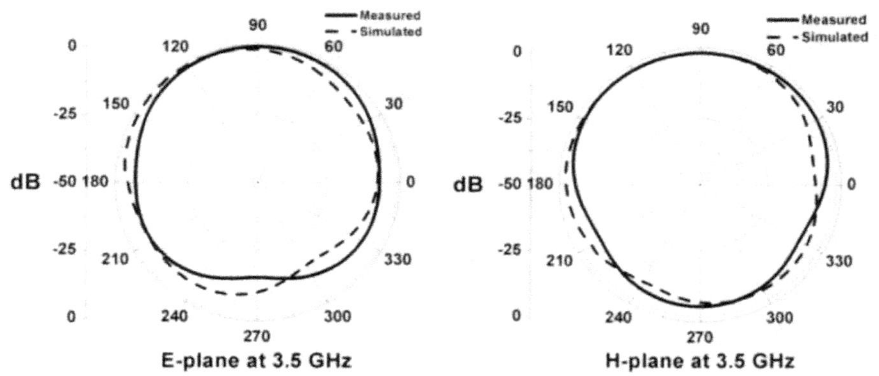

(a)

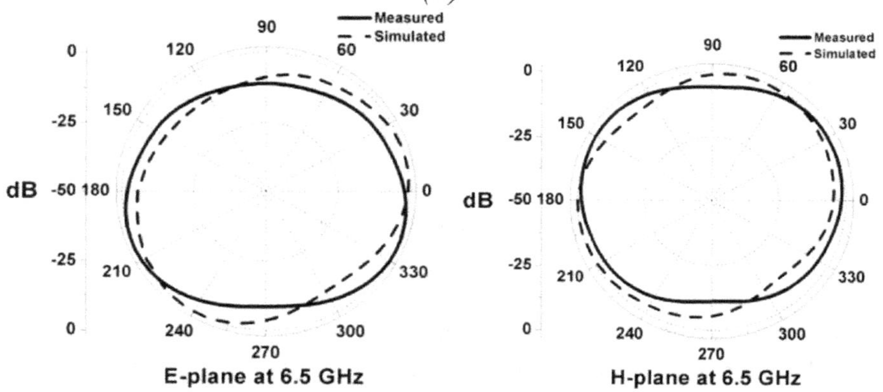

(b)

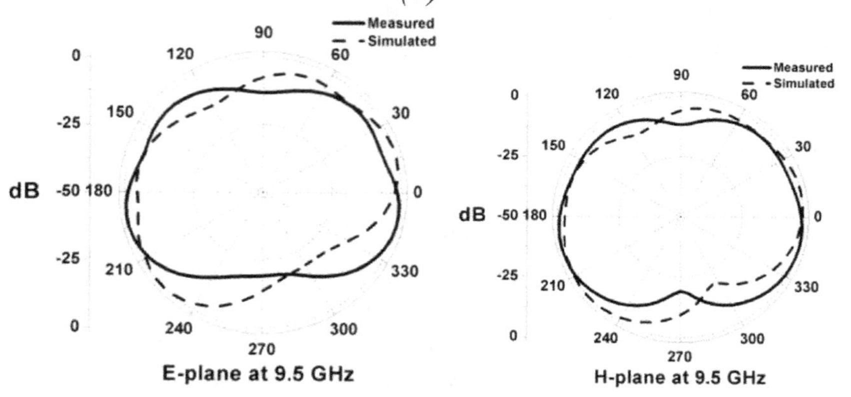

(c)

have also depicted that the proposed antenna has high isolation of >17 dB over the entire UWB bandwidth. The focus of this design is to improve isolation between the radiating elements using fractal shaped structures. The ECC (Envelope Correlation Coefficient) is less than 0.002 for the entire UWB band with an average realized gain of 4 dBi and radiation efficiency greater than 75% throughout the operating band making it an exceptional candidate for various UWB MIMO communication systems.

REFERENCES

Abed, A. T. (2018). Highly compact size serpentine-shaped multiple-input–multiple-output fractal antenna with CP diversity. *IET Microwaves, Antennas & Propagation*, *12*(4), 636–640. doi:10.1049/iet-map.2017.0770

Sharawi, M. (2017, April). Current Misuses and Future Prospects for Printed Multiple-Input, Multiple-Output Antenna Systems. *IEEE Antennas & Propagation Magazine*, 162–170.

Debnath, P., Karmakar, A., Saha, A., & Huda, S. (2018). UWB MIMO Slot Antenna with Minkowski Fractal Shaped Isolators for Isolation Enhancement. *Progress In Electromagnetics Research M*, *75*, 69–78. doi:10.2528/PIERM18090506

Federal Communications Commission. (2002). *Federal communications commission revision of Part 15 of the Commission's rules regarding ultra-wideband (UWB) transmission system from 3.1 to 10.6 GHz*. Washington, DC: ET-Docket.

Gallo, M., Antonino-Daviu, E., Ferrando-Bataller, M., Bozzetti, M., Molina-Garcia-Pardo, J. M., & Juan-Llacer, L. (2012). A broadband pattern diversity annular slot antenna. *IEEE Transactions on Antennas and Propagation*, *60*(3), 1596–1600. doi:10.1109/TAP.2011.2180314

Gao, P., He, S., & Wei, X. (2014). Compact printed UWB diversity slot antenna with 5.5-GHz band-notched characteristics. *IEEE Antennas and Wireless Propagation Letters*, *13*, 376–379. doi:10.1109/LAWP.2014.2305772

Ghosh, J., Ghosal, S., Mitra, D., & Chaudhuri, S. R. B. (2016). Mutual coupling reduction between closely placed microstrip patch antenna using meander line resonator. *Prog. Electromagn. Res. Lett*, *59*, 115–122. doi:10.2528/PIERL16012202

Iqbal, A., Saraereh, O. A., Ahmad, A. W., & Bashir, S. (2018). Mutual Coupling Reduction Using F-Shaped Stubs in UWB-MIMO Antenna. *IEEE Access: Practical Innovations, Open Solutions*, *6*, 2755–2759. doi:10.1109/ACCESS.2017.2785232

Kaiser, T., Zheng, F., & Dimitrov, E. (2009). An overview of ultra-wide-band systems with MIMO. *Proceedings of the IEEE, 97*(2), 285–312. doi:10.1109/JPROC.2008.2008784

Kang, L., Li, H., Wang, X., & Shi, X. (2015). Compact offset microstrip-fed MIMO antenna for band-notched UWB applications. *IEEE Antennas and Wireless Propagation Letters*, *14*, 1754–1757. doi:10.1109/LAWP.2015.2422571

Khan, B. D. B. M. S., Capobianco, A. D., Shafique, M. F., Ijaz, B., Naqvi, A., & Braaten, B. D. (2015). Isolation Enhancement of a Wideband MIMO Antenna using Floating Parasitic Elements. *Microwave and Optical Technology Letters*, *57*(7), 1677–1682. doi:10.1002/mop.29162

Kumar & Nath. (2017). *Introducing Multiband and Wideband Microstrip Patch Antennas Using Fractal Geometries: Development in Last Decade*. Springer.

Li, L., Zhou, Z. L., Hong, J. S., & Wang, B. Z. (2011). Compact dual-band-notched UWB planar monopole antenna with modified SRR. *Electronics Letters*, *47*(17), 950–951. doi:10.1049/el.2011.1874

Li, Q., Feresidis, A. P., Mavridou, M., & Hall, P. S. (2015). Miniaturized double-layer EBG structures for broadband mutual coupling reduction between UWB monopoles. *IEEE Transactions on Antennas and Propagation*, *63*(3), 1168–1171. doi:10.1109/TAP.2014.2387871

Li, Z., Du, Z., Takahashi, M., Saito, K., & Ito, K. (2012). Reducing Mutual Coupling of MIMO Antennas With Parasitic Elements for Mobile Terminals. *IEEE Transactions on Antennas and Propagation*, *60*(2), 473–481. doi:10.1109/TAP.2011.2173432

Liu, Cheung, & Yuk. (2013). Compact MIMO antenna for portable devices in UWB applications. *IEEE Transactions on Antennas and Propagation, 8*, 4257–4264.

Ren, J., Hu, W., Yin, Y., & Fan, R. (2014). Compact printed MIMO antenna for UWB applications. *IEEE Antennas and Wireless Propagation Letters*, *13*, 1517–1520. doi:10.1109/LAWP.2014.2343454

Sharawi, M. S. (2017). Current Misuses and Future Prospects for Printed Multiple-Input, Multiple-Output Antenna Systems. *IEEE Antennas & Propagation Magazine*, *59*(2), 162–170. doi:10.1109/MAP.2017.2658346

Tripathi, S., Mohan, A., & Yadav, S. (2015). A compact Koch fractal UWB MIMO antenna with WLAN band-rejection. *IEEE Antennas and Wireless Propagation Letters*, *14*, 1565–1568. doi:10.1109/LAWP.2015.2412659

Zhang, J.-Y., Zhang, F., Tian, W.-P., & Luo, Y.-L. (2015). Acs-fed UWB-MIMO antenna with shared radiator. *Electronics Letters*, *51*(7), 1301–1302. doi:10.1049/el.2015.1327

Zhang, S., Lau, B. K., Sunesson, A., & He, S. (2012). Closely-packed UWB MIMO/diversity antenna with different patterns and polarizations for USB dongle applications. *IEEE Transactions on Antennas and Propagation*, *60*(9), 4372–4380. doi:10.1109/TAP.2012.2207049

Zhang, S., Ying, Z., Xiong, J., & He, S. (2009). Ultrawideband MIMO/diversity antennas with a tree-like structure to enhance wideband isolation. *IEEE Antennas and Wireless Propagation Letters*, *8*, 1279–1282. doi:10.1109/LAWP.2009.2037027

KEY TERMS AND DEFINITIONS

Band Notch: To avoid the electromagnetic interference between the UWB system and the wireless system, band notch functionality is necessary. It is achieved by etching different notches in UWB mono-poles. The length of the band notch element is equal to half wave length at notch frequency.

Fractal Curve: Recursive structures describe many real-world objects like clouds, mountains, trees, coastlines and Brownian motion, which do not correspond to simple geometric shapes. Fractal geometry, which was introduced in 1975 by B.B. Maldelbrot, mathematically defined these extremely intricate self-similar shapes. His work inspired interest and has made fractal geometry and its application a very popular field of study. A fractal is a rough or fragmented geometric shape that can be subdivided in parts, each of which is a reduced copy of the whole. The amazing realization towards combination of fractal geometry with electro-magnetic theory has led to the development of novel class of antennas called fractal shaped antennas which have received significant attention in applied electromagnetic research community. Fractals in antenna engineering are based on analysis and design of fractal radiating elements and incorporation of fractal shaped antenna arrays. In antenna community, it is known as fractal antenna engineering. As fractal geometry is an extension of classical geometry, its appearance provides engineers with the unprecedented opportunity to explore number of configurations for possible use in the development of new and innovative antenna designs.

Mandelbrot: Mr. Mandelbrot was the first person who realized that it is very often impossible to describe nature using only Euclidian geometry in terms of straight lines, circles and cubes. He coined the term 'fractal' about 20 years ago in his book named "The fractal geometry of nature." He proposed that fractals and fractal geometry could be used to describe real objects such as trees, lightening, river meanders, and coastline and so forth.

MIMO (Multiple-Input Multiple-Output) Antenna: The design of MIMO antenna systems has been driven by the wireless industry as this technology is a major enabling one in 4G wireless standards and will continue at the upcoming 5G standards. Using MIMO, channel capacity can be drastically increased which in turn produces higher bit rate transmission.

Minkowski Curve: The geometrical construction of fractal Minkowski curve can be described as where each side of the basic structure is divided into three equal line segments and the middle line segment of each side is replaced by an inward projection with two vertical and two horizontal line segments of equal length. This process is repeated for successive iterations while maintaining the basic geometry of the antenna.

Monopole Antenna: The monopole is a resonant antenna where the radiator acts as a resonator for radio waves, oscillating with standing waves of voltage and current along its length. Therefore, the length of the antenna is determined by the wavelength of the radio waves it is used with.

UWB (Ultra-Wideband): February 2002 witnessed the allocation of the frequency band between 3.1 GHz to 10.6 GHz with an effective isotropic radiated power (EIRP) less than -41.3 dBm/MHz for the unlicensed indoor UWB wireless communication system by the Federal Communication Commission (FCC), USA which initiated research in component development. According to the released regulation, UWB technology which is based on transmitting ultra short pulses with duration of only a few nano-seconds or less, has recently received great attention in various fields for the short-distance (< 10 m) wireless communications.

Chapter 6
Design of Spiral Square Patch Antenna for Wireless Communications

Ketavath Kumar Naik
https://orcid.org/0000-0002-1338-0239
Koneru Lakshmaiah Education Foundation, India

ABSTRACT

The kapton polyimide material is considered to design conformal antenna with spiral square for radio frequency identification (RFID) and wireless local area network (WLAN) applications. In this chapter, the analysis and investigation has been carried out with spiral square techniques using coplanar waveguide (CPW) feed. The proposed antenna operates at 5.8 GHz with impedance bandwidth of 170 MHz (5.73 - 5.9 GHz) with return loss -25.6 dB and gain is 2.4 dBi. The proposed antenna has considered with different bending angles for investigating the conformal characteristics due to flexibility of the material. These results are presented for omni-directional radiation patterns.

INTRODUCTION

Rapid growth in wireless communications makes RFID and WLAN technology more popular. RFID is an emerging technology and many service oriented, manufacturing industries, logistics, material flow systems and bio-medical systems use to identify and provide information about goods, products, animals and people (Finkenzeller, Wiley & Sons, 2010). Operating frequency is the most important criterion for RFID systems, federal communication commission (FCC) has allocated 5.8 GHz

DOI: 10.4018/978-1-5225-8531-2.ch006

Copyright © 2019, IGI Global. Copying or distributing in print or electronic forms without written permission of IGI Global is prohibited.

band for RFID, In bio-medical RFID is used for patient identification (Santiago, Costa, & Fernandes, 2013), (Wang et al., 2005). There are so many RFID systems commercially available with S-band operating frequencies, and an active research is going on for the next higher frequency of RFID systems (Kossel, Kung, Benedickter, & B¨achtold, 1999).

RFID antenna design becomes more complicated and critical for microwave frequency region (Tseng, Huang, & Hsu, 2006). So far, several RFID systems have been proposed, and a capacitive coupled CPW-fed folded-slot structure (Chen & Hsu, 2004) is proposed with single band frequency to obtain miniaturization. Multipurpose novel CPW-fed folded slot monopole antenna (Li et al., 2014) is proposed for RFID/WLAN applications, here the antenna is designed with U, F, L-shape slots to get additional resonances. In (Liu & Hu, 2005) a miniature folded slot CPW-fed monopole antenna is designed for RFID applications. However the above proposed antennas are FR-4 substrate based, rigid ones and not suitable to mount on curved objects. WiMAX is allocated with the 2.5/3.5/5.5-GHz bands, and WLAN is allocated with the 2.4/5.2/5.8 GHz bands. The proposed antenna also works for the WLAN applications. Many designs have investigated and proposed to support WLAN/WiMAX standards. Design of RFID/WLAN antenna systems for microwave frequencies with compact, flexible, conformal considerations are very important (Liu, Ku, & Yang, 2010). So, design of low profile flexible patch antenna with conformability is needful to fulfill the requirements of RFID, bio-medical, industrial, scientific, medical (ISM) and wireless wearable devices. A dual band CPW-fed kapton polyimide substrate based antenna is proposed with two deformations (Mesquita, Pires, & Moreira, 2013) to study the performance characteristics, only single bent configuration is used to characterize this antenna. A CPW-fed patch antenna is designed with rectangular slots in (Niranjan, Saxena, & Srivastava, 2012), a polyflon polyguide substrate is used and it is a rigid one. Paper antennas are flexible antennas, most of the antenna designs are either complex or not compact. In (Kumar Naik & Sri, 2017; Kumar Naik & Sri, 2018) circular patch antenna has described for satellite communication.

Modern day communications require demand of compact antenna with larger bandwidth applications and multifunctional wireless communication systems. The patch antennas are used in many applications because they offer light weight, compact size and economically efficient to fabricate (Garg, BahlI, & Ittipiboon, 2001). However the patch antennas are limited to narrow bandwidth. To overcome this there are numerous techniques and methods to enhance the bandwidth. Use of increase in thickness of substrate, low dielectric substrate (Schaubert, Pozar, & Adrian, 1989), various feeding and matching techniques (Pues & Van De Capelle, 1989) and several multi resonators (Lee, Lee, & Bobinchak, 1987) are used to enhance the bandwidth. In (Mandal & Prtim Sarkar, 2013) using U- shaped patch with defected ground, they achieved wideband with return loss of -18dB at 7.8GHz frequency, gain of 4.1dBi. Similarly (Wi, Lee, & Yook, 2007) used two U-shaped structures; they placed it as mirror image to other with co-axial feed and enhance two operating frequencies of 5.32GHz and 6.11GHz with impedance bandwidth of 1.5GHz at centre frequency of 5.5GHz, gain at the operating frequencies are 5.4dB and 6.2dB. To achieve large bandwidth U- shaped patch is considered.

In (Qu & Xue, 2007) Y-shaped structure is used as stub to enhance the bandwidth and gain in proximity feeding of the antenna. Y-shaped feed line is used as barrier for current flow to rhombus shaped patch for gain enhancement and reduce the losses in patch antenna (Lee & Sung, 2015).

In this chapter a flexible, compact spiral square CPW-fed monopole patch antenna is proposed and it consists of four spiral shaped rectangular patches symmetrically to achieve compact antenna size, better radiation patterns for RFID and WLAN applications.

ANTENNA DESIGN

The design and parametric study of the proposed spiral square CPW-fed flexible compact patch antenna have been carried out using method of moment (MOM) based computer simulation technology (CST). It is a demanding task to match the impedance in patch antennas, various factors like thickness of dielectric substrate and dielectric properties of substrate materials have to be considered in the design targets of impedance matching. The detailed dimensional layout of the proposed antenna is presented in Figure 1. The optimized design parameters are tableted in Table 1. The proposed configuration consists of monopole structure connected with microstrip line to the four spiral square structures. The rectangular shape of the ground planes designed to get the desired resonant frequencies of the antenna. For conformability and flexibility dielectric material are polyimide-based materials used as a substrate with a loss tangent of 0.002, dielectric constant of 3.5, and compact size of 24 × 20 mm^2.

The dimensions of spiral shaped CPW-fed flexible compact patch antenna is denoted by $L_1 \times W_1$, the spiral square shape strip size is denoted by L_4. The four spiral shaped patches are connected to the single pole of width W_4 and is fed through CPW transmission line impedance is a 50Ω. Each side of microstrp line is considered as ($L_2 \times W_2$). The radiating patch on only one side of the substrate (W × L). There are other parameters like open end feed line gap (L_3), feed line width (W_4), and gap between feed line to spiral square (W_5), in the given structure (Figure 1). Parametric analysis has been carried out with different bent angles (Figure 2).

Figure 1. Geometry of spiral square CPW-fed flexible compact patch antenna

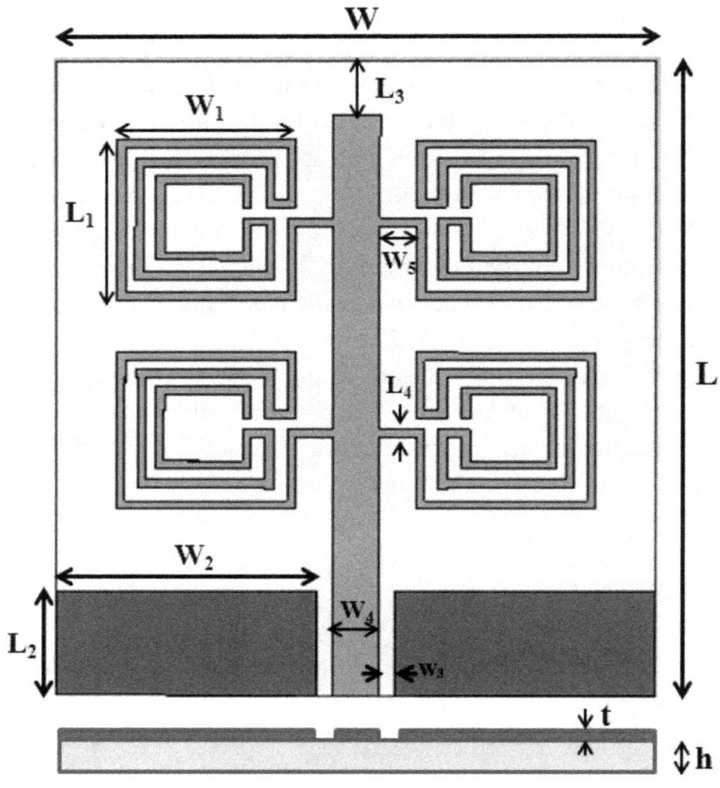

RESULTS AND DISCUSSIONS

The simulated return loss response of the spiral square CPW-fed flexible patch antenna is presented in Figure 3 for plane and bent configurations. As seen from the simulated results it is observed that the plane antenna resonates at 5.8 GHz frequency with -25.6 dB return loss and a bandwidth of 170 MHz (5730-5900 MHz) (Figure 3).

As the bending angle increases there is a frequency shift with a change in return loss. The resonant frequencies are 5.65 GHz, 5.75 GHz, 5.72GHz, 5.72 GHz, and 5.69 GHz with –20dB, –18dB, –7dB, –12dB, –24.7dB return losses for 10^0, 20^0, 30^0, 40^0, 50^0 bending configurations (Figure 3).

The gain plots of spiral square shaped CPW-fed flexible compact patch antenna for plane and bent configurations are shown in Figure 4. For plane configuration the gain is –2.4 dBi for 5.8 GHz frequency and for 10^0, 20^0, 30^0, 40^0, 50^0 bent configurations the gains are 2.3 dBi, 2.6 dBi, 2.9 dBi, 2.5 dBi, 2.7 dBi respectively.

Table 1. Optimized Values of Spiral Square CPW-fed Antenna

Antenna Parameters	Values (mm)
L	24
L_1	6
L_2	4
L_3	2
L_4	0.3
W	20
W_1	6
W_2	8.7
W_3	0.3
W_4	2
W_5	1.2
t	0.03
h	0.07

Figure 2. Cylindrical bent configuration of spiral square CPW-fed flexible compact patch antenna

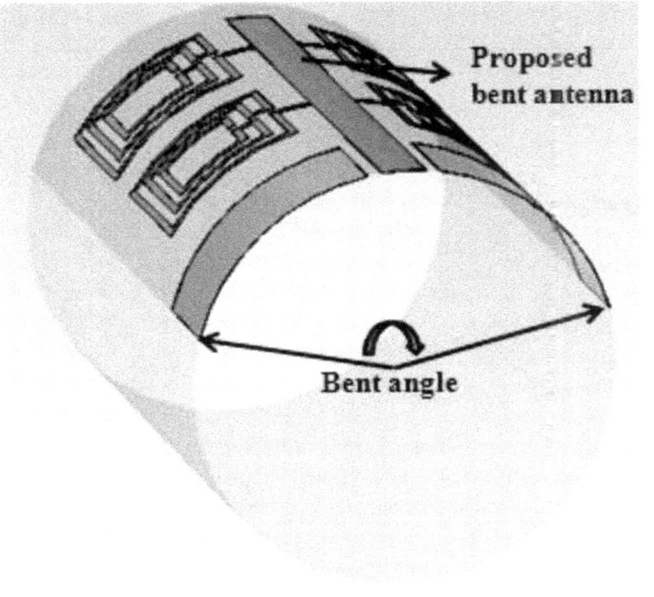

Figure 3. Return Loss plot of spiral square CPW-fed flexible compact patch antenna when plane and bent states with different angels

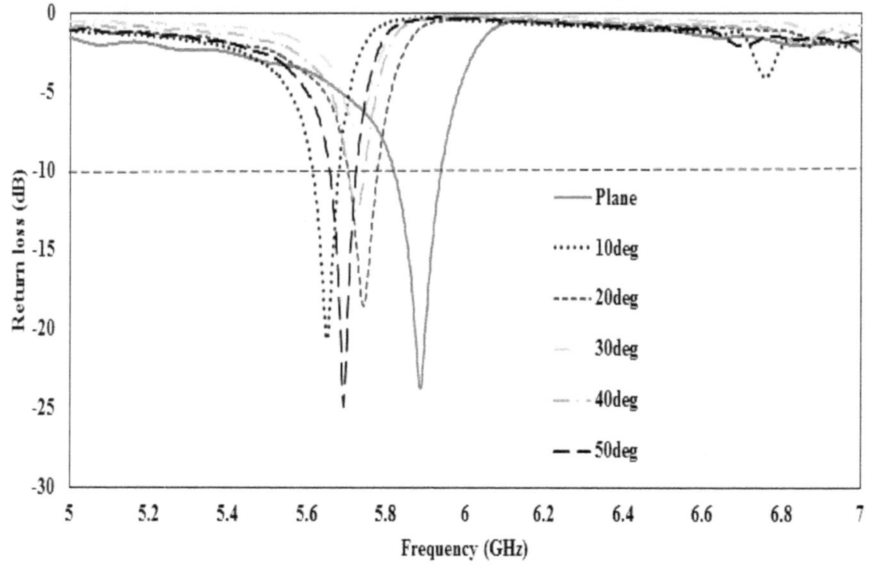

Figure 4. Gain plot of spiral square CPW-fed flexible compact patch antenna for plane and bent states with different angels

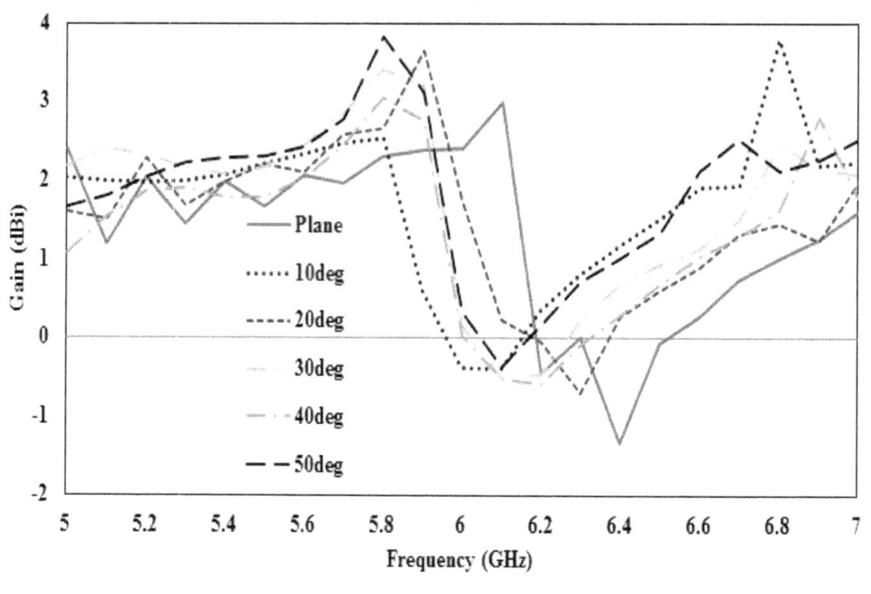

The simulated radiation patterns in E-plane, H-plane for planar, bent configurations are shown in Figure 5. The radiation pattern for planar antenna is almost omnidirectional in both E, H-planes. For 50^0(Figure 5(b)) bend the radiation patterns are similar to omnidirectional patterns.

The performance of the spiral square CPW-fed flexible compact patch antenna was also understandable through surface current distributions of the correspond resonant frequencies. The surface current distributions are shown in Figure 6.

Figure 5. E-field, H-field radiation patterns of spiral square CPW-fed flexible compact patch antenna at frequencies (a) 5.8 GHz (Planar), (b) 5.69 GHz (50 deg bent).

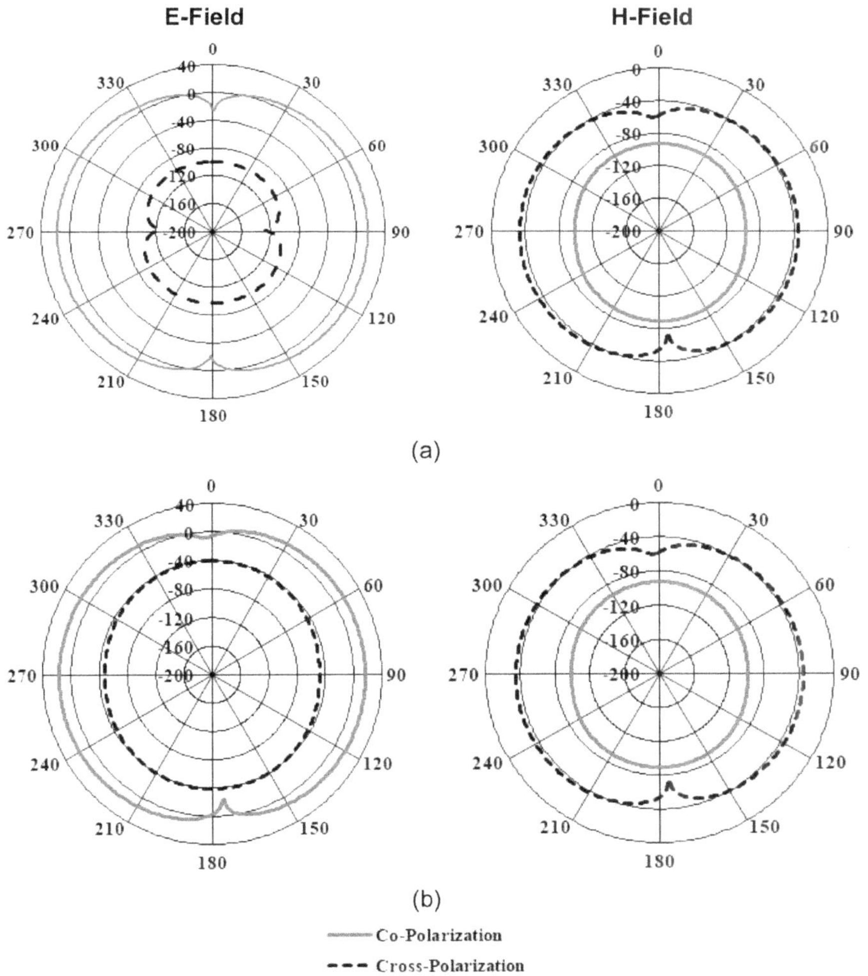

Figure 6. E, H-Field, surface current distributions of Spiral square CPW-fed flexible compact patch antenna at frequencies (a) 5.8 GHz (Planar), (b) 5.69 GHz (50 deg bent).

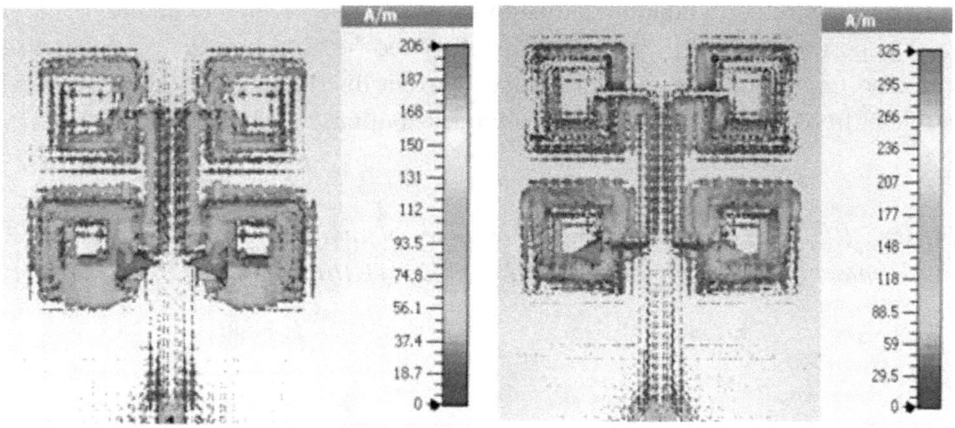

(a) (b)

**For a more accurate representation see the electronic version.*

CONCLUSION

Flexible compact spiral square CPW-fed patch antenna is designed and presented with simulated results. The proposed antenna analyzed with different bent configurations and resonates at 5.8 GHz frequency with a bandwidth of 19.4% for plane configuration and it is enough bandwidth to fulfill the requirements of the RFID or WiMAX/WLAN standards. The gain is 2.4 dBi with omnidirectional radiation pattern is observed for the proposed design. The proposed antenna is flexibility and compactness is suitable for RFID and WLAN/WiMAX at wireless applications.

ACKNOWLEDGMENT

This work was supported by the SERB, New Delhi, India. (File no1: EEQ/2016/000754, File no2: SB/FTP/ETA-0179/2014).

REFERENCES

Chen, S.-Y., & Hsu, P. (2004). CPW-fed folded-slot antenna for 5.8 GHz RFID tags. *Electronics Letters*, *40*(24), 1516–1517. doi:10.1049/el:20046780

Finkenzeller, K. (2010). Fundamentals and applications in contactless smart cards, radio frequency identification and near-field communication. In *RFID handbook*. John Wiley & Sons.

Garg, R., Bhartia, P., Bahl, I., & Ittipiboon, A. (2001). *Microstrip Antenna Design Handbook*. Artech House.

Kossel, M. A., Kung, R., Benedickter, H., & B¨achtold, W. (1999). An active tagging system using circular-polarization modulation. *IEEE Transactions on Microwave Theory and Techniques*, *47*(12), 2242–2248. doi:10.1109/22.808966

Kumar Naik, K., & Vijaya Sri, A. (2017). Design of Concentric Circular Ring Patch with DGS for Dual-Band at Satellite Communication and Radar Applications. In *Wireless Personal Communications*. Springer.

Kumar Naik, K., & Amala Vijaya Sri, P. (2018). Design of hexadecagon circular patch antenna with dgs at ku band for satellite communications. *Progress In Electromagnetics Research M*, *63*, 163–173. doi:10.2528/PIERM17092205

Kumar Naik, K., & Amala Vijaya Sri, P. (2018). Design of hexadecagon circular patch antenna with DGS at Ku band for satellite communications. *Progress In Electromagnetics Research M*, *63*, 163–173. doi:10.2528/PIERM17092205

Lee, R. Q., Lee, K. F., & Bobinchak, J. (1987, September). Characteristics of a two-layer electro-magnetically coupled rectangular patch antenna. *Electronics Letters*, *23*(20), 1070–1072. doi:10.1049/el:19870748

Lee, S. W., & Sung, Y. J. (2015). Reconfigurable Rhombus-Shaped Patch Antenna With Y-Shaped Feed for Polarization Diversity. *IEEE Antennas and Wireless Propagation Letters*, *14*, 163–166. doi:10.1109/LAWP.2014.2358651

Li, H., Zhou, Y., Mou, X., Ji, Z., Yu, H., & Wang, L. (2014). Miniature four-band CPW-fed antenna for RFID/WiMAX/WLAN applications. *IEEE Antennas and Wireless Propagation Letters*, *13*, 1684–1688. doi:10.1109/LAWP.2014.2345564

Liu, H.-W., Ku, C.-H., & Yang, C.-F. (2010). Novel CPW-fed planar monopole antenna for WiMAX/WLAN applications. *IEEE Antennas and Wireless Propagation Letters*, *9*, 240–243. doi:10.1109/LAWP.2010.2044860

Liu, W. C., & Hu, Z.-K. (2005). Broadband CPW-fed folded-slot monopole antenna for 5.8 GHz RFID application. *Electronics Letters*, *41*(17), 937–939. doi:10.1049/el:20052222

Mandal, K., & Sarkar, P. P. (2013, April). High Gain Wide-Band U-Shaped Patch Antennas with Modified Ground Planes. *IEEE Transactions and Antenna Propagations*, *61*(4), 2279–2282. doi:10.1109/TAP.2012.2233455

Mesquita, J. M., Pires, N., & Moreira, A. A. (2013). Influence of deformations on the matching of a flexible dual-band antenna. *Antennas and Propagation (EuCAP), 2013 7th European Conference on*, 3283-3285.

Niranjan, V., Saxena, A. K., & Srivastava, K. V. (2012). CPW-fed slot patch antenna for 5.2/5.8 ghz wlan application. *PIERS Proceedings,* 1350-1352.

Pues, H. F., & Van De Capelle, A. R. (1989, November). An impedance-matching technique for increasing the bandwidth of microstrip antennas. *IEEE Transactions and Antenna Propagations*, *AP-37*(11), 1345–1354. doi:10.1109/8.43553

Qu, S.-W., & Xue, Q. (2007). A Y-Shaped Stub Proximity Coupled V-Slot Microstrip Patch Antenna. *IEEE Antennas and Wireless Propagation Letters*, *6*, 40–42. doi:10.1109/LAWP.2007.891515

Santiago, A. G., Costa, J. R., & Fernandes, C. A. (2013). Broadband UHF RFID passive tag antenna for near-body applications. *IEEE Antennas and Wireless Propagation Letters*, *12*, 136–139. doi:10.1109/LAWP.2013.2243400

Schaubert, D. H., Pozar, D. M., & Adrian, A. (1989, June). Effect of microstrip antenna substrate thickness and permittivity: Comparison of theories and experiment. *IEEE Transactions and Antennas Propagations*, *AP-37*(6), 677–682. doi:10.1109/8.29353

Tseng, C.-F., Huang, C.-L., & Hsu, C.-H. (2006). A wideband planar inverted-F dielectric resonator antenna for RFID system applications. *Microwave and Optical Technology Letters*, *48*(7), 1302–1305. doi:10.1002/mop.21616

Wang, L., Johannessen, E. A., Hammond, P. A., Cui, L., Reid, S. W. J., Cooper, J. M., & Cumming, D. R. S. (2005). A programmable microsystem using system-on-chip for real-time biotelemetry. *IEEE Transactions on Biomedical Engineering*, *52*(7), 1251–1260. doi:10.1109/TBME.2005.847562 PMID:16041988

Wi, S.-H., Lee, Y.-S., & Yook, J.-G. (2007, April). Wideband Microstrip Patch Antenna With U-Shaped Parasitic Elements. *IEEE Transactions and Antenna Propagations*, *55*(4), 1196–1199. doi:10.1109/TAP.2007.893427

Chapter 7
Simulative Approach to Realize All Optical-Frequency-Encoded Dibit-Based Integrated Logic Gates:
Controlled AND/OR Logic Gates by Optical Switches

Bitan Ghosh
University of Burdwan, India

Partha Pratim Sarkar
University of Burdwan, India

ABSTRACT

Optics is considered a potential candidate for the realization of logic devices, digital optical systems for communication, and computation exploiting its super-fast speed. Optical logic gates also can act on the basis of frequency conversion process of some nonlinear materials. Further, in this chapter, the authors have mentioned the dibit representation technique for reducing bit error problem at the input and output terminals of all optical digital logic circuits and a control input for selecting particular logic operation. Here the authors have proposed frequency encoded all optical dibit-based integrated AND and OR logic gates with control input, where a single circuit acts as both AND logic gate and OR logic gate using the optical switches like reflected semiconductor optical amplifier and add/drop multiplexer.

DOI: 10.4018/978-1-5225-8531-2.ch007

Copyright © 2019, IGI Global. Copying or distributing in print or electronic forms without written permission of IGI Global is prohibited.

INTRODUCTION

For tremendous swift information processing (Ghosh et al., 2012), photon can be treated as better information carrier than electron. Many all optical logic gates (Dey & Mukhopadhyay, 2017), flip-flops (Gayen et al., 2015), bi-stable multivibrators, latches etc. are developed with optics (Amiri et al., 2015). Semiconductor optical amplifier (SOA) (Sarkar et al, 2017) is established as a very promising optical device for conducting many all- optical (Sarkar & Mukhopadhyay, 2017) logical operations (Dutta & Mukhopadhyay, *All optical* …., 2011). SOA is an optoelectronic device that under suitable operating conditions can amplify an input light signal (Cakmak et al., 2017). The active region in the device imparts gain to an input signal. Three radiative mechanisms are possible in the semiconductor. Incident light of photon of adequate energy can excite a carrier from the valance band to the conduction band in stimulated absorption. This type of process is a lossy process as the incident photon is extinguished. Now, if a photon with appropriate energy is incident on the semiconductor, it causes stimulated recombination between a conduction band carrier and a valance band hole. After this process the carrier loses its energy in the form of a photon. This new stimulated photon will be identical in all aspects to the previous inducing photon with identical phase (Chandra & Mukhopadhyay, 2012), frequency and direction that means a coherent interaction happens. Now, both the incident photon and stimulated photon can go up to additional stimulated transitions. For the sufficiently high injected current a population inversion is formed when the carrier population in the conduction band exceeds population in the valance band. For this time the likelihood of stimulated emission is greater than stimulated absorption. So semiconductor will show the optical gain. In the spontaneous emission process, there is a non-zero probability per unit time that a carrier of conduction band will spontaneously recombine with a hole of valance band and thus produce a photon (Wang & Chen, 2015) with random phase and direction. The spontaneous emissions of photons have a wide range of frequencies. These spontaneously emitted photons are basically a noise and also take part in reducing the carrier population available for the optical gain. This type of spontaneous emission of photons is a direct consequence of the amplification process.

SOA can be used as a super-fast non-linear medium. When photo-excitation increases, absorption reduces and becomes negative for high excitation levels. This causes the changing in refractive index. Slow response associated with carrier recombination is one of the most inconvenient features of the SOA for their ultra-fast signal processing (Dey & Mukhopadhyay, *Implementation* …., 2018).

For implementing (Dey & Mukhopadhyay, *A method of generation* …., 2018) the all optical logic and arithmetic devices different types of encoding principle like, spatial encoding (Saha et al., 2017), polarization encoding, frequency encoding

(Dutta & Mukhopadhyay, *Alternating* …., 2011), intensity encoding and phase encoding (Hu et al., 2017) are needed. As the frequency of light beam (Sarkar et al, 2016) remains unaffected and constant after reflection, refraction, absorption etc, the frequency encoding principle (Ghosh et al., 2018) is the most suitable one among all other encoding principles (Zhu et al., 2016).

BACKGROUND

Now, in frequency encoding or decoding technique (Garai & Mukhopadhyay, *A method of*…., 2010) two different states of information can be represented by two different frequencies at the time of data computation (Garai & Mukhopadhyay, *All-optical*…., 2010). Mostly, the presence of a specific frequency of light is treated as '1' logic state and other specific frequency represents as '0' logic state. All optical schemes are designed by all optical switches (Sarkar et al, 2013) like add/drop multiplexer (ADM) (Chen et al., 2016) and reflected semiconductor optical amplifier (RSOA). Add/drop multiplexer (Mansoor et al., 2015) (ADM) is a frequency choosy optical switching device. If it is biased by a light beam of particular frequency then the particular frequency of light beam is reflected by ADM (Rashed, 2012) from input and passes all other frequency of light beam. Opposite incident happens if the biasing input changes. Now, reflected semiconductor optical amplifier (RSOA) (Sarkar & Mukhopadhyay, 2014) is another optical switching device (Kim et al., 2011), where a weak probe beam with particular frequency (υ_1 frequency) and a strong pump beam with another frequency (say, υ_2 frequency) are inserted to the input terminals of RSOA (Wang & Foster, 2012). Then, with the power of the pump beam this switch (Singh et al., 2007) will give light beam of υ_1 frequency. Therefore they should be recognized as extremely useful optical switches for implementing several all optical arithmetical and logical operations (Singh et al., 2016).

The dibit representation (Ghosh & Mukhopadhyay, 2013) has the advantage of the possibility of implementing ternary, quaternary, octal and decimal operations with the help of optics. This technique (Mukhopadhyay, 1992) can be done to represent a digit by two consecutive bits. These two bits may be considered like $\upsilon_1 = 195$ THz and $\upsilon_2 = 193$ THz frequencies. Now if we denote the optical frequency of light beam υ_1 = digital logic state '0' and υ_2 = digital logic state '1' then by the dibit system this digital logic state '0' is represented as <0><1> and digital logic state '1' is denoted as <1><0> respectively. Thus, the two frequencies $<\upsilon_1><\upsilon_2>$ represents logic state '0' and similarly $<\upsilon_2><\upsilon_1>$ represents logic state '1'. Here, the authors have proposed an alternative approach to realize all optical frequency encoded dibit based integrated AND and OR logic gates with a control input for selecting particular logic operation with the help of RSOA and ADM. This scheme is successfully implemented on integrated AND and OR logic operations with a single circuit just altering one switch. Also the truth tables have been verified with the simulation result.

REALIZATION OF THE INTEGRATED AND OR LOGIC GATES

At first, the scheme will be divided into three units for realization of all optical frequency encoded dibit based integrated two input AND and OR logic gates with a control input using optical switches. They are controlling dibit input unit, controlling logic unit and controlling dibit output unit.

Controlling Dibit Input Unit

Here, the two input terminals are ‘A’ and ‘B’. These two channels are divided further as ‘A′’ and ‘A″’, ‘B′’ and ‘B″’ respectively to provide dibit inputs. In this single bit transformation one optical light beam of frequency υ_1 or digital ‘0’ is treated as $<\upsilon_1><\upsilon_2>$ or <0><1> in case of dibit representation technique. Similarly, optical light beam of frequency υ_2 or digital logic state ‘1’ is considered as $<\upsilon_2><\upsilon_1>$ or <1><0> for dibit system. The block diagram of controlling dibit input unit is shown in Figure 1. Here, the authors have designed the controlling dibit input unit, which offers the actual or effective dibit, i. e. this system allows the input in form of dibit $<\upsilon_1><\upsilon_2>$ or <0><1> or digital ‘0’ and dibit $<\upsilon_2><\upsilon_1>$ or <1><0> or digital ‘1’. The system can reject to operate for any other formation.

Now for this reason, if we apply $<\upsilon_1><\upsilon_1>$ (or, <0><0>) to ‘A′’ and ‘A″’, then, from ‘A′’ terminal, one part of light beam of υ_1 frequency acts as a biasing of ADM_1 and other part acts as probe beam of $RSOA_1$. Now, from ‘A″’ terminal, light beam of υ_1 frequency goes to the input of ADM_1, so ADM_1 restricts to penetrate the light beam of υ_1 frequency, since υ_1 frequency acts as a biasing of ADM_1. Therefore, $RSOA_1$ has no pump beam, as output of ADM_1 connected to pump beam terminal of $RSOA_1$. So, there is no light beam at the ‘A’ terminal. This is shown in the first combination for the channel ‘A’ in Table 1. Again, if we apply $<\upsilon_2><\upsilon_2>$ (or, <1><1>) to ‘A′’ and ‘A″’ then, from ‘A′’ terminal, one part of light beam of υ_2 frequency acts as a biasing of ADM_1 and other part acts as probe beam of $RSOA_1$. Now, from ‘A″’ terminal, light beam of υ_2 frequency goes to the input of ADM_1, so ADM_1 restricts to penetrate the light beam of υ_2 frequency as υ_2 frequency acts as a biasing of ADM_1. Therefore, $RSOA_1$ has no pump beam. So, there is no light beam at the ‘A’ terminal. Also this is shown in the fourth combination for the channel ‘A’ in Table 1.This principle of operation will be applicable for ‘B’ terminal also. When we apply $<\upsilon_1><\upsilon_1>$ (or, <0><0>) to ‘B′’ and ‘B″’, then, one part of light

Figure 1. Block diagram of controlling dibit input unit

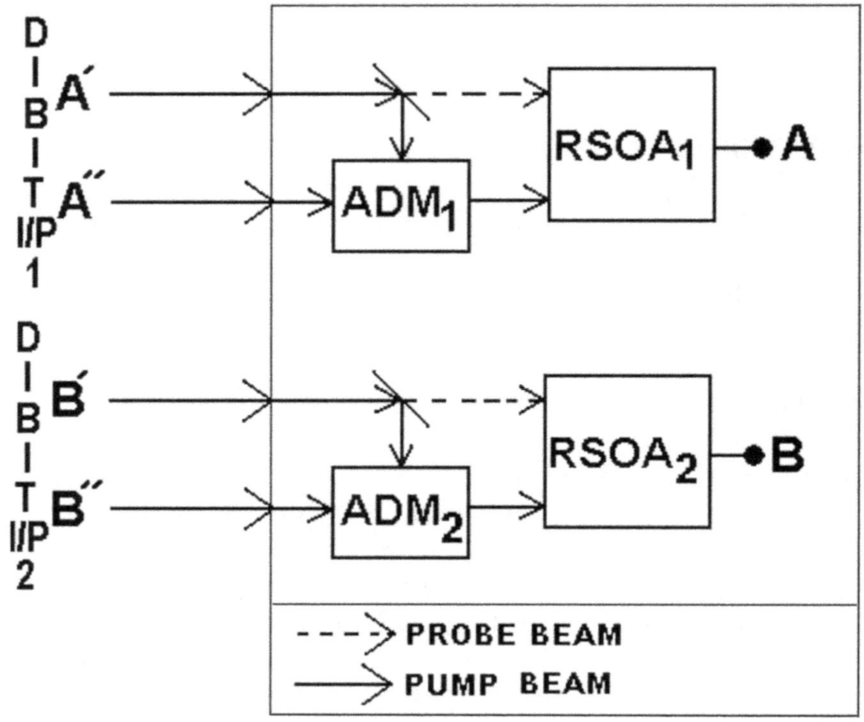

Table 1. Truth table of controlling dibit input unit

1st Dibit Input		Controlled single bit form for Channel A	2nd Dibit Input		Controlled single bit form for Channel B
A′	**A′′**	**A**	**B′**	**B′′**	**B**
υ_1 (0)	υ_1 (0)	×	υ_1 (0)	υ_1 (0)	×
υ_1 (0)	υ_2 (1)	υ_1 (0)	υ_1 (0)	υ_2 (1)	υ_1 (0)
υ_2 (1)	υ_1 (0)	υ_2 (1)	υ_2 (1)	υ_1 (0)	υ_2 (1)
υ_2 (1)	υ_2 (1)	×	υ_2 (1)	υ_2 (1)	×

beam of υ_1 frequency from 'B′′ terminal, acts as a biasing of ADM_2 and other part acts as probe beam of $RSOA_2$. Now, the light beam of υ_1 frequency from 'B′′' terminal goes to the input of ADM_2, so ADM_2 blocks to penetrate the light beam of υ_1 frequency, since υ_1 frequency acts as a biasing of ADM_2. Therefore, $RSOA_2$ has no pump beam, as output of ADM_2 connected to pump beam terminal of $RSOA_2$.

So, there is no light beam at the 'B' terminal. This is shown in the first combination for the channel 'B' in Table 1. Again, if we apply $<\upsilon_2><\upsilon_2>$ (or, $<1><1>$) to 'B′' and 'B″' then, from 'B′' terminal, one part of light beam of υ_2 frequency acts as a biasing of ADM_2 and other part acts as probe beam of $RSOA_2$. Now, from 'B″' terminal, light beam of υ_2 frequency goes to the input of ADM_2, so ADM_2 restricts to penetrate the light beam of υ_2 frequency as υ_2 frequency acts as a biasing of ADM_2. Therefore, $RSOA_2$ has no pump beam. So, there is no light beam at the 'B' terminal. Also it is shown in the fourth combination for the channel 'B' in Table 1.

But, if we apply $<\upsilon_1><\upsilon_2>$ (or, $<0><1>$) to 'A′' and 'A″', then, from 'A′' terminal, one part of light beam of υ_1 frequency act as a biasing of ADM_1 and other part acts as probe beam of $RSOA_1$. Now, from 'A″' terminal, light beam of υ_2 frequency goes to the input of ADM_1, so ADM_1 passes the light beam of υ_2 frequency. So, $RSOA_1$ has the pump beam of υ_2 frequency. As we know that, frequency of the output beam of RSOA is same as the frequency of the probe beam when pump beam is present. So, there is a light beam of υ_1 frequency (or, digital '0') at 'A'. This is shown in the second combination for the channel 'A' in Table 1. This principle of operation will be applicable for 'B' terminal also. When we apply $<\upsilon_1><\upsilon_2>$ (or, $<0><1>$) to 'B′' and 'B″', then, from 'B′' terminal, one part of light beam of υ_1 frequency act as a biasing of ADM_2 and other part acts as probe beam of $RSOA_2$. Now, from 'B″' terminal, light beam of υ_2 frequency goes to the input of ADM_2, so ADM_2 passes the light beam of υ_2 frequency. So, $RSOA_2$ has the pump beam of υ_2 frequency. So, there is a light beam of υ_1 frequency (or, digital state '0') at the 'B' terminal. Also it is shown in the second combination for the channel 'B' in Table 1. Now, if we apply $<\upsilon_2><\upsilon_1>$ (or, $<1><0>$) to 'A′' and 'A″' then, from 'A′' terminal, one part of light beam of υ_2 frequency act as a biasing of ADM_1 and other part acts as probe beam of $RSOA_1$. Now, from 'A″' terminal, light beam of υ_1 frequency goes to the input of ADM_1, so ADM_1 passes the light beam of υ_1 frequency. So, $RSOA_1$ has the pump beam of υ_1 frequency. So, there is a light beam of υ_2 frequency (or, digital state '1') at the 'A' terminal. It is shown in the third combination for the channel 'A' in Table 1. Similarly, this process will be applicable for 'B' terminal also. When we apply $<\upsilon_2><\upsilon_1>$ (or, $<1><0>$) to 'B′' and 'B″' then, from 'B′' terminal, one part of light beam of υ_2 frequency act as a biasing of ADM_1 and other part acts as probe beam of $RSOA_1$. Now, from 'B″' terminal, light beam of υ_1 frequency goes to the input of ADM_2, so ADM_2 passes the light beam of υ_1 frequency. So, $RSOA_2$ has the pump beam of υ_1 frequency. So, there is a light beam of υ_2 frequency (digital '1') at 'B' channel. Also it is shown in the third combination for the channel 'B' in Table 1

Therefore, this unit restricts those unwanted inputs, like $<\upsilon_1><\upsilon_1>$ (or, $<0><0>$) and $<\upsilon_2><\upsilon_2>$ (or, $<1><1>$) to perform, because they are not the proper inputs as per logical representations of dibit and it checks the real dibit input combinations, like $<\upsilon_1><\upsilon_2>$ (or, $<0><1>$) and $<\upsilon_2><\upsilon_1>$ (or, $<1><0>$), which are shown in Table 1. So it can be said that the inputs are controlled before entering to the system and reduces bit error problem at input terminals, coming from the channels.

Controlling Logic Unit

Two input terminals 'A' and 'B', control input terminal 'C' and one controlled output terminal 'Y' are used to implement the controlling logic unit, which is shown in Figure 2. Now, this scheme of unit provides output as AND as well as OR logic depending upon the control input states of light beam of υ_1 (or digital state '0') and υ_2 (or digital state '1') frequencies.

Controlling Logic Unit Acts as AND Logic Gate

Considering control input terminal 'C' as light beam of υ_1 frequency (or digital state '0'), at first, if light beam of υ_1 frequency (or digital state '0') is given to both the

Figure 2. Block diagram of controlling logic unit

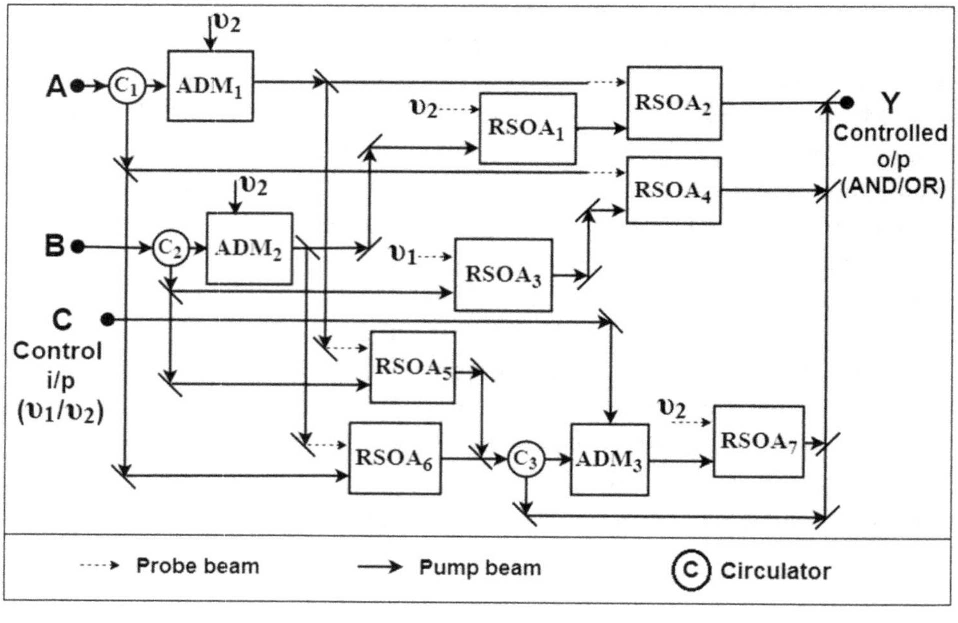

input terminals of 'A' and 'B' at the same time for realizing this scheme. So, light beam of υ_1 frequency, coming from terminal 'A' is passed by ADM_1 because ADM_1 is biased by υ_2 frequency and this light beam of υ_1 frequency goes as probe beam of $RSOA_2$ and one part of this beam is coming as probe beam of $RSOA_5$. Again, light beam of υ_1 frequency, coming from terminal 'B' is passed by ADM_2 as it is biased by υ_2 frequency and the first part of this light beam goes as probe beam of $RSOA_6$ but other part acts as pump beam of $RSOA_1$, which has constant probe beam of υ_2 frequency. Therefore, $RSOA_1$ provides light beam of υ_2 frequency, which will act as pump beam of $RSOA_2$. So, $RSOA_2$ will give light beam of υ_1 frequency (or digital state '0') to the controlled output terminal 'Y' because light beam of υ_1 frequency is present as probe beam of $RSOA_2$. But $RSOA_5$ does not work as there is no pump beam coming from circulator 'C_2'. Similarly, $RSOA_3$ (with constant probe beam of υ_1 frequency) does not work as there is no pump beam coming from circulator 'C_2'. Since $RSOA_3$ does not work then $RSOA_4$ does not work. Again $RSOA_6$ does not have any pump beam coming from circulator 'C_1'. So $RSOA_6$ does not work. Though light beam of υ_1 frequency coming from control input terminal 'C', acts as a biasing of ADM_3, ADM_3 does not provide any light beam because $RSOA_5$ and $RSOA_6$ do not work. As there is no pump beam of $RSOA_7$ (with constant probe beam of υ_2 frequency) coming from ADM_3, $RSOA_7$ does not work. So, only $RSOA_2$ will give light beam of υ_1 frequency (or digital state '0') to the controlled output terminal 'Y'. It is theoretically satisfied by the first input combination of truth table of AND logic gate, which is given in the Table 2.

Now, light beam of υ_1 (or digital state '0') frequency is given to the input terminal of 'A' and υ_2 (or digital state '1') frequency is given to the input terminal of 'B'. By the similar way, light beam of υ_1 frequency, coming from terminal 'A' is passed by ADM_1(biased by υ_2 frequency) and the light beam goes as probe beam of $RSOA_2$ and other part of this beam is coming as probe beam of $RSOA_5$. Now, $RSOA_1$ (with constant probe beam of υ_2 frequency) has no pump beam coming from ADM_2. So,

Table 2. Truth table of controlling logic unit acts as AND logic gate

Controlled single bit (or Digital) form input at channel 'A'	Controlled single bit (or Digital) form input at channel 'B'	Control input single bit (or Digital) form at terminal 'C'	Controlled output terminal 'Y'
A	B	C	Y
$\upsilon_1(0)$	$\upsilon_1(0)$	$\upsilon_1(0)$	$\upsilon_1(0)$
$\upsilon_1(0)$	$\upsilon_2(1)$		$\upsilon_1(0)$
$\upsilon_2(1)$	$\upsilon_1(0)$		$\upsilon_1(0)$
$\upsilon_2(1)$	$\upsilon_2(1)$		$\upsilon_2(1)$

$RSOA_1$ does not provide pump beam of $RSOA_2$. Therefore, $RSOA_2$ does not work. But, light beam of υ_2 frequency, coming from terminal 'B' is reflected by ADM_2 because ADM_2 is biased by υ_2 frequency and one part of the reflected beam comes as pump beam of $RSOA_3$, which has constant probe beam of υ_1 frequency. So, $RSOA_3$ gives light beam of υ_1 frequency, which will act as pump beam of $RSOA_4$. But $RSOA_4$ doesn't work as the lack of probe beam of $RSOA_4$ coming from circulator 'C_1'. Now, other part of the reflected beam of ADM_2 comes as pump beam of $RSOA_5$. Thus, $RSOA_5$ provides light beam of υ_1 frequency because light beam of υ_1 frequency (coming from ADM_1) is present as probe beam of $RSOA_5$. But $RSOA_6$ neither has probe beam coming from ADM_2 nor has pump beam coming from circulator 'C_1'. Therefore, $RSOA_6$ does not work. Now, the light beam coming from $RSOA_5$ is entering to the ADM_3 which is biased by υ_1 frequency, coming from control input terminal 'C'. So, light beam of υ_1 frequency coming from $RSOA_5$ is reflected by ADM_3. As there is no pump beam of $RSOA_7$ (with constant probe beam of υ_2 frequency) coming from ADM_3, $RSOA_7$ does not work. Only the reflected light beam of υ_1 frequency (or digital state '0') coming from circulator 'C_3' goes to the controlled output terminal 'Y', which is theoretically satisfied by the second input combination of AND logic gate (truth table is given in the Table 2).

Next, light beam of υ_2 (or digital state '1') and υ_1 (or digital state '0') frequencies are inserted to the input channels of 'A' and 'B' respectively. Therefore, light beam of υ_2 frequency, coming from terminal 'A' is reflected by ADM_1 (biased by υ_2 frequency) and this beam goes as probe beam of $RSOA_4$ and other part of the reflected beam from circulator 'C_1' comes as pump beam of $RSOA_6$. Now, light beam of υ_1 frequency, coming from terminal 'B' is passed by ADM_2 (biased by υ_2 frequency) and one part of this light beam propagates as pump beam of $RSOA_1$, which has constant probe beam of υ_2 frequency. Thus, $RSOA_1$ provides light beam of υ_1 frequency as pump beam of $RSOA_2$. Here, $RSOA_2$ has no probe beam coming from ADM_1. So, $RSOA_2$ does not work. Now, the other part of the light beam of υ_1 frequency, coming from ADM_2, goes as probe beam of $RSOA_6$. Since light beam of υ_2 frequency is present as pump beam, $RSOA_6$ gives light beam of υ_1 frequency. Whereas, $RSOA_3$ (with constant light beam of υ_1 frequency) has no pump beam coming from circulator 'C_2', $RSOA_3$ does not work. So, $RSOA_4$ does not work as there is no pump beam coming from $RSOA_3$. Again, $RSOA_5$ neither has probe beam coming from ADM_1 nor has pump beam coming from circulator 'C_2'. Therefore, $RSOA_5$ does not work. So, the light beam of υ_1 frequency from $RSOA_6$ goes to the ADM_3 (biased by υ_1 frequency, coming from control input terminal 'C'). As a result, light beam of υ_1 frequency is reflected by ADM_3. Since there is no pump beam of $RSOA_7$ coming from ADM_3, $RSOA_7$ does not work. But reflected light beam of υ_1 frequency (or digital state '0') coming from ADM_3 via circulator 'C_3' goes to the controlled output terminal 'Y'. It is satisfied by the third input combination of AND logic gate truth table (shown in the Table 2).

Lastly, light beam of υ_2 frequency (or digital state '1') is inserted to both the input terminals of 'A' and 'B' at the same time. So, light beam of υ_2 frequency, coming from terminal 'A' is reflected by ADM_1 which is biased by υ_2 frequency and the first part of this light beam goes as probe beam of $RSOA_4$ and the other part of the reflected beam comes as pump beam of $RSOA_6$. Again, light beam of υ_2 frequency, coming from terminal 'B' is reflected by ADM_2 and one part of the reflected beam comes as pump beam of $RSOA_5$. Whereas the other part of this reflected light comes as pump beam of $RSOA_3$, which has constant probe beam of υ_1 frequency. So, $RSOA_3$ provides light beam of υ_1 frequency, which will proceed as pump beam of $RSOA_4$. As a result, $RSOA_4$ provides light beam of υ_2 frequency. As there is no pump beam of $RSOA_1$ (with constant light beam of υ_2 frequency) $RSOA_1$ does not work. So $RSOA_2$ has no pump beam (no probe beam also) thus it does not work. Again, $RSOA_5$ does not have probe beam coming from ADM_1 and $RSOA_6$ does not have probe beam coming from ADM_2. Therefore, both the $RSOA_5$ and $RSOA_6$ do not work. So there is no input to the ADM_3 and obviously ADM_3 does not work though light beam of υ_1 frequency coming from control input terminal 'C', acts as a biasing of ADM_3. Thus, with constant light beam of υ_2 frequency $RSOA_7$ does not work as there is no pump beam coming from ADM_3. So the light beam of υ_2 frequency (or digital state '1') coming from $RSOA_4$ goes to the controlled output terminal 'Y'. It is satisfied by the fourth input combination of truth table of AND logic gate (given in the Table 2).

By this way, after observing the logic output from the controlled output terminal 'Y' for these four input combinations, we can say that, the logic output fully satisfies the truth table of AND logic gate, for control input terminal 'C' as constant light beam of υ_1 frequency (or digital state '0'). This is shown in Table 2.

Controlling Logic Unit Acts as OR Logic Gate

Here, at first, light beam of υ_1 frequency (or digital state '0') is inserted to both the inputs of 'A' and 'B' terminals simultaneously, maintaining control input terminal 'C' of constant light beam of υ_2 frequency (or digital state '1'). Thus, light beam of υ_1 frequency, coming from terminal 'A' is passed by ADM_1 (biased by υ_2 frequency) and goes as probe beam of $RSOA_2$ and another part of this beam is coming as probe beam of $RSOA_5$. Similarly, light beam of υ_1 frequency, coming from terminal 'B' is passed by ADM_2 (biased by υ_2 frequency) and the first part of this light beam goes as probe beam of $RSOA_6$. But other part acts as pump beam of $RSOA_1$, which has constant probe beam of υ_2 frequency. Therefore, $RSOA_1$ provides light beam of υ_2 frequency, which will act as pump beam of $RSOA_2$. So, $RSOA_2$ will give light beam of υ_1 frequency. $RSOA_5$ does not work due to absent of pump beam coming from circulator 'C_2'. And $RSOA_6$ is stop for the absence of pump beam coming from circulator 'C_1'. So, there is no light beam for the input of ADM_3 even if light beam of υ_2 frequency coming from control input terminal 'C', which acts as a biasing of ADM_3. Therefore, ADM_3 does not provide any light beam for the pump beam of $RSOA_7$ (with constant light beam of υ_2 frequency). So, $RSOA_7$ does not work. Again, $RSOA_3$ (with constant probe beam of υ_1 frequency) does not work as there is no pump beam coming from circulator 'C_2'. Since $RSOA_3$ does not work then $RSOA_4$ does not work. Therefore, light beam of υ_1 frequency (or digital state '0') from $RSOA_2$ goes to the controlled output port 'Y'. It is satisfied by the first input combination of truth table of OR logic gate, which is given in the Table 3.

Now, light beam of υ_1 (or digital '0') and υ_2 (or digital '1') frequencies are inserted to the inputs of 'A′' and 'B′' respectively. By the similar way, light beam of υ_1 frequency, coming from terminal 'A' is passed by ADM_1 and one part goes as probe beam of $RSOA_2$ and other part of this beam is coming as probe beam of $RSOA_5$. But, light beam of υ_2 frequency, coming from terminal 'B' is reflected by ADM_2 which is biased by υ_2 frequency and the part of this reflected light beam goes as pump beam of $RSOA_3$, which has constant probe beam of υ_1 frequency. So, $RSOA_3$ gives light beam of υ_1 frequency, which will act as pump beam of $RSOA_4$. But $RSOA_4$ doesn't work as the absence of the probe beam coming from circulator 'C_1'. Now, other part of the reflected beam of ADM_2 comes as pump beam of $RSOA_5$. Thus, $RSOA_5$ provides light beam of υ_1 frequency because light beam of υ_1 frequency is present as probe beam of $RSOA_5$. Now, the beam, coming from $RSOA_5$, proceeds to the ADM_3 which is biased by υ_2 frequency, coming from control input terminal 'C'. So, light beam of υ_1 frequency is passed by ADM_3 and goes as pump beam of $RSOA_7$, which has constant probe beam of υ_2 frequency. Therefore, $RSOA_7$ provides light beam of υ_2 frequency. As $RSOA_1$ (with constant light beam of υ_2 frequency) does not have any pump beam, So $RSOA_1$ does not generate any light beam as pump beam of $RSOA_2$. Therefore, $RSOA_2$ does not work. Only light beam of υ_2 frequency (or digital state '1') coming from $RSOA_7$ propagates to the controlled output channel 'Y'. This is satisfied by the second input combination of truth table of OR logic gate, which is given in the Table 3.

Next, light beam of υ_2 (or digital logic state '1') and υ_1 (or digital logic state '0') frequencies are inserted to the inputs of 'A' and 'B' respectively. Therefore, light beam of υ_2 frequency, coming from terminal 'A' is reflected by ADM_1 which is biased by υ_2 frequency and the part of this light beam comes as probe beam of $RSOA_4$ and other part of the reflected beam comes as pump beam of $RSOA_6$. Now, light beam of υ_1 frequency, coming from terminal 'B' is passed by ADM_2 which is biased by υ_2 frequency and the part of this light beam propagates as pump beam of $RSOA_1$, which has constant probe beam of υ_2 frequency. Thus, $RSOA_1$ provides light beam of υ_1 frequency as pump beam of $RSOA_2$. As there is no probe beam of $RSOA_2$ coming from ADM_1, it does not work. Again, as $RSOA_3$ (with constant light beam of υ_1 frequency) has no pump beam coming from circulator 'C_2', $RSOA_3$ does not work. So, $RSOA_4$ does not work as there is no pump beam coming from $RSOA_3$. Whereas, $RSOA_3$ (with constant light beam of υ_1 frequency) has no pump beam coming from circulator 'C_2', $RSOA_3$ does not work. So, $RSOA_4$ does not work as there is no pump beam coming from $RSOA_3$. Again, $RSOA_5$ neither has probe beam coming from ADM_1 nor has pump beam coming from circulator 'C_2'. Therefore, $RSOA_5$ does not work. Now, other part of the beam, coming from ADM_2, goes as probe beam of $RSOA_6$. Since there is a light beam of υ_2 frequency is present as pump beam coming from circulator 'C_1', $RSOA_6$ gives light beam of υ_1 frequency to the ADM_3 (biased by υ_2 frequency, coming from control input terminal 'C'). As a result, light beam of υ_1 frequency is passed by ADM_3 and proceeds as pump beam of $RSOA_7$, which has constant probe beam of υ_2 frequency. Therefore, $RSOA_7$ provides light beam of υ_2 frequency (or digital state '1') to the controlled output terminal 'Y'. It is satisfied by the third input combination of truth table of OR logic gate (shown in Table 3).

Finally, light beam of υ_2 frequency (or digital state '1') is set to both the input channels of 'A' and 'B' simultaneously. Thus, light beam of υ_2 frequency, coming from terminal 'A' is reflected by ADM_1 (biased by υ_2 frequency) and the part of this beam comes as probe beam of $RSOA_4$ and other part of the reflected beam comes as pump beam of $RSOA_6$. Again, light beam of υ_2 frequency, coming from terminal 'B' is reflected by ADM_2 and one part of the reflected beam comes as pump beam of $RSOA_5$. While another part of the reflected light beam enters as pump beam of $RSOA_3$, which has constant probe beam of υ_1 frequency. So, $RSOA_3$ provides light beam of υ_1 frequency, which will proceed as pump beam of $RSOA_4$. As a result, $RSOA_4$ gives the light beam of υ_2 frequency. Since there is no pump beam of $RSOA_1$ (with constant light beam of υ_2 frequency) $RSOA_1$ does not work. Therefore $RSOA_2$

Table 3. Truth table of controlling logic unit acts as OR logic gate

Controlled single bit (or Digital) form input at channel 'A'	Controlled single bit (or Digital) form input at channel 'B'	Control input single bit (or Digital) form at terminal 'C'	Controlled output terminal 'Y'
A	B	C	Y
υ_1 (0)	υ_1 (0)	υ_2 (1)	υ_1 (0)
υ_1 (0)	υ_2 (1)		υ_2 (1)
υ_2 (1)	υ_1 (0)		υ_2 (1)
υ_2 (1)	υ_2 (1)		υ_2 (1)

has no pump beam as well as no probe beam therefore it does not work. Again, $RSOA_5$ does not have probe beam coming from ADM_1 so $RSOA_5$ does not work. And $RSOA_6$ does not have probe beam coming from ADM_2. Therefore, $RSOA_6$ does not work. So there is no input to the ADM_3 and obviously ADM_3 does not work though light beam of υ_1 frequency coming from control input terminal 'C', acts as a biasing of ADM_3. So, $RSOA_7$ (with constant light beam of υ_2 frequency) does not work as there is no pump beam coming from ADM_3. So the light beam of υ_2 frequency (or digital state '1') coming from $RSOA_4$ goes to the controlled output terminal 'Y'. It is fully satisfied by the fourth input combination of truth table of OR logic gate (given in the Table 3).

In this way, we can say that, the logic output coming from the controlled output terminal 'Y' satisfies the truth table of OR logic gate after analyzing four input combinations for control input terminal 'C' as constant light beam of υ_2 frequency (or digital state '1'). And this is shown in Table 3.

Controlling Dibit Output Unit

After getting the output from controlling logic unit at 'Y' terminal, we can proceed to the controlling dibit output unit. The block diagram of controlling dibit output unit is shown in Figure 3.

Figure 3. Block diagram of controlling dibit output unit

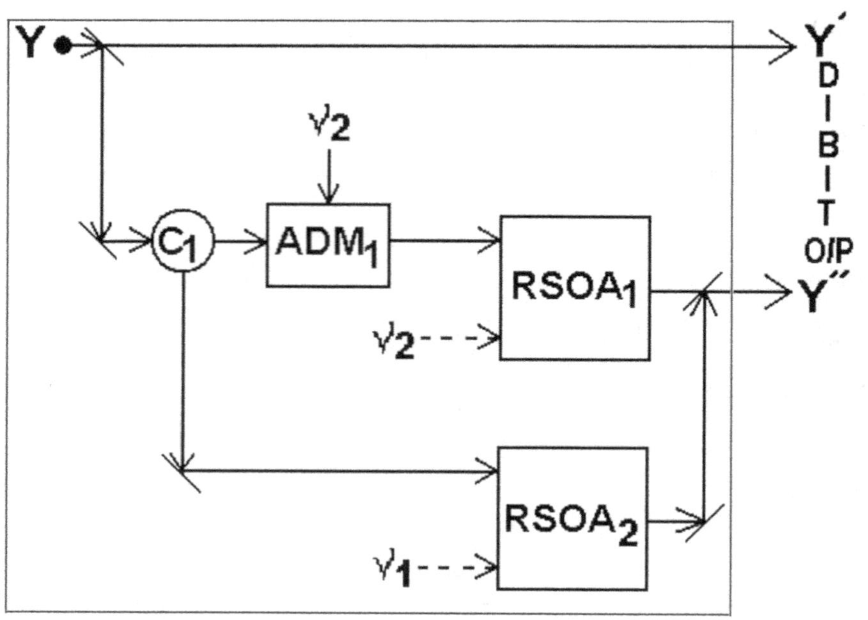

Now, if light beam of υ_1 frequency (or, digital state ‘0’) is present at ‘Y’ terminal, then this beam goes directly to the ‘Y′’ and we can get light beam of υ_1 frequency (or, digital logic state ‘0’) at ‘Y′’ output terminal. And the other part of the light beam of υ_1 frequency from ‘Y’ terminal is passed by ADM_1, which is biased by υ_2 frequency and goes as pump beam of $RSOA_1$. But a constant probe beam of υ_2 frequency is present in $RSOA_1$. So, we can get light beam of υ_2 frequency (digital ‘1’) at ‘Y″’ output terminal. $RSOA_2$ (with constant light beam of υ_1 frequency) does not work, since there is no pump beam of $RSOA_2$ coming from circulator ‘C_2’. This is shown in second last row in Table 4.

By the same way, if light beam of υ_2 frequency (or, digital state ‘1’) is present at ‘Y’ terminal, then this light beam goes directly to the ‘Y′’ output terminal and we can get light beam of υ_2 frequency (digital ‘1’) at ‘Y′’ terminal. Other part of the light beam of υ_2 frequency from ‘Y’ terminal is reflected by ADM_1, which is biased by υ_2 frequency and goes as pump beam of $RSOA_2$. But we can get light beam of υ_1 frequency (or, digital logic state ‘0’) at ‘Y″’ output terminal as there is a constant probe beam of υ_1 frequency. $RSOA_1$ (with constant light beam of υ_2 frequency) does not work, since there is no pump beam of $RSOA_1$ coming from ADM_1. It is shown in last row in Table 4.

Therefore, if light beam of υ_1 frequency (or, digital state ‘0’) and υ_2 frequency (or, digital state ‘1’) are present at ‘Y’ terminal then we get output as $<\upsilon_1><\upsilon_2>$ (or, $<0><1>$) and $<\upsilon_2><\upsilon_1>$ (or, $<1><0>$) at ‘Y′’ and ‘Y″’ terminals respectively. This is shown in Table 4.

Finally, the overall scheme of the frequency encoded all optical dibit based integrated AND and OR logic gates with control input is shown in the Figure 4.

SIMULATION OF THE INTEGRATED AND & OR LOGIC GATES

MATLAB (R2008a) simulink tools have been used for simulation of all optical frequency encoded dibit based integrated two input AND and OR logic gates with a control input using optical switches (RSOA and ADM). “DIBIT I/P 1”, “DIBIT I/P 2”, “Control i/p” and “DIBIT O/P” terminals are used for applying proper inputs

Table 4. Truth table of controlling dibit output unit

Single bit (or Digital) form at terminal ‘Y’	Controlled dibit output	
Y	Y′	Y″
υ_1 (0)	υ_1 (0)	υ_2 (1)
υ_2 (1)	υ_2 (1)	υ_1 (0)

Figure 4. Block diagram of dibit based integrated two input AND and OR logic gates

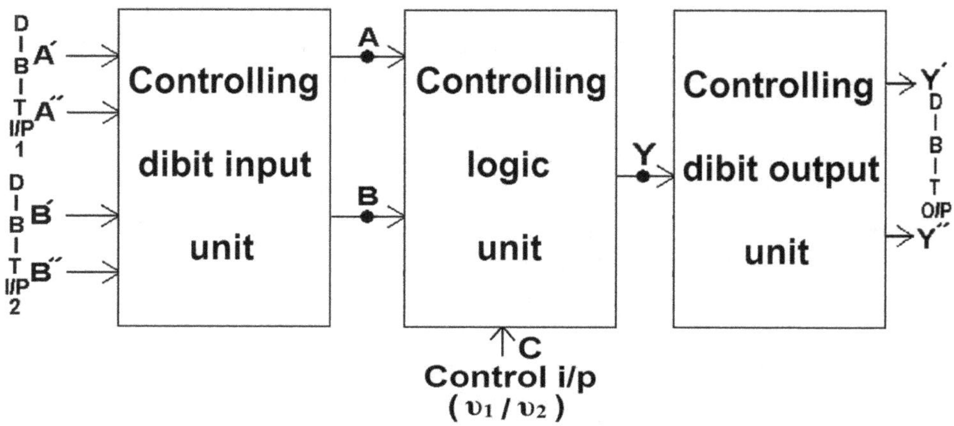

and getting proper output. In simulation process, for providing the two frequency states, υ_1 frequency (or, digital state '0') and υ_2 frequency (or, digital state '1') are represented by '3' and '8' respectively. Maintaining the similarity of the block diagram (shown in Figure 4) simulative model of the integrated scheme when acts as AND logic gate is shown in Figure 5 keeping control input as υ_1 frequency or '3'. Another simulative model of the integrated scheme when acts as OR logic gate is also shown in Figure 6 keeping control input as υ_2 frequency or '8'.

Now, each of the RSOA blocks is properly programmed according to the RSOA mechanism with two terminals "pr" for probe beam and "pm" for pump beam and

Figure 5. Simulative model of the integrated scheme when acts as AND logic gate

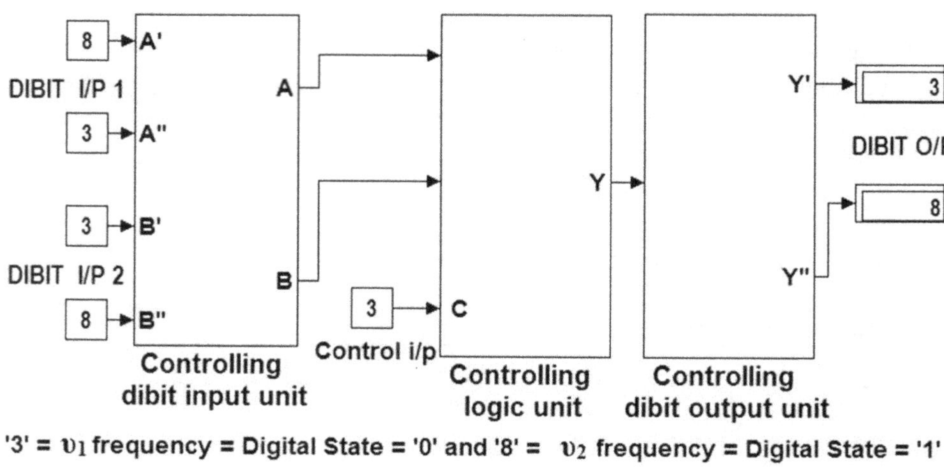

Figure 6. Simulative model of the integrated scheme when acts as OR logic gate

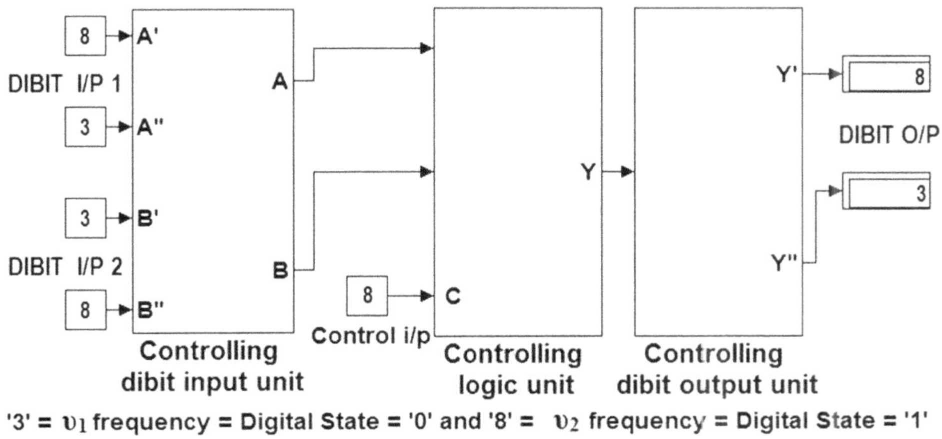

"y" as output line in RSOA blocks. According to the theory of RSOA, if pump beam and probe beam are treated as say "3 peta Hz" = υ_1 = digital logic state "0" and "8 peta Hz" = υ_2 = digital logic state "1" respectively, then we get "8 peta Hz" = υ_2 = digital "1" at the output line of these blocks. The state of the output alters if pump beam and probe beam are changed consequently.

Also, ADM block has been simulated such a way that, it obeys the proper mechanism of ADM. Now, if ADM is biased by a particular light beam frequency, say "3 peta Hz" = υ_1 = digital logic state "0" at "bias" terminal, then this block drops the frequency coming from "input" terminal to "drop" terminal and passes other frequency, say "8 peta Hz" = υ_2 = digital logic state "1" to "out" terminal. If the biasing is changed, opposite incident happens. Internally, RSOA and ADM blocks have been connected in proper way for modeling three internal units (subsystems) of the 'controlling dibit input unit', 'controlling logic unit' and 'controlling dibit output unit' of the simulative model, which are shown in Figure 7, Figure 8 and Figure 9 respectively, maintaining the similarity of the theoretical block diagrams in Figure 1, Figure 2 and Figure 3 respectively.

RESULTS AND DISCUSSION

Now, any input combination must be taken into consideration for satisfying the above integrated scheme of all optical digital circuit. For example, if we apply <8><3> and <3><8> are inserted at the two dibit input lines and '3' is inserted as control

Figure 7. Internal simulative model of the controlling dibit input unit (subsystem)

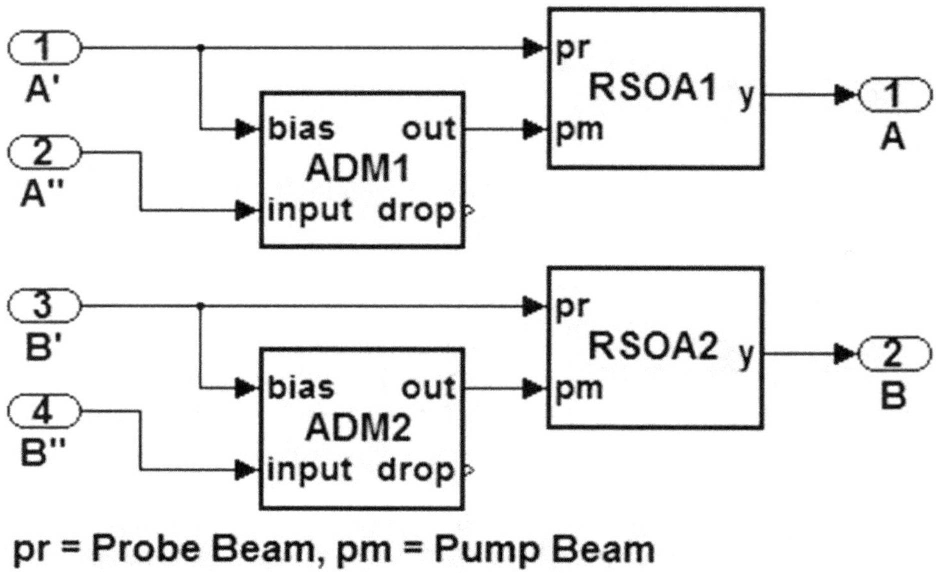

Figure 8. Internal simulative model of the controlling logic unit (subsystem)

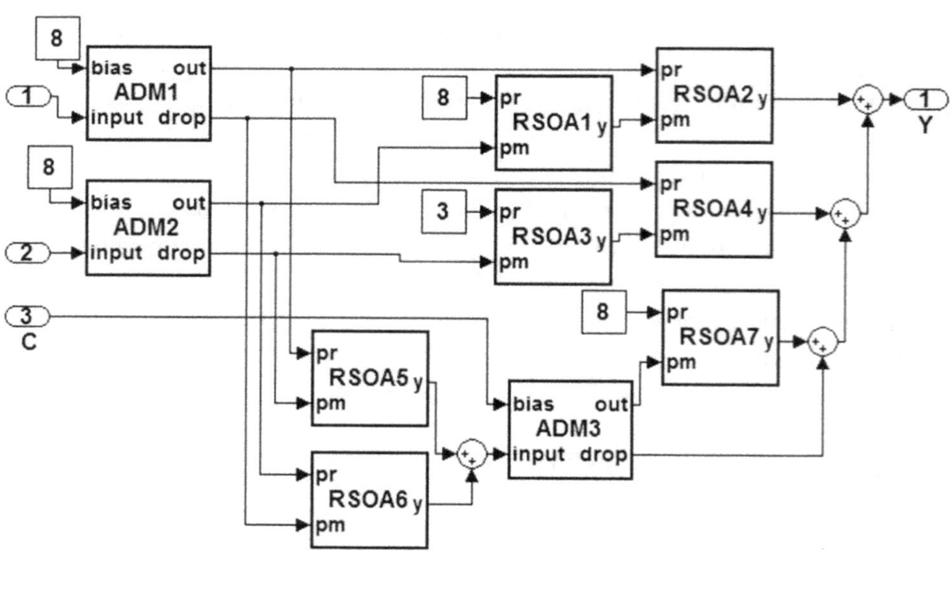

Figure 9. Internal simulative model of the controlling dibit output unit (subsystem)

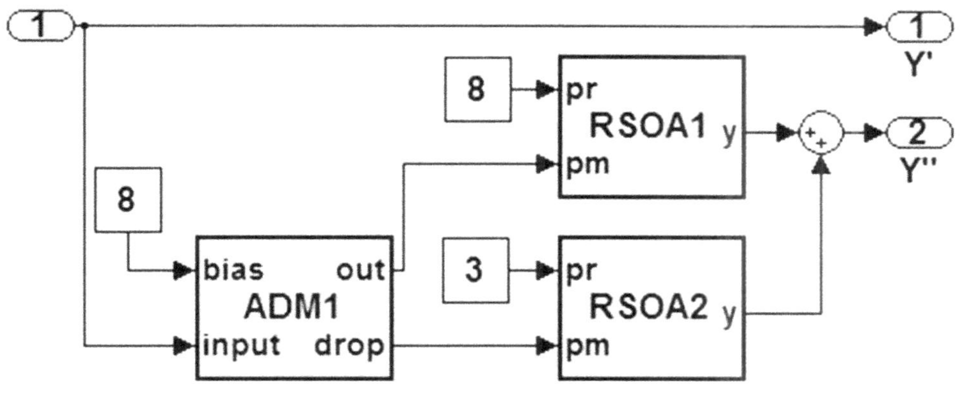

input then we can get <3><8> at the dibit output terminal (shown in Figure 5). By this way, we can see that, this model satisfies truth table of AND logic which is shown in Table 2. But, when '8' is inserted as control input then we can get <8><3> at the dibit output terminal (shown in Figure 6). So, we can say that, this model satisfies truth table of OR logic which is shown in Table 3. This integrated scheme can also be satisfied by the other input combinations of the truth tables of respective AND and OR logic gates.

FUTURE RESEARCH DIRECTIONS

The simulative verification satisfies these integrated AND and OR logic gates with control input. Therefore its performance can directly be utilized for developing and verifying the performances of other arithmetic and logic devices based on frequency encoding principle. Also, one can implement various schemes of all optical integrated logic and arithmetic circuits using this technique for optical computation. Furthermore, it may be easy to construct advanced optical computer by combining several all optical digital devices or circuits in near future.

CONCLUSION

This dibit representation technique is very much accurate and reliable one because it supports for decline of bit error problem by increasing high signal to noise ratio. It can also expect high degree of parallelism. Now, if the dibit output is interchanged between 'Y'' and 'Y''' terminals, then AND logic output can be treated as NAND logic output and OR logic output can act as NOR logic output with the advantage of dibit representation technique. Again, the control input provides the opportunity for selecting particular logic operation of the integrated scheme of all optical logic circuits.

REFERENCES

Amiri, I. S., Alavi, S. E., Bahadoran, M., Afroozeh, A., & Ahmad, H. (2015). Nanometer bandwidth soliton generation and experimental transmission within nonlinear fiber optics using an add-drop filter system. *Journal of Computational and Theoretical Nanoscience*, *12*(2), 221–225. doi:10.1166/jctn.2015.3721

Çakmak, V., Ufuk, F., & Karabulut, N. (2017). Diffusion-weighted MRI of pulmonary lesions: Comparison of apparent diffusion coefficient and lesion-to-spinal cord signal intensity ratio in lesion characterization. *Journal of Magnetic Resonance Imaging*, *45*(3), 845–854. doi:10.1002/jmri.25426 PMID:27519160

Chandra, S. K., & Mukhopadhyay, S. (2012). All optical alternative approach of conducting NAND and NOR logic gates with phase encoding principle. *Optik-International Journal for Light and Electron Optics*, *123*(11), 1022–1025. doi:10.1016/j.ijleo.2011.07.022

Chen, S., Shi, Y., He, S., & Dai, D. (2016). Compact eight-channel thermally reconfigurable optical add/drop multiplexers on silicon. *IEEE Photonics Technology Letters*, *28*(17), 1874–1877. doi:10.1109/LPT.2016.2574459

Dey, S., & Mukhopadhyay, S. (2017). Approach of implementing phase encoded quantum square root of NOT gate. *Electronics Letters*, *53*(20), 1375–1377. doi:10.1049/el.2017.2500

Dey, S., & Mukhopadhyay, S. (2018). A method of generation of sharp peaked parabolic and hyper parabolic optical pulses by the use of Pockels effect for long distance communication. *Journal of Optics*, *47*(3), 272–277. doi:10.100712596-018-0461-2

Dey, S., & Mukhopadhyay, S. (2018). Implementation of all-optical Pauli-Y gate by the integrated phase and polarisation encoding. *IET Optoelectronics*, *12*(4), 176–179. doi:10.1049/iet-opt.2017.0138

Dutta, S., & Mukhopadhyay, S. (2011). All optical frequency encoding method for converting a decimal number to its equivalent binary number using tree architecture. *Optik - International Journal for Light and Electron Optics, 122*(2), 125-127.

Dutta, S., & Mukhopadhyay, S. (2011). Alternating approach of implementing frequency encoded all-optical logic gates and flip-flop using semiconductor optical amplifier. *Optik - International Journal for Light and Electron Optics, 122*(12), 1088-1094.

Garai, S. K., & Mukhopadhyay, S. (2010). A method of optical implementation of frequency encoded different logic operations using second harmonic & difference frequency generation techniques in non-linear material. *Optik-International Journal for Light and Electron Optics*, *121*(8), 715–721. doi:10.1016/j.ijleo.2008.10.011

Garai, S. K., & Mukhopadhyay, S. (2010). All-optical frequency encoded binary half subtractor using periodically poled lithium niobate waveguide and reflecting semiconductor optical amplifier. *Optics and Photonics Letters*, *3*(1), 15–22. doi:10.1142/S1793528810000074

Gayen, D. K., Chattopadhyay, T., & Zoiros, K. E. (2015). All-optical D flip-flop using single quantum-dot semiconductor optical amplifier assisted Mach–Zehnder interferometer. *Journal of Computational Electronics*, *14*(1), 129–138. doi:10.100710825-014-0632-6

Ghosh, A. K., Bhattacharya, A., Raul, M., & Basuray, A. (2012). Trinary arithmetic and logic unit (TALU) using savart plate and spatial light modulator (SLM) suitable for optical computation in multivalued logic. *Optics & Laser Technology*, *44*(5), 1583–1592. doi:10.1016/j.optlastec.2011.11.044

Ghosh, B., Hazra, S., Haldar, N., Roy, D., Patra, S. N., Swarnakar, J., ... Mukhopadhyay, S. (2018). A Novel Approach to Realize of All Optical Frequency Encoded Dibit Based XOR and XNOR Logic Gates Using Optical Switches with Simulated Verification. *Optics and Spectroscopy*, *124*(3), 337–342. doi:10.1134/S0030400X1803013X

Ghosh, B., & Mukhopadhyay, S. (2013). A novel realization of all-optical dibit represented frequency encoded Boolean and quaternary inverters without switching device. *Optik-International Journal for Light and Electron Optics*, *124*(21), 4813–4815. doi:10.1016/j.ijleo.2013.02.035

Hu, G., Xiao, D., Wang, Y., Xiang, T., & Zhou, Q. (2017). Securing image information using double random phase encoding and parallel compressive sensing with updated sampling processes. *Optics and Lasers in Engineering*, *98*, 123–133. doi:10.1016/j.optlaseng.2017.06.013

Kim, H., Kwon, B., Suh, M., Kang, D. S., Kim, Y., & Jeon, D. Y. (2011). Degradation characteristics of red light-emitting CuInS2/ZnS quantum dots as a wavelength converter for LEDs. *Electrochemical and Solid-State Letters*, *14*(10), K55–K57. doi:10.1149/1.3610644

Mansoor, R., Koziel, S., Sasse, H., & Duffy, A. (2015). Crosstalk suppression bandwidth optimisation of a vertically coupled ring resonator add/drop filter. *IET Optoelectronics*, *9*(2), 30–36. doi:10.1049/iet-opt.2014.0059

Mukhopadhyay, S. (1992). Binary optical data subtraction by using a ternary dibit representation technique in optical arithmetic problems. *Applied Optics*, *31*(23), 4622–4623. doi:10.1364/AO.31.004622 PMID:20725467

Rashed, A. N. Z. (2012). Optical Add Drop Multiplexer (OADM) Based on Dense Wavelength Division Multiplexing Technology in Next Generation Optical Networks. *International Journal of Multimedia & Ubiquitous Engineering*, *7*(1), 1–14.

Saha, S., Biswas, S., & Mukhopadhyay, S. (2017). Optical scheme of conversion of a positionally encoded decimal digit to frequency encoded Boolean form using Mach–Zehnder interferometer-based semiconductor optical amplifier. *IET Optoelectronics*, *11*(5), 201–207. doi:10.1049/iet-opt.2016.0078

Sarkar, B., & Mukhopadhyay, S. (2017). An all optical scheme for implementing an integrated Pauli's X, Y and Z quantum gates with optical switches. *Journal of Optics*, *46*(2), 143–148. doi:10.100712596-017-0398-x

Sarkar, P. P., Ghosh, B., & Patra, S. N. (2016). Simulative study of all optical frequency encoded dibit based universal NAND and NOR logic gates using a reflective semiconductor optical amplifier and an add/drop multiplexer. *Journal of Optical Technology*, *83*(4), 257–262. doi:10.1364/JOT.83.000257

Sarkar, P. P., Ghosh, B., Patra, S. N., & Mukhopadhyay, S. (2017). A new scheme of an all optical frequency encoded dibit based latch with its simulated result. *Journal of Optical Technology*, *84*(9), 631–634. doi:10.1364/JOT.84.000631

Sarkar, P. P., & Mukhopadhyay, S. (2014). All optical frequency encoded NAND logic operation along with the simulated result. *Journal of Optics*, *43*(3), 177–182. doi:10.100712596-014-0197-6

Sarkar, P. P., Satpati, B., & Mukhopadhyay, S. (2013). New simulative studies on performance of semiconductor optical amplifier based optical switches like frequency converter and add-drop multiplexer for optical data processors. *Journal of Optics*, *42*(4), 360–366. doi:10.100712596-013-0133-1

Singh, S., Kaur, S., Kaur, R., & Kaler, R. S. (2016). Photonic processing of all-optical Johnson counter using semiconductor optical amplifiers. *IET Optoelectronics*, *11*(1), 8–14. doi:10.1049/iet-opt.2015.0106

Singh, S. P., Kar, S., & Jain, V. K. (2007). Performance of all-optical WDM network in presence of four-wave mixing, optical amplifier noise, and wavelength converter noise. *Fiber and Integrated Optics*, *26*(2), 79–97. doi:10.1080/01468030601131477

Wang, J., & Chen, L. R. (2015). Low crosstalk Bragg grating/Mach-Zehnder interferometer optical add-drop multiplexer in silicon photonics. *Optics Express*, *23*(20), 26450–26459. doi:10.1364/OE.23.026450 PMID:26480158

Wang, K. Y., & Foster, A. C. (2012). Ultralow power continuous-wave frequency conversion in hydrogenated amorphous silicon waveguides. *Optics Letters*, *37*(8), 1331–1333. doi:10.1364/OL.37.001331 PMID:22513676

Zhu, J., Zhou, P., Su, X., & You, Z. (2016). Accurate and fast 3D surface measurement with temporal-spatial binary encoding structured illumination. *Optics Express*, *24*(25), 28549–28560. doi:10.1364/OE.24.028549 PMID:27958499

Chapter 8
Optical Switching in Next-Generation Data Centers:
Architectures Based on Optical Switching

Vaibhav Shukla
Allenhouse Institute of Technology, India

Rajiv Srivastava
Independent Researcher, India

Dilip Kumar Choubey
National Institute of Technology Patna, India

ABSTRACT

The leading content provider companies like Google, Yahoo, and Amazon installed mega-data centers that contain hundreds of thousands of servers in very large scale. The current data center systems are organized in the form of the hierarchal tree structure based on bandwidth-limited electronic switches. Modern data center systems face a number of issues like high power consumption, limited bandwidth availability, server connectivity, energy and cost efficiency, traffic complexity, etc. One of the most feasible solution of these issues is the use of optical switching technologies in the core of data center systems. In this chapter a brief description about the modern data center system is presented, and some prominent optical packet switch architectures are also presented in this chapter with their pros and cons.

DOI: 10.4018/978-1-5225-8531-2.ch008

Copyright © 2019, IGI Global. Copying or distributing in print or electronic forms without written permission of IGI Global is prohibited.

INTRODUCTION

Over the past few years, data-centric applications like high-definition video streaming (Hadbib, 2012) social networking (Koley et al, 2010) and cloud computing (Sato et al, 2009) in combination with the large diffusion of mobile devices such as smart phones and tablets (Jinno et al, 2009) have contributed to the rapid growth of the Internet traffic. Data centres form a key set-up of today's Internet and providing a myriad of services from social networking to large-scale scientific calculations. They can be defined as mega-centres of computing and storage resources that communicate extensively with each other to serve the ever-increasing demands of customers (Habib et al, 2012). In the past two decades, the high-performance computing section has achieved a performance improvement trend of 85-90% of the compound annual growth rate (CAGR) (Taubenblatt et al, 2012). Due to the advancement in multicore processors, accelerators, and increasing levels of parallelism across all layers of the hierarchal system, this trend of performance improvement is expected to continue in the near future also. Besides the rapidly increasing demand of bandwidth, that requires massive-scale interconnection solutions, the ICT (Information and Communication Technology) industry reported a contribution of 2 to 3 percent of global greenhouse gas emissions, which will rapidly increase soon (Despins et al, 2011). Data Centres, being ICT infrastructure, thus needs to be revisited in terms of interconnection design to support very large scales data centres while leading to ultimate footprint and power savings. In recent years, many researchers focused on examining the various approaches to address the scalability, speed and footprint requirements of Next-Generation data centres (NGDCs).

One of the main objectives of this chapter is to develop an effective networking solution which is good enough for the Next Generation Data Centre. Our definition of Next Generation data centres, contains massive-scale data centres that can host millions of processing cores. Also referred to as cloud data centres, these entities will be employed by large online service providers to offer Internet-facing services and are expected to host data-intensive applications. In this chapter, we are trying to report the requirements of these data centres in terms of footprint, scale, energy efficiency and flexibility. Inspired by the ongoing efforts in data centre network design, we aim at deploying the most appropriate technologies in the upcoming massive-scale data centres.

FUTURE SCENARIO OF DATA CENTRE

The rise of cloud computing and other emerging web applications have created the need of more powerful warehouse-scale data centres. These data centres comprise hundreds of thousands of servers that need to communicate with each other via high performance and low latency interconnection networks (Kachris et al, 2013). With the rapid growth of Internet applications, data centres have witnessed demands for more and more storage, computation power and communication bandwidth. In present days telecommunication environment thousands of servers, are very common in heavy data centre systems. In the recent survey report given by Cisco ("Cisco Global index", 2011), the yearly worldwide data centre traffic is expected to reach several zetta-bytes by the end of 2017.In near future, the data traffic which is generated between the data centres and within the data centre is expected to grow extensively. This increase in the traffic, generates the need of high performance network technologies and architectures for intra and inters data centre networks. In the present electrical data centre network, the major challenge in the design of the data centre is the power consumption of the infrastructure, mainly due to the associated operational expenditure (OPEX) costs. According to relevant studies, data centre networks consume around 10–20 percent of the total IT power consumption of data centre sites, and this is expected to increase soon in near future (Kachris et al, 2013).

CHALLENGES OF DATA CENTRE NETWORK

As our focused area in this project is on data centre interconnects, we are primarily focusing on the current challenges associated with the current data centre architectures. Figure 1 shows a hierarchal system of current data centre system (Greenberg et al, 2009; Greenberg et al, 2009; Benson eta l, 2010; Kachirs et al, 2012). In this figure the servers present at the lowest level of hierarchy are organized in server racks (usually consist of 40-80 blades). Each of the blade server is connected to the Top of the Rack (TOR) switch with a 1 Gbps ethernet link. For redundancy, each of the TOR switches is connected with more than one aggregation switch and the traffic from these switches is further aggregated through routers. At the top level, the core routers are used to provide the connectivity between data centre with the internet (multi routed tree topology). We note that all links use Ethernet as the physical-layer protocol, with a mix of copper and multi-mode fiber cabling. Switches below each access router pair form a single layer-2 domain, typically connecting thousands of servers.

In this hierarchal system there are some disadvantages which makes it an inappropriate choice in the case of large data centre system. Some of the major disadvantages are as follows: -

Figure 1. Hierarchal architecture of current data centres

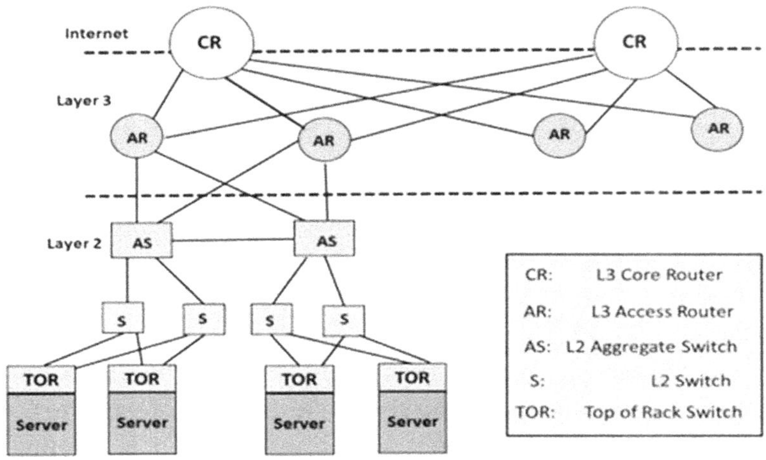

Poor Server to Server Connectivity

In this system, while moving up through the layers of switches and routers, the oversubscription ratio increases. Oversubscription is a technique through which the total cost of design goes down. The oversubscription is defined as the ratio between the worst case achievable aggregate bandwidth and the end host to the total bisection bandwidth of a particular communication topology (Al-Fares et al, 2008). For example, in the same rack the servers typically have *1:1* oversubscription with the other servers. The uplinks from TORs, however, are typically *5:1* to *20:1* oversubscribed (*i.e.,* 2 to 8 Gbps of uplink for *40* servers) (Greenberg et al, 2012). The paths which are through the highest layer of the tree may be oversubscribed with the ratio *240:1* (Kachris et al, 2013; Kachris et al, 2012). The large oversubscription introduced in a present electronic data centre network designs fragments that serve the server pool by preventing idle servers from being assigned to services.

Fragmentation of Server Pool and Limited Flexibility

In data centre systems, the cost and performance of communication depend upon the distance along with the hierarchy. In terms of layer 2 domains, the hierarchal architecture forms clusters of servers. Further, as spreading a service outside a single layer 2 domain leads to the overhead of reconfiguring the IP addresses and

VLAN trunks, spare capacity throughout the data centre which is overprovisioned per individual service so that each service can scale out to nearby servers to respond rapidly to the demand spikes or to failures. Preventing resource sharing, this approach can suffer significant disturbance in case of increasing resource needs.

Limited Scalability

As the traffic flow within the datacentre network continues to grow, so it is very difficult to manage the traffic through the switches based on ethernet links. According to the recent survey report given by Cisco (2014), the global data centre IP traffic will triple during 2012-17. This measures the compound annual growth rate of about 25% (up from 220 exabytes per month in 2013 to 648 exabytes per month in year 2017). In the next several years we can expect the doubling of transistor density on chips in every 18-24 months. This stems from the possibility of integrating multiple cores on a single die, the multicore processor technology (Geer et al, 2005). The hierarchal design cannot support the growing traffic heavy data centre systems that will be equipped with more numbers of servers and also more microprocessor cores on each server.

Power Consumption

One of the most challenging issues in the deployment of data centre systems is power consumption. According to the recent Greenpeace survey report, the global demand of electricity in data centre system is expected to triple from 340 billion kWh in 2006 to 1000 billion kWh in 2020 (“Greenpeace”, 2010; “GreenDataProject”, 2008). Studies also suggest that in data centre system, around 40%, 37% and 23% of the IT power is consumed in servers, storage and in networking equipment’s respectively (“GreenDataProject”, 2008). That’s why the savings in the power consumption of network elements make a positive impact on the overall power consumption of data centre sites. Electronic links and switches in the data centre design do not leave much space for improvement in this area as they are power hungry devices especially as the bit rate increases. The other shortcoming of current designs in terms of energy efficiency arises from the fact that the power consumption of servers is not necessarily low when they are idle. They can consume as much as 60 percent of their peak power when idle. Measurements on current deployments have recorded an average server utilization as low as 30% (Zhang et al, 2012; Meisner et al, 2009) indicating a huge energy wastage due to idling hardware starved for data.

The challenges discussed above provide us with a list of requirements for networking in NGDCs that are pointed out as follows.

Scalability

Due to the rapid growth of network traffic and the fixed growth in the processing power of multicore servers, future data centres should enable the support for millions of microprocessor cores. We expect NGDCs of more than 10 million server cores, which is one order of magnitude larger than the numbers supported by the available designs and proposals.

Flexibility

NGDCs are expected to support the requirements of various applications by providing non-blocking connectivity among clusters of computing racks. This is the key requirement of cloud computing. A uniform high capacity and non-blocking network design provides flexibility in slicing the data centre enabling the efficient mapping of application requirements to virtualized resources (Lam et al, 2010).

Space Management

Providing a nonblocking data centre connectivity would require a massive amount of Ethernet cables when the data centre accommodate millions of server cores, resulting in severe implementation, management and maintenance problems. So, an alternative technology is required that reduces the cabling requirements in the future data centre systems.

Energy Efficiency

In the information and communication technology, the data centres are the major power consumers. The key areas that have become the major concern for the research community are cost associated with fuelling data centres as well as greenhouse gas emissions. In heavy data centre systems, the electronic links and switches consume a large amount of power and it is difficult to perform power savings by turning off the under-utilized machines. An energy efficient interconnect is required for significant energy saving in the future data centre.

PROPOSED SOLUTION: OPTICAL SWITCHING BASED ON SCALABLE ARCHITECTURE

Due to the limited scalability and high cabling complexity, the ethernet switching is not suitable for the next generation data centre system (NGDCs). In designing of interconnects for NGDCs, the first step is to select an appropriate switching technology. The best solution of the problems faced by NGDCs is the use of optical technology in data centres. It has been widely deployed in long haul communication networks (Lam et al, 20120). The major driving forces behind large capacity optical network deployments are the content providers and large-scale data centre operators. The properties like ultra-high capacity, small footprint and low power consumption makes the optical transmission best suitable option for data centre network. These three properties are discussed in detail:

Ultra-High Capacity

In present days, in telecommunication system, the optical fiber technology is considered as the best communication channel in the sense that it provides THz bandwidth (Shacham et al, 2007).. For effective utilization of this huge amount of bandwidth the WDM (Wavelength Division Multiplexing) technology has been deployed to launch multiple channel parallel in the same fiber. In a DWDM (Dense Wavelength Division Multiplexing) 100 MHz, 50 MHz, or even smaller channel spacing can be achieved, so it can be possible to set up tens of channels in a fiber. In NGDs large number of server cores are reported so, it is predictable that the amount of traffic to be transmitted over various links will far exceed the capacity of available electronic links. The primary solution of this problem is to increase the cabling in the system significantly through this necessary amount of bandwidth between the nodes is guaranteed. This solution reported a serious shortage of space and power dissipation problem. However, by using optical fiber the same capacity can be achieved at a very low transmission loss.

Low Power Consumption

One of the major characters of optical links and switches is its low power consumption (Shacham et al, 2007; Yoo et al, 2006). By using optical fiber this is feasible to set up very long links without the need of regeneration as because the loss of optical fiber is too low (around .2 db/km). Also, a low bit rate makes optical switching a power-efficient technology. In other words, an optical switch consumes power independent of the bandwidth of the signal it switches. Hence, in optical switches, the power per unit of bandwidth could be extremely low in comparison to electronic switches.

Small Footprint

The other key property of optical devices is a small footprint, the property of which align well with the requirements of data centre (Yoo et al, 2011). Despite of significant capacity, fiber has a pitch of only 125 μm. Optical packet switching components could be compactly built through photonic integration. Large scale photonic integration is expected to improve the reliability, power consumption, manufacturing cost and it reduces the overall size of the system (Nicholes et al, 2010).

The features discussed above makes the optical technology the best suitable technology in next generation data centre systems. Despite these properties, we should not disregard its shortcomings. The processing capabilities in the electronic domain are better than the optical domain. One of the major hurdles in the optical domain is the unavailability of optical buffers, while the electronic RAMs can be used as a buffer in the electronic domain (Yoo et al, 2006; Hunter et al, 1998; Chang et al, 2006). Hence our aim is to consolidate the benefits of the optical and electronic domain to improve the overall design of a switch.

The introduction of optical switching in data centre follows an evolutionary trend. In data centre currently, optical transmission is used for some point to point links on multimode fibers and optical switching is totally absent. With the use of hybrid approach which uses optical links and electronic switches, the main drawback is the power-hungry O/E (optical to electrical) and E/O (electrical to optical) conversion that is required on switch ports. We expect the proliferation of optical switching in NGDCs to follow a trend as represented in Figure 2. As in figure (2 (a)), in multi stage data centre architecture, optical switching will start to emerge in NGDCs at the top tier. In other words, we can expect that inside the server racks the data traffic is not large enough to transmit through optical links. However, as we aggregate more in the hierarchy, long-lived flows become more common and an optical switching network can be beneficial. As in figure 2 (a) at the top tier, electronic packet switching is also present with the optical switching, so electronic switching can be used for transferring the short-lived flows in a cost-effective fashion.

The boundaries of the optical switch network in NGDCs can still evolve in 2 dimensions over the time. The optical switching will mature until there is no need of electronic switching network at the top tier as shown in figure 2 (b). Moreover, as traffic grows up, the boundaries of optical fabric move down in multi stage data centre architecture. In future, as the advancement in multicore processor technology is reported, we predict the use of optics to become ubiquitous in data centres, penetrating into servers, server racks and even in multicore chips (figure 2 (c)). At this stage the electronics will face a serious problem in addressing bandwidth requirements even inside the chips.

The above discussion shows that the use of optical packet switching technology in data centre systems improves the overall performance of DC systems. Now literature about the optical packet switching is discussed in the further sections of this chapter.

Figure 2. Evolution of Optical Switching in NGDCs

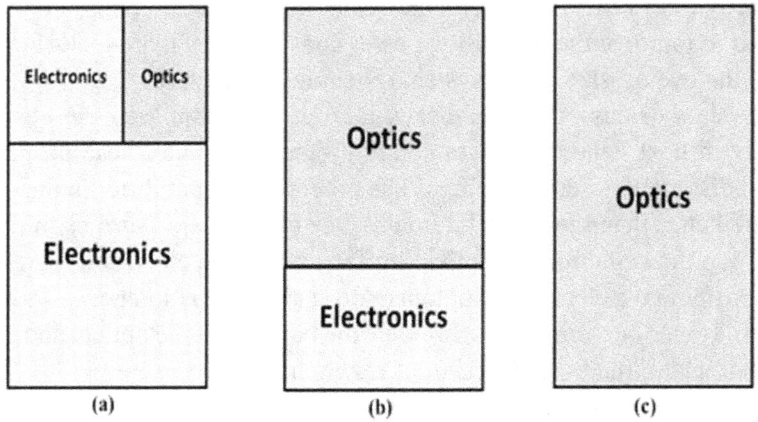

OVERVIEW OF PACKET SWITCHING

In 1960, during his research for US Air force, RAND corporation scientist Paul Baran was the first one to introduce the concept of packet switching in 1960. In packet switching, the information is transferred into small size packets. Each packet contains small chunks of data which is to be transmitted between various nodes into the network. Inside the network, at each node the packets are either buffered or it can be directly transmitted towards the output port which results into variable queuing delay among packets. This feature makes a difference between circuit and packet switching as in circuit switching a dedicated path is established in advance and continue till communication takes place. For optimizing available network channel capacity, packet switching technologies are used. By using packet switching, the transmission latency can be minimized (i.e. the time taken by data to pass across the network), and the robustness of network can be increases.

Optical packet switching (OPS) is a switching technology with fine granularity. OPS technology transits data in the optical domain without O/E and E/O conversion (Yoo et al, 2006; Xu eta l, 2001). The key issue involved in the OPS is the design of the switch/router which can perform switching of the data at a very high data rate.

An arrayed waveguide is a device, it consists of two slab waveguides and a star coupler, connected by a dispersive waveguide array. An interesting device is obtained if the AWG is designed with ‘N’ input and ‘N’ output waveguides and free spectral range equaling ‘N’ times the channel spacing. With such an arrangement the device behaves cyclical: a signal disappearing from output ‘N’ will reappear at output 1, if the frequency is increased by an amount equal to the channel spacing. The arrayed waveguide grating has very less insertion loss. Therefore, many optical switch designs using AWGR have emerged. In this chapter, some recently published AWGR based optical packet switch routers are discussed with their pros and cons.

WDM Fiber Delay Line and AWG Based Optical Switch

In 2009 R. Srivastava et.al (Srivastava, n.d.; Shukla et al, 2015) proposed a WDM arrayed waveguide-based optical packet switch architecture as shown in figure 3. The whole switch is divided into two sections, scheduling and switching sections. In the scheduling section, a set of TWC is present at each input of switch, which converts the wavelength of incoming packets as per the routing pattern of the first AWG router. The upper N ports of the scheduling AWG is connected with the set of fiber delay lines these delay lines are used as the optical storage.

First when packet arrives at the input of switch the TWC converts the wavelength of incoming packets as per the routing pattern of scheduling section AWG after this if the situation of contention among packets arises at the input of switch then the AWG forwarded the contended packets towards the fiber delay lines and packet stores there till the contention removes after that packet forwarded towards the switching section where the TWC converts the wavelength of packets as per the routing pattern of switching section AWG and after that AWG delivered the packet towards the corresponding output port.

Advantages

1. The design is very simple in nature.
2. By using wavelength division multiplexing technique an effective buffer pool is designed.

Disadvantage

1. In this switch the directly transmitted packets have better signal quality in comparison to buffered packets.
2. The AWG used into the design of switch introduces an undesired crosstalk factor which reduces the signal quality.

Petabit Optical Switch

In 2013, the realization of petabit optical switch is based on the Tunable Wavelength Converter (TWC), Arrayed waveguide router (AWG) and Tunable lasers (TL) as shown in figure 4. In the petabit model Top of The Rack (TOR) switches are connected to the optical switch this creates three stage clos networks, having input,

Figure 3. WDM, AWG and FDL based switch (S1)

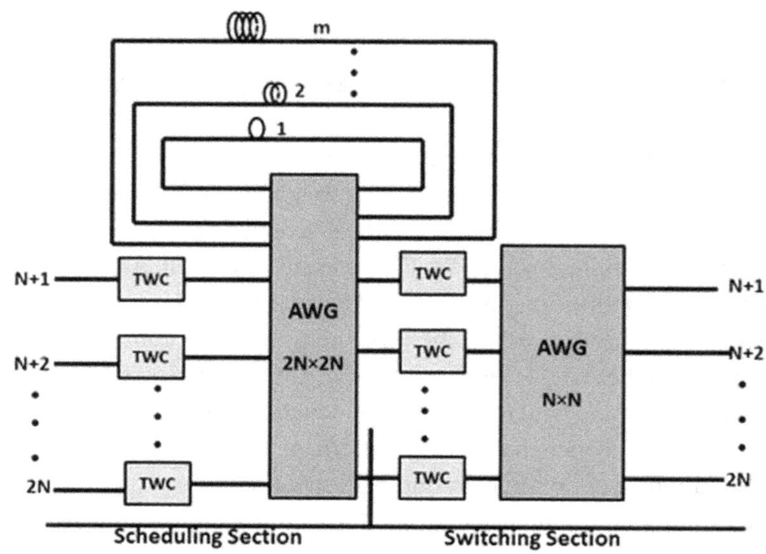

Figure 4. The petabit optical switch architecture (S2)

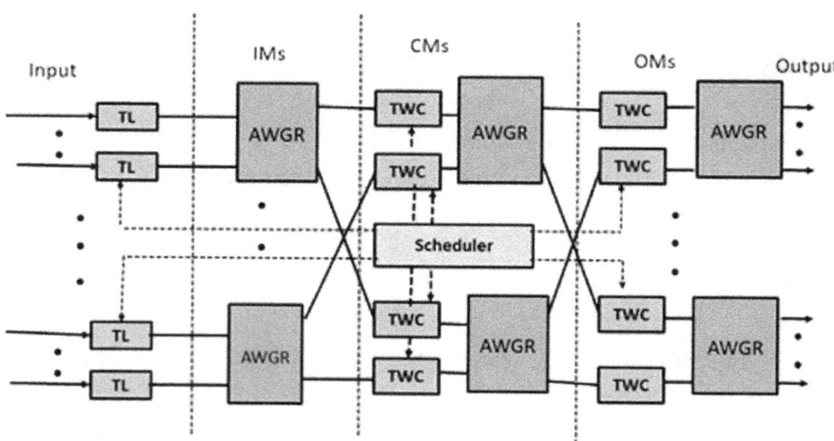

output and central modules (Xi et al, 2013). In each module AWG router is used as the core. A scheduler is used to control the SOA based TWC as well as the TLs, present in the line cards. One of the prominent features of the switch is that the packets are buffered only at the line cards, while input output and control modules do not require buffers. By this feature the implementation complexity reduces and it achieves low latency.

Advantages

1. The prominent feature of this switch is that it uses line cards for buffering of packets this helps to reduce implementation complexity and to achieve low latency.
2. A very high throughput is achieved through petabit switch.

Disadvantages

1. A bufferless design is used in the switch so the storage of packets is an important concern in the switch.

Recirculating Loop Buffer-Based Switch

In 2013 H. Rastegarfar et.al. (2013) presented a recirculating loop buffer-based switch. In this design if the packet feels contention it can be stored in to the loop buffer until contention resolve and when contention is resolved, it is forwarded towards the appropriate output port. In this switch design AWG router is used for directing the packets towards appropriate output port.

Advantages

1. The buffer used in this switch is recirculating in nature it increases the storage duration of packets into the buffer
2. A very low packet loss probability is achieved through this switch.

Disadvantages

1. The buffer is made recirculating in nature so the advantages of WDM is loosed in the switch.
2. As AWG router is used in the core of switch so more circulation means more crosstalk in nature and we know that the crosstalk is an undesired factor introduced by AWG router.

Data Center Optical Switch

Figure 6 shows a general LIONs switch architecture (Yawei et al,2 013). The switch uses as AWG router, Tunable wavelength converters, an electrical control plane (CP), electrical loopback buffer, fiber delay lines and label extractor (LE) in the core of the switch. As a media interface an optical channel adaptor is used between the

Figure 5. AWGR based recirculating loop buffer-based switch (S3)

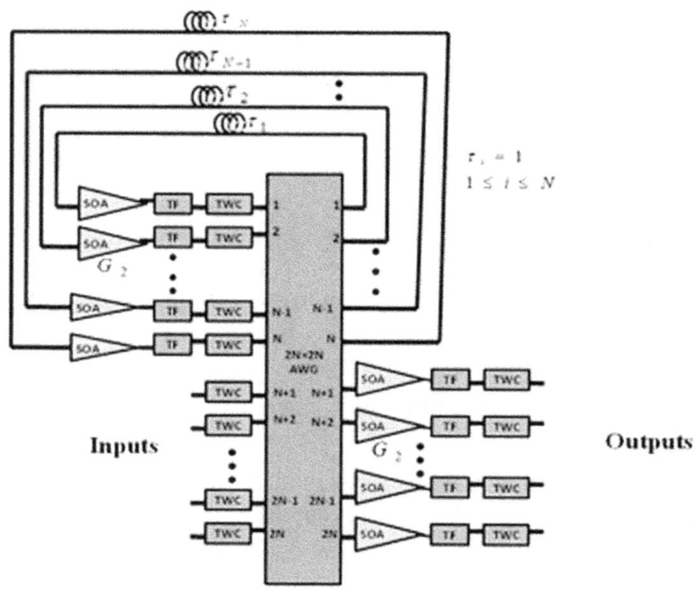

Figure 6. LIONS architecture (S4)

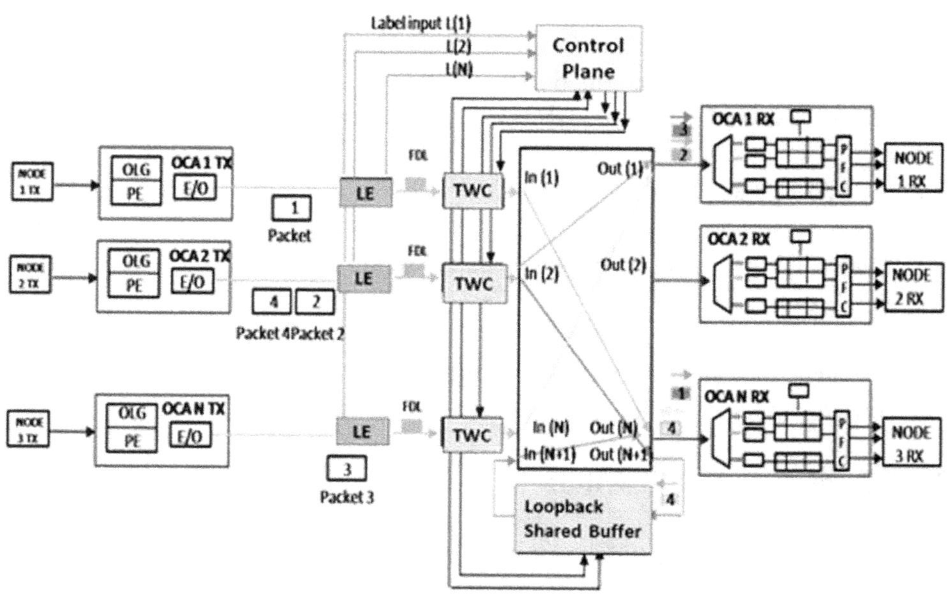

switch and each end nodes. An output queue structure is easily realized by using an AWG based switching fabric, provided that *N* receivers and a *1:N* optical DEMUX is used at each AWGR output.

In an optical switch, the store-and-forward mechanism cannot be applied due to the lack of feasible optical buffers. In this design the electronic buffer is used in the loopback path called loopback buffer. In the situation of contention this loopback buffer switch play an important role. The one of the proposed loopback buffer architecture is represented by SLB in the figure.

A shared buffer is needed for contention resolution when the number of nodes is more than the number of output receivers. It is used to store temporarily for the transmitted packets, which cannot reach the desired outputs, so that they can try it later. The figure shows a shared loopback SDRAM. For realizing this SDRAM, a *1:N* optical demultiplexer and N receivers are required because switch may receive delayed packets with different wavelengths from different input concurrently (Yawei et al, 2013). For sending delayed packets on different outputs on different wavelengths a *N:1* Mux and *N* transmitters are used in the design. In this design optical Mux and Demux both are realized by a *1:N* AWG router. In this design the packet received on different wavelength will be copied to the shared memory parallelly. The shared memory can also be used as the buffer for delayed packets and the packets are stored there before final transmission. The queuing structure in the shared memory is organized based on outputs, since the packet destined for the same output should be sent out serially. The transmission for a delayed packet can start before the entire packet arrives at the input of switch.

Advantages

1. The DOS architecture has quite low latency which also stays independent of the number of inputs. Because the ToR packets only travel through optical switches, no delay from the buffer of electrical switches.
2. The TWC has a rapid reconfiguring time of a few nanoseconds, which is useful to meet the demand of bursty traffic fluctuation.

Disadvantages

1. By using an electronic buffer, a heavy Optical to electrical and electrical to optical conversions are required and this increases the packet latency.
2. A set of TWC is used in the core of switch and this increases the overall cost of the switch.

Figure 7. SLB (Shared loopback buffer)

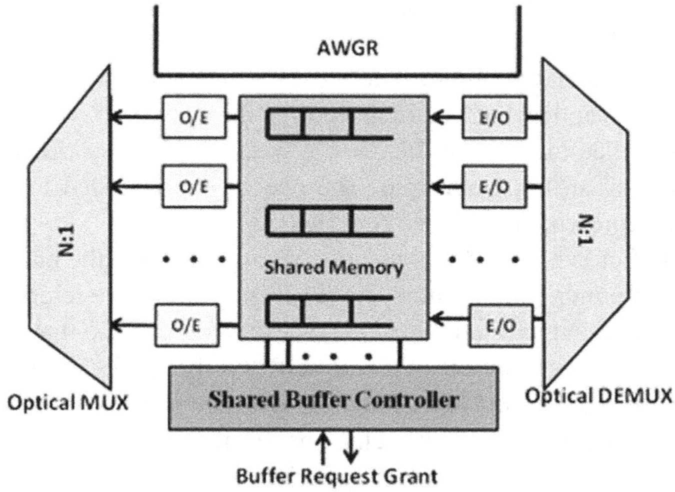

AO-NACK Based Optical Switch

In 2010Y. Yin et.al. (2012) proposed an AWGR based optical switch here in this design in the case of blocking or dropping the packets a negative acknowledgment is sent to the sender to notify and after receiving this negative acknowledgment sender resend the dropped or blocked packet again as shown in the figure. In this design a buffer is not used. In the figure, the path between circulators C_2 and C_3 is used for the label extractor where an old label is extracted and after O/E conversion fed to the electronic controller, and a new label is inserted and fed with payload at the output of the switch. The figure shows an AO-NACK based optical packet switch architecture. Here T_1 $T_{2\ldots\ldots}$ $T_{i\ldots\ldots}$ T_N represents the N transmitters and R_1 $R_{2\ldots}$ $R_{i\ldots\ldots}$ R_N represents N receivers. The information generated through transmitter T_i first passes through circulator C_1 and then passes through fiber delay lines and then again passes through circulator C_2. Over here, label is extracted and as per the destination wavelength of the incoming packet is tuned to direct them to the appropriate output port of the AWG. For blocking the appropriate branch, the tunable wavelength convertors convert the wavelength of the packet in such a way that it is blocked by AWG, the AWG used here acts as a reflecting grating and reflected information chosen other path it appears at circulator C_1 and it is transferred to appropriate transmitter where information is detected with associated Rx, and after receiving a negative acknowledgment the transmitter retransmits the blocked or non-acknowledged packets.

Advantages

1. In this switch packet will not be dropped
2. The control unit design is simple in nature.
3. WDM can be easily implemented in the switch

Disadvantages

1. As the buffer is not used in the design so a large number of packets will be retransmitted.
2. The situation of congestion of network arises as because retransmission of a large number of packets are required.

Modified WDM Fiber Delay Lines and AWG Based Switch

The major drawback of the switch design proposed by R. Srivastava (n.d.) is its signal quality and huge power requirement, for the correct operation of switch. To overcome these limitations V. Shukla (2014) proposed a modification in the design of switch (Shukla et al, 2015; Shukla et al, 2014). In this modification a semiconductor optical amplifier is placed at each input of switch just after the input TWC. This modification improves the overall performance of switch. In this modification by

Figure 8. AO-NACK based switch (S5)

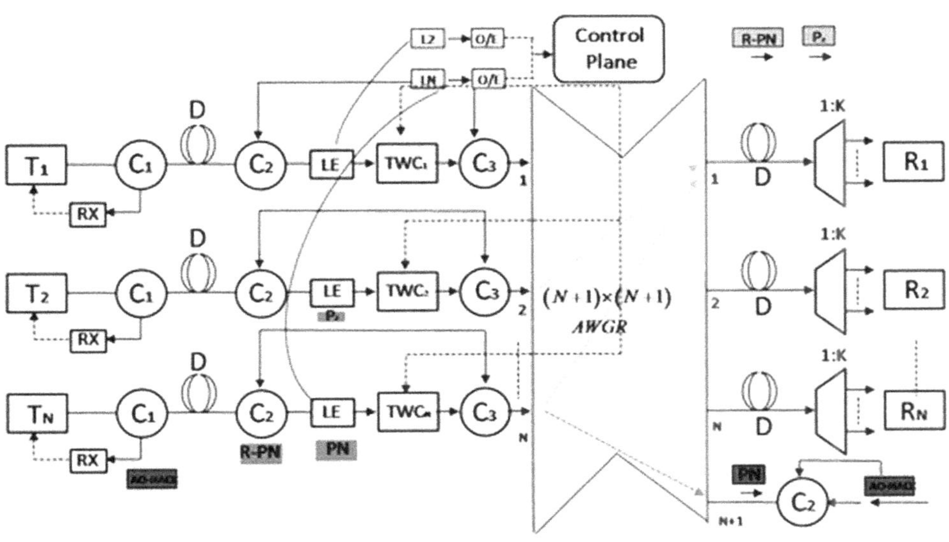

using SOA we compensate the overall packet loss of the switch as SOA amplify the signals.

Advantages

1. As WDM technique is used in the switch so effective buffer utilization is achieved through the switch.
2. In this design at each input of switch a SOA is used which amplify the signals to improve the quality of transmitted packets.

Disadvantages

1. The packets cannot be stored for longer duration.
2. The AWG introduces an undesired factor crosstalk.

Design of Optical Switch Having AWGR With Output Buffer and AO-NACK Mechanism

In 2013 Y. Yin et.al. proposed a bufferless NACK-based optical switch design in which due to absence of effective buffering mechanism a large number of packets will be re-transmitted. In 2017 P. Bhattacharya et.al (2017) proposed a design

Figure 9. Modified WDM AWG based switch (S6)

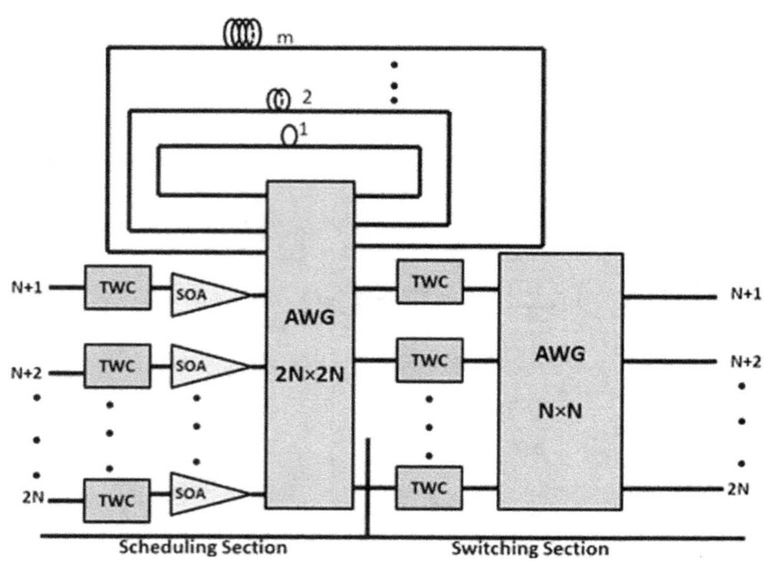

having a N×N AWG router and N input / output lines used as the core of switch. The data receives at the input lines are in the form of packets. In design each input is connected with a TWC which tunes the wavelength of incoming packets as per the routing pattern of AWG. Each packet will be assigned a specific wavelength from the set of wavelengths, depending upon the output destined port. A packet destined for the port j will be assigned a wavelength λ_j using routing pattern of AWGR. After assigning wavelength packet passes the AWGR to specific output and stored at the output buffer at destination. In the switch to accommodate the contending packets the size of output buffer is as large as possible as because any packet from all N input can be sent to any output.

In the design, the decision of AWGR is crucial. If the packet is not assigned a wavelength from the set of TWC set T, we pick a wavelength called reflector wavelength which is normally out of range of T to reflect the packet back to sender as AO-NACK.

Advantages

1. WDM can be implemented easily.
2. Packet will not be dropped.

Disadvantages

1. The situation of congestion of network arises as because retransmission of large number of packets is required.

WDM Hybrid Buffer Based Optical Switch

One of the major drawbacks of the switch design proposed by V. Shukla et.al. is that in this switch the packet cannot be stored for longer duration and the limited storage is achieved through the switch. To overcome this limitation researcher have proposed buffering in electronic domain through O/E conversion. In this context in 2018 A. Singh et.al. (2018) proposed a switch design presented in figure 11 which uses a hybrid optical buffer that merges the advantages of both fast-electronic buffering and fiber optical buffering. In this design numerous sets of fiber loops and electronic buffering modules are joined and develop the hybrid optical buffer. Initially the packets are stored in to the optical buffer, in a temporary basis. In the event that packet cannot be sent out after a particular interval, the conversion of packet will be done and it will be stored in electronic memory. The major advantage of this switch is that we can use big electronic memory in order to decreases the loss of packets.

Figure 10. AWGR with output buffer and AO-NACK based switch(S7)

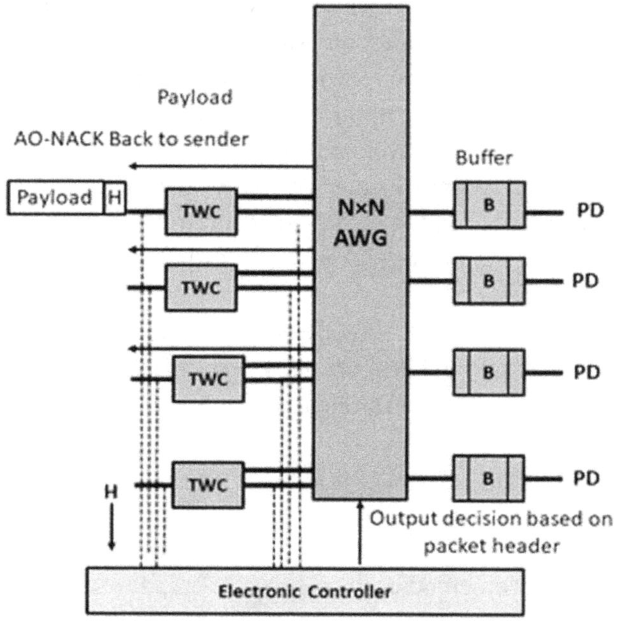

Advantages

1. The buffer is designed only through a piece of fiber so signal quality degradation is very less in comparison to other switches.
2. The switch performs well when longer duration buffering is desirable.

Disadvantages

1. For storing the packets into the buffer a heavy O/E and E/O conversions are required.
2. Energy consumption in the buffer is higher in comparison to pure optical buffer.

Dual Loop Buffer Based Optical Packet Switch Architecture

The major limitations of the switch design proposed by V. Shukla et.al is its limited storage capacity and a packet cannot be stored in to buffer for longer duration. To overcome these limitations the A. Singh et.al. have proposed a hybrid optical buffer-based switch but in this design a heavy Optical/ Electrical conversion is required.

Figure 11. Hybrid O/E Buffer based optical switch (S8)

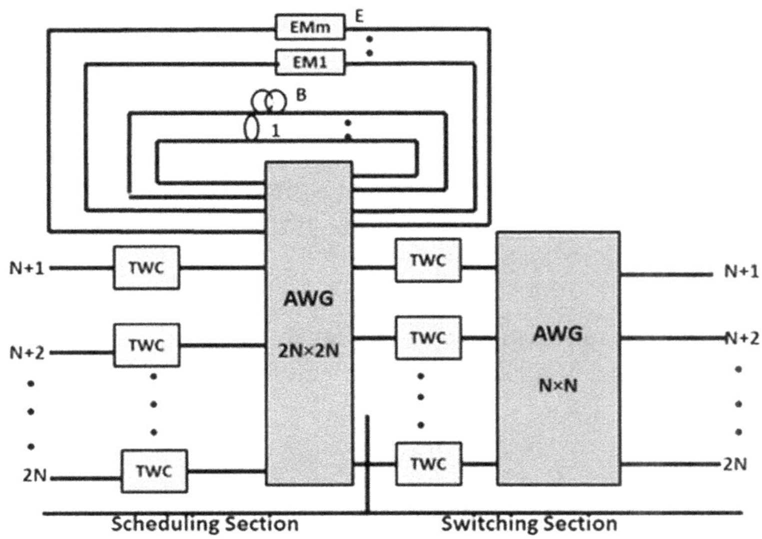

For making design effective P. Bhattacharya et.al. (In press) (n.d.) proposes a switch design as shown in figure 12. In this design the two AWG based scheduling sections are combined. In the first AWG section buffering of delay lines varying from '0' to '9' slots with the increment of unit slots is considered, while in second section buffering of delay lines of 10 to 100 slots increment of 10 slots is used. The size of the first scheduling AWG is $2N \times 2N$, while actual switch size is $N \times N$, out of rest input N ports only 9 ports are chosen for connecting buffer FDLs rest N-9 ports are left free. Similarly, in the second scheduling AWG, N-10, ports remain vacant. The principle of operation of switch is similar to switch (S_1) except the additional stage for buffering of larger delay buffer.

The TWCs placed at each input port of the scheduling and switching AWGs are used to convert the wavelength of the incoming packets as per control unit decision, either to switch them in the appropriate FDL in buffer or to direct them towards the correct output ports. Considering a situation that for a particular output port j, k-1 packets are stored in buffer, and at the input for the same output j two packets arrive, then one packet from the buffer will leave and incoming packets will be placed in the buffer module which provides a delay of k slots. Hence, for a packet arriving at input 'i' will be assigned a wavelength using the relation

$$\lambda(i,k) = \lambda_{[1+(i+k-2)\bmod N]}$$

After the delay of '*k*' time slots the buffered packet will re-appear at the input port of scheduling and due to cyclic nature of AWG it gets directed to the output port '*i*' of the scheduling AWG, here either it will be placed in second buffer of longer delays or it will be forwarded towards the appropriate output '*j*' using TWC of switching section.

The SOAs placed next to input TWC, fully compensate the losses of different components and SOAs also adjust their gain for buffered and straight through packets.

Advantages

1. The huge buffer space is achieved through this switch.
2. A very low packet loss probability is achieved through this switch.

Disadvantages

1. The switch is bulky in design
2. As the design of switch is bulky and many optical components are used in the core of switch, this increases overall cost of this switch.

IRIS

In IRIS (Gripp et al, 2012) project a photonic packet switch router is developed that scales to hundreds of terabit/s capacity. The whole architecture is divided into three stages: the first stage that is space switch; the second one which is time switch; and the last one which is again a space switch. A load balanced multistage architecture is used as shown in the figure 13. At the first stage for connecting each node to an input port N WDM wavelength channels each carrying synchronous fixed-length data packets is used. To set the wavelength routing, an array of all optical SOA based wavelength converters are used in this switch. The time-based switch is used in the second stage of design in which N time buffers consisting of shared ODLs is used. The wavelengths of incoming packets are configured in such a way that the packets entering into the time buffer always exited on their corresponding output ports. Then finally the third stage is used to forward the packet to their desired destination port.

Advantages

1. It is easy to scale the architecture. A 40Gb/s wavelength converter and 80x80 AWGRs allows the system to scale to 256Tb/s.

Figure 12. Dual Buffer based optical packet switch architecture (S9)

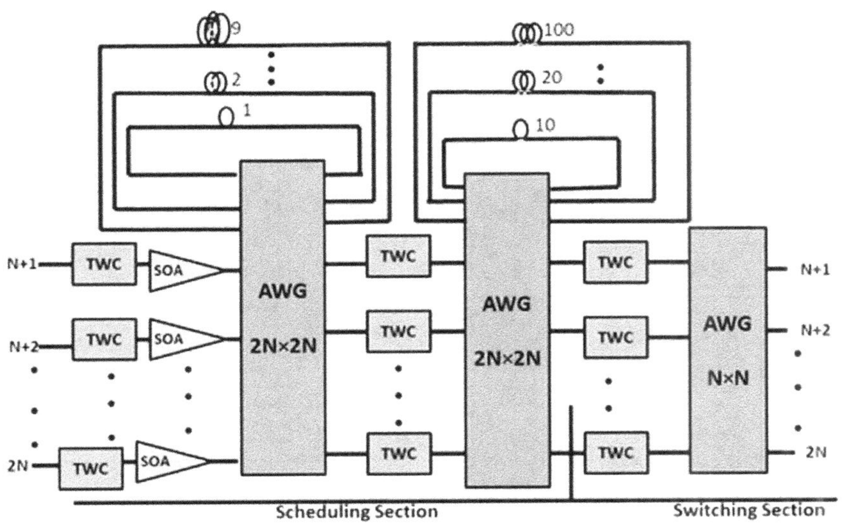

2. The router employs a load-balanced multi-stage architecture with an all-optical data plane consisting of space switches and shared time buffers based on fast wavelength converters and large $N \times N$ AWGs.

Disadvantage

1. Speed is a limitation in the design due to the use of space switch.
2. Cost is high.

Comparison of Switches Under Different Parameters

In this section table 1 presents a comparison among all switches under different parameters like cost, power requirement, buffer type, WDM feature (Shukla et al, 2018). Through this table it is clear that it is impossible to design an efficient switch which covers all attributes.

CONCLUSION

In the present day's telecommunication environment, the use of optical technology is not limited to switching technology but it can also be used in the data center

Figure 13. IRIS Photonic terabit switch (S10)

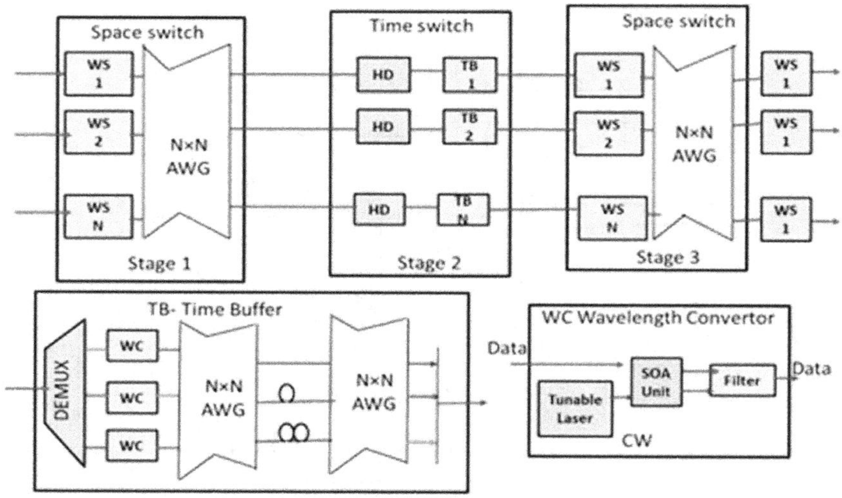

Table 1. Switch comparison table

S.No.	Switch	Year	Technology	WDM	Contention Resolution	Power Efficiency	Cost
1	WDM, AWG based switch (S1)	2009	OPS	Yes	FDL	Low	Low
2	Petabit (S2)	2013	OPS	Yes	Electronic Buffer	High	Moderate
3	Recirculating loop Buffer Switch (S3)	2013	OPS	Partial	FDL	Moderate	Moderate
4	LIONS (S4)	2013	OPS	Yes	Electronic Buffer	Moderate	Low
5	AO-NACK Based Switch (S5)	2013	OCS	Yes	Buffer less	Moderate	Moderate
6	Modified WDM, AWG based switch (S6)	2014	OPS	Yes	FDL	Low	Low
7	AWGR AO-NACK based switch (S7)	2017	OPS	Yes	Electronic Buffer	Moderate	Moderate
8	WDM Hybrid Buffer Switch (S8)	2018	OPS	Yes	Electronic RAM+FDL	Low	Low
9	Dual loop buffer-based switch (S9)	2019	OPS	Yes	FDL	High	High
10	IRIS(S10)	2010	OPS	Yes	FDL	Moderate	Moderate

applications. In this chapter the present status of the data center environment is discussed with their deployment constraints and solutions. The best feasible solutions of all the problems faced by data center network is the use of fast processing optical switching technologies in the core of the designs. In this chapter some notable switch designs are discussed in the field of optical communication and the data center designs. Full detail description of each switch architecture is presented in the chapter with advantages and disadvantages of each design. The following conclusions are made after critical evaluation of switch design:

1. Both optical and electrical technologies coexist in next few years.
2. The use of an optical packet switching system provides a feasible solution for the problems faced by current data center system.
3. For high-speed networks, optical packet switching will remain as next-generation technology.
4. From the discussion in this chapter, it is clear that AWGR will be an integral part of the optical switch and data centers design in the future.
5. The commercialization of tunable wavelength converters will be break-through in the design of optical data center system.

REFERENCES

Al-Fares, Loukissas, & Vahdat. (2008). A scalable, commodity data centre network architecture. *ACM SIGCOMM'08*, 63–74.

Benson, T., Akella, A., & Maltz, D. A. (2010). Network traffic characteristics of data centres in the wild. *10th ACM SIGCOMM Conference on Internet Measurement*, 267–280.

Bhattacharya, P. (2017). Comparative Study for Proposed Algorithm for All-Optical Network with Negative Acknowledgment (AO-NACK). In *Proceedings of the 7th International Conference on Computer and Communication Technology*. ACM. 10.1145/3154979.3154981

Bhattacharya, P., Tiwari, K. A., & Srivastava, R. (in press). Dual Buffers Optical Based Packet Switch incorporating Arrayed Waveguide Gratings. *Journal of Engineering Research.*

Chang, G.-K., Yu, J., Yeo, Y.-K., Chowdhury, A., & Jia, Z. (2006, May). Enabling technologies for next generation optical packet-switching networks. *Proceedings of the IEEE, 94*(5), 892–910. doi:10.1109/JPROC.2006.873433

Cisco Global Cloud Index. (2011). *Forecast and Methodology, 2011–2016*. Cisco White Paper.

Cisco global cloud index: forecast and methodology, 2012-2017. (n.d.). Available: http://www.cisco.com/

Despins, C., Labeau, F., Le-Ngoc, T., Labelle, R., Cheriet, M., Thibeault, C., & Gagnon, F. (2011, August). Leveraging green communications for carbon emission reductions: Techniques, testbeds, and emerging carbon footprint standards. *IEEE Communications Magazine*, *49*(8), 101–109. doi:10.1109/MCOM.2011.5978422

Geer, D. (2005, May). Chip makers turn to multicore processors. *IEEE Computer*, *38*(5), 11–13. doi:10.1109/MC.2005.160

Greenberg, Hamilton, Jain, Kandula, Kim, Lahiri, … Sengupta. (2009). VL2: a scalable and flexible data centre network. *ACM SIGCOMM'09, 39*, 51–62.

Greenberg, A., Hamilton, J., Maltz, D. A., & Patel, P. (2009, January). The cost of a cloud: Research problems in data centre networks. *Computer Communication Review*, *39*(1), 68–73. doi:10.1145/1496091.1496103

GreenDataProject. (2008). *Where does power go?* Available: http://www.greendataproject.org

Greenpeace International. (2010). *Make IT green: cloud computing and its contribution to climate change*. Available: http://www.greenpeace.org/

Gripp, Simsarian, LeGrange, Bernasconi, & Neilson. (2012). Photonic terabit routers: the IRIS project. *Optical Fiber Communication Conference*.

Habib, M. F., Tornatore, M., Leenheer, M. D., Dikbiyik, F., & Mukherjee, B. (2012, August). Design of disaster resilient optical data centre networks. *Journal of Lightwave Technology*, *30*(16), 2563–2573. doi:10.1109/JLT.2012.2201696

Hunter, D. K., Chia, M. C., & Andonovic, I. (1998, December). Buffering in optical packet switches. *Journal of Lightwave Technology*, *16*(12), 2081–2094. doi:10.1109/50.736577

Jinno, M., Takara, H., Kozicki, B., Tsukishima, Y., Sone, Y., & Matsuoka, S. (2009, November). Spectrum-efficient and scalable elastic optical path network: Architecture, benefits, and enabling technologies. *IEEE Communications Magazine*, *47*(11), 66–73. doi:10.1109/MCOM.2009.5307468

Kachris, C., Kanonakis, K., & Tomkos, I. (2013). Optical Interconnection Networks in Data Centres: Recent Trends and Future Challenges. *IEEE Communications Magazine, 51*(9), 39-45.

Kachris, C., & Tomkos, I. (2012). A survey on optical interconnects for data centres. *IEEE Communications Surveys and Tutorials*, *14*(4), 1021–1036. doi:10.1109/SURV.2011.122111.00069

Katz, R. H. (2009, February). Tech titans building boom. *IEEE Spectrum*, *46*(2), 40–54. doi:10.1109/MSPEC.2009.4768855

Koley, B., Vusirikala, V., Lam, C., & Gill, V. (2010). 100GbE and beyond for warehouse scale computing. *15th IEEE Opto electronics and Communications Conference (OECC),* 106–107.

Lam. (2010). Optical network technologies for data centre networks. *OFC/NFOEC 2010*, NWA3.

Lam, C. F., Liu, H., Koley, B., Zhao, X., Kamalov, V., & Gill, V. (2010, July). Fiber optic communication technologies: What's needed for datacentre network operations? *IEEE Communications Magazine*, *48*(7), 32–39. doi:10.1109/MCOM.2010.5496876

Meisner, D., Gold, B. T., & Wenisch, T. F. (2009). PowerNap: eliminating server idle power. *14th International Conference on Architectural Support for Programming Languages and Operating Systems*, 205–216.

Miller, D. A. B. (2009, July). Device requirements for optical interconnects to silicon chips. *Proceedings of the IEEE*, *97*(7), 1166–1185. doi:10.1109/JPROC.2009.2014298

Nagaraj, K., Khandelwal, H., Killian, C., & Kompella, R. R. (2012). Hierarchy-aware distributed overlays in data centres using DC2. *Fourth International Conference on Communication Systems and Networks (COMSNETS)*, 1–10.

Nicholes, S. C. (2010, February). An 8×8 InP monolithic tunable optical router (MOTOR) packet forwarding chip. *Journal of Lightwave Technology*, *28*(4), 641–650. doi:10.1109/JLT.2009.2030145

Proietti, Yin, Yu, Ye, Nitta, Akella, & Yoo. (2012). All-Optical Physical Layer NACK in AWGR-Based Optical Interconnects. *IEEE Photonics Technology Letters, 24*(5), 410-412.

Rastegarfar, Garcia, Laroachelle, & Rusch. (n.d.). Cross Layer Performance Analysis of Recirculation Buffers for Optical Data Centres. *Journal of Lightwave Technology*, *31*(3), 432–445. doi:10.1109/JLT.2012.2226865

Sato, K.-I., & Hasegawa, H. (2009, July). Optical networking technologies that will create future bandwidth abundant networks. *Journal of Optical Communications and Networking*, *1*(2), A81–A93. doi:10.1364/JOCN.1.000A81

Shacham, A., & Bergman, K. (2007, July/August). Building ultralow-latency interconnection networks using photonic integration. *IEEE Micro*, *27*(4), 6–20. doi:10.1109/MM.2007.64

Shacham, A., Bergman, K., & Carloni, L. P. (2008, September). Photonic networks-on-chip for future generations of chip multiprocessors. *IEEE Transactions on Computers*, *57*(9), 1246–1260. doi:10.1109/TC.2008.78

Shukla, Jain, & Srivastava. (2015). Performance Evaluation of an AWG based Optical Router. *Optical and Quantum Electronics, 48*(1).

Shukla, Jain, & Srivastava. (2014). Physical Layer Analysis of Arrayed Waveguide Based Optical Switch. *International Journal of Applied Engineering Research, 9*(21), 10035-10050.

Shukla, Jain, & Srivastava. (2016). Design of an Arrayed Waveguide Gratings based Optical Packet Switch. *Journal of Engineering Science and Technology, 11*(12), 1705-1721.

Shukla, V. (2016). Design of AWG based Optical Switch for High Speed Optical Networks. *IJE Transactions A: Basics*, *29*(7), 909–915.

Shukla, V., & Jain, A. (2018). Design and analysis of high-speed optical routers for next generation data centre network. *Journal of Engineering Research*, *6*(2), 122–137.

Shukla, V., & Singh, O. P., Khare, M., Sharma, S., & Srivastava, R. (2018). Estimation of burst length in optical burst switching networks based on early release of control packet. *Journal of Engineering Research*, *6*(1).

Shukla, V., & Srivastava, R. (2015). WDM fiber delay lines and AWG based optical packet switch architecture. *Proceedings of National Conference on Innovative Trends in Computer Science Engineering (ITCSE-2015).*

Singh, Tiwari, & Srivastava. (2018). Design and analysis of hybrid optical and electronic buffer based optical packet switch. *Sādhanā, 43*(2), 19.

Srivastava, R. (n.d.). *WDM fiber delay lines based optical packet switch architectures* (Ph.D. dissertation). IIT Kanpur.

Taubenblatt, M. A. (2012, February). Optical interconnects for high-performance computing. *Journal of Lightwave Technology, 30*(4), 448–457. doi:10.1109/JLT.2011.2172989

Tucker, R. S. (2006, December). The role of optics and electronics in high-capacity routers. *Journal of Lightwave Technology, 24*(12), 4655–4673. doi:10.1109/JLT.2006.885774

Wong, E. W. M., & Zukerman, M. (2006, December). Bandwidth and buffer trade offs in optical packet switching. *Journal of Lightwave Technology, 24*(12), 4790–4798. doi:10.1109/JLT.2006.885249

Xi, K., Kao, Y.-H., & Chao, H. J. (2013). *A Petabit buffer less optical switch for data center networks. In Optical Interconnects for Future Data Center Networks* (pp. 135–154). Springer New York.

Xu, L., Perros, H. G., & Rouskas, G. (2001, January). Techniques for optical packet switching and optical burst switching. *IEEE Communications Magazine, 39*(1), 136–142. doi:10.1109/35.894388

Yin, Y. (2013, March). LIONS: An AWGR-based low-latency optical switch for high-performance computing and data centres. *Selected Topics in Quantum Electronics. IEEE Journal of, 19*(2), 3600409–3600409.

Yoo, S. J. B. (2006, December). Optical packet and burst switching technologies for the future photonic Internet. *Journal of Lightwave Technology, 24*(12), 4468–4492. doi:10.1109/JLT.2006.886060

Yoo, S. J. B. (2006, December). Optical Packet and Burst Switching Technologies for the Future Photonic Internet. *Journal of Lightwave Technology, 24*(12), 4468–4492. doi:10.1109/JLT.2006.886060

Yoo, S. J. B. (2011, March/April). Energy efficiency in the future Internet: The role of optical packet switching and optical-label switching. *IEEE Journal of Selected Topics in Quantum Electronics*, *17*(2), 406–418. doi:10.1109/JSTQE.2010.2076793

Zhang, W., Wang, H., & Bergman, K. (2012, December). Next-generation optically-interconnected high performance data centres. *Journal of Lightwave Technology*, *30*(24), 3836–3844. doi:10.1109/JLT.2012.2212696

APPENDIX

Table 2.

Acronyms and Abbreviations	
AWG	Arrayed Waveguide Grating
DCN	Data Centre Network
DWDM	Dense Wavelength Division Multiplexing
CAGR	Compound annual growth rate
CMOS	complementary metal-oxide-semiconductor
EDFA	Erbium Doped Fiber amplifier
FDLs	Fiber Delay Lines
FF	Fixed Filter
KEOPS	Keys to Optical Packet Switching
NGDC	Next Generation Data Centre
OEO	Optical Electrical Optical
OLB	Optical Loop Buffer
OBS	Optical Burst Switching
OPS	Optical Packet Switching
OCS	Optical circuit Switching
RAM	Random Access Memory
SLOB	Switch with Large Optical Buffer
SOA	Semiconductor Optical Amplifier
TDM	Time Division Multiplexing
TF	Tunable Filter
TWC	Tunable Wavelength Convertor
WASPNET	Wavelength Switched Optical Network
WDM	Wavelength Division Multiplexing

Chapter 9
Applications of a Birefringent Lens as an Optical Image Processing Device

Surajit Mandal
B. P. Poddar Institute of Management and Technology, India

ABSTRACT

A uniaxial birefringent crystal lens with its optic axis perpendicular to the system axis and sandwiched between two properly oriented linear polarizers behaves as an isotropic lens with a radially varying complex mask on its pupil plane. The proposed system may be adapted for both apodization and enhanced resolution just by rotating one of the two linear polarizers even when it is illuminated with a polychromatic source of light. Hence, the system may find applications in the fields of spectroscopy and astronomy. In general, it behaves as a double focus lens. However, by varying the birefringent lens parameters, it is possible to obtain a noticeably large depth-of-focus compared to an identical isotropic lens. An optical imaging system with large depth-of-focus is a prerequisite for many fields of applications particularly for microscope imagery, medical imaging, as well as for automatic inspection in microelectronics industry.

INTRODUCTION

The pupil function plays an important role in the final image formation by an optical system. The fact can be understood as some of the image assessment parameters such as the point spread function (PSF), the optical transfer function (OTF) etc. are directly related to the pupil function. Many researchers reported that an imaging

DOI: 10.4018/978-1-5225-8531-2.ch009

Copyright © 2019, IGI Global. Copying or distributing in print or electronic forms without written permission of IGI Global is prohibited.

system may exhibit an appropriate behavior by modifying either the amplitude or phase or both of the pupil function. An appropriate coating of metal or dielectrics on the pupil may reduce the size of the central spot of the point spread function which in turn may increase the resolving power of the system (Luneberg, 1944). Suppression of the secondary maxima in the diffraction pattern using suitable apodizing masks also gives some advantages which may be utilized in spectroscopy and astronomy (Jacquinot & Roizen-Dossier, 1964). Some researchers reported the use of an appropriate phase filter to improve the resolving power (Osterberg & Wilkins, 1949, 1950; Lit, 1971). A special type of phase filter, known as Walsh filter was also reported for increasing resolution in microscopic imaging (Hazra, 2006).

In some applications such as automatic inspection in microelectronics industry, medical imaging and microscope imagery, imaging systems with high focal depth are required. The use of annular apertures and apodizers for increasing the focal depth as well as reducing spherical aberration was studied by some researchers (Castaneda et al., 1985, 1986, 1988, 1989, 1990). Use of phase filters to decrease the defocus and spherical aberrations was also reported by some scientists (Mezouari & Harvey, 2003).

In almost all of the above cases, the masks used are not versatile and have been utilized for some particular applications. Moreover, on-line modification of these masks and hence, imaging characteristics of the optical systems is not possible. In this connection it may be mentioned that the imaging behavior of such systems has been considered to be independent of the polarization properties of the incident light and only scalar wave theory of light has been employed to study the imaging behavior. Some scientists, however, realized that the polarization properties of light may be used as a parametric variable to modify the complex amplitude of the pupil function. Two orthogonal states of polarization control the nature of the pupil function in this case and their relative contributions can allow real-time modification of the imaging behavior of polarization-based optical systems. Kubota and Inoue (1959), and Kubota and Saito (1960) first demonstrated the application of polarization properties of light in investigating the diffraction images of a pinhole produced by a polarizing microscope with parallel and crossed nicols. Marathay (1969) suggested a versatile analytical technique to realize complex spatial filters using polarizaton properties of light and Polaroid Vectograph film to obtain the desired image. Whiting (2003) utilized polarization properties of light for achieving superresolution for an optical system. The advantage of polarization-based optical systems is that the rotation of any of the two polarizers can modify the imaging properties of the system. This makes the system more versatile for different applications.

However, polarization-based optical devices have certain inherent disadvantages. Manufacturing and proper positioning of polarization masks are very difficult and an appreciable amount of light is always absorbed and/or scattered by the masks. So, research has been carried out to address these problems. Optical imaging systems with their components made from birefringent materials may reduce the associated problems to a certain extent.

An optical lens made up of birefringent material may alter the imaging properties of the system. However, birefringent optical elements have already been used in interferometry (Dyson, 1957; Yang, Xiang & Zhang, 1993; Davis, Evans, Crabtree & Moreno 2004) as well as in many other applications because of their some unique features. Ray, Sarkar, Chakraborty and Basuray (2008) reported a technique of birefringence measurement with the aid of a confocal microscopic setup. Chou, Shyu, Huang and Yuan (1998) suggested a birefringent element based common-path optical heterodyne profilometer for measurement of the profiles of flat surfaces. The design of birefringent bifocal lenses to be used for contact lenses and intra-ocular lenses was also demonstrated some researchers (Fiala, 1990, 1992; Fiala, Finkelmann & Gfeller1993).

In most of the above cases applications of a uniaxial crystal lens as a bifocal lens was taken into consideration. However, a birefringent lens may also be used as an imaging device which needed further research. Any vector quantity may be resolved into two mutually independent orthogonal components. The electric vector of a polarized beam of light also vibrates in a particular direction. Likewise, it may be assumed to consist of two orthogonal components (conventionally called the *p*-polarization and the *s*-polarization) which are also independent to each other and they do not interfere. Because of this reason, a polarized light wave may be termed as a "vector wave". Thus, vector algebra will add a new dimension in the analysis of the imaging behavior of optical elements made up of birefringent material.

THE PROPOSED SYSTEM AND ITS WORKING PRINCIPLE

Figure 1 shows our proposed system followed by its working principle.

When a beam of polarized light is incident on the birefringent lens, it splits into two rays: ordinary ray and extraordinary ray. As the refractive indices of the birefringent material for these two rays are different, corresponding spherical wavefronts have different curvatures. When these two spherical wavefronts pass through the second polarizer, they produce two foci on the axis of the proposed system. At the same time, when these two wavefronts interfere with each other, they produce circular fringes on the pupil plane of the proposed system. Therefore proposed system behaves like a normal lens with its pupil plane bearing a complex mask. The complex mask is found to vary in the radial direction.

Figure 1. Proposed system

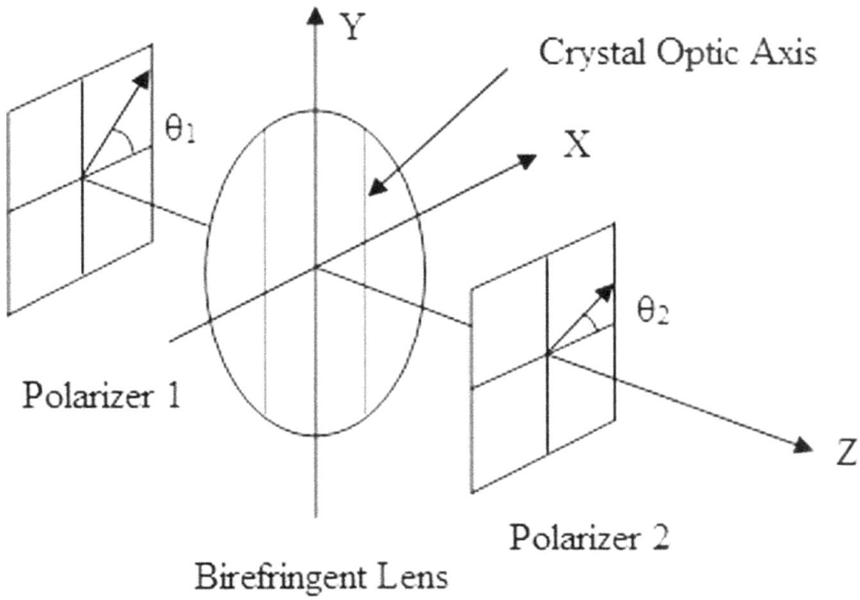

The direction of the pass axes of the linear polarizers together with the design parameters of the uniaxial crystal lens determines the nature of this virtual mask. Therefore, if any of the polarizers is rotated, the virtual mask will also be modified, which, in turn, will modify the properties of the imaging system. As two spherical wavefronts with different curvatures due to extraordinary ray and ordinary ray interfere with each other to produce the mask on the pupil plane of the said system, it is self-oriented. However, a considerable amount of light is absorbed by the polarizers and the uniaxial crystal, so a laser source may be appropriate for illuminating such a system. Care should also be taken for proper alignment of the crystal optic axis during manufacturing process of the uniaxial crystal lens. Sanyal *et al.* (1998, 1999, 2000, 2002, 2003, 2004) studied the behavior of the proposed system when illuminated with a source of monochromatic light. Here the properties of such a system are investigated under broadband illumination.

PERFORMANCE OF THE PROPOSED SYSTEM UNDER BROADBAND ILLUMINATION

It is well-known that the phenomena of reflection, transmission, linear and circular birefringence, dichroism etc. in general affect the state of polarization of a beam of light and the effects are dependent on the wavelength of the input light beam. In other words, polarization phenomena are wavelength-sensitive. The overall behavior of the polychromatic beam that results from the superposition of these individual monochromatic components obviously leads to a partial depolarization of the output beam, even though the individual monochromatic components are completely polarized. Thus, in order to study the interaction of a polychromatic beam with any polarization device one must consider the effect of each of the wavelength components separately and finally appropriately add up these individual components over the given wavelength range to get the resultant effect. We intend to study the effect of a polychromatic input beam on the imaging characteristics of the proposed system. The input beam is considered to be uniform for all wavelengths.

In order to investigate the focusing characteristics of the proposed system under polychromatic beam illumination, the axial irradiance distribution function is chosen as the image assessment parameter. The imaging performance of the proposed system under diffraction-limited condition when illuminated with a broadband source is studied in terms of the polychromatic intensity point spread function (IPSF). Since its introduction, the monochromatic OTF has emerged as a merit function and powerful tool in characterizing and assessing the performance of an optical system, because of the fact that experimental evaluation of the OTF is possible and measuring instruments are available for this purpose. It is also useful for addressing issues that may be difficult to judge from the intensity distributions. Because the OTF is suitable for a wide range of systems it has become an obvious choice for extension to the polychromatic case. Evaluation of the performance of optical systems by means of polychromatic OTF is now a standard operating procedure. The imaging behavior of the proposed system is also studied using polychromatic OTF as the image assessment parameter.

EXPRESSIONS OF VARIOUS IMAGE ASSESSMENT PARAMETERS

The effective pupil function of the proposed system (for parallel- and crossed-polarizers configurations) under defocused condition is given by (Sanyal, Bandyopadhyay & Ghosh, 1998; Sanyal & Ghosh, 1999)

$$\begin{aligned} \Omega_{-}^{+}\left(r,\alpha,\bar{W}_{20}\right) &= \begin{matrix}\cos\\ i\sin\end{matrix}\left\{k\alpha\left(1-r^{2}\right)\right\}\exp(i\,k\,\bar{W}_{20}\,r^{2})\,; && \text{for } r\le 1 \\ &= 0; && \text{otherwise} \end{aligned} \tag{1}$$

where $k = 2\pi / \lambda$ is the propagation constant, $\bar{W}_{20} = W_{20} - \sigma$ is the effective defocus, W_{20} is the defocus coefficient, $\alpha = \dfrac{\delta n\ \Delta}{2}$ and

$$\sigma = \left(\bar{n} - 1\right)\Delta = \frac{2(\bar{n} - 1)}{\delta n}\alpha$$

are two design parameters of the birefringent lens, n_o and n_e are the ordinary and extraordinary refractive indices respectively, $\delta n = n_e - n_o$ is the birefringence of the lens material, $\bar{n} = \dfrac{n_o + n_e}{2}$ is the mean refractive index of the lens material and Δ is the central thickness of the lens.

Upon simplification and identifying the wavelength dependent terms, the axial irradiance distribution function for the proposed system comes out as (Mandal, Sanyal & Ghosh, 2007)

$$I_{\lambda-}^{+}\left(W_{20}\right) = \frac{1}{4}\left[\mathrm{sinc}^2\left\{W_{20} - \left(\alpha_\lambda + \sigma_\lambda\right)\right\} + \mathrm{sinc}^2\left\{W_{20} + \left(\alpha_\lambda - \sigma_\lambda\right)\right\}\right]$$

$$\pm \frac{1}{2}\mathrm{sinc}\left\{W_{20} - \left(\alpha_\lambda + \sigma_\lambda\right)\right\}\ \mathrm{sinc}\left\{W_{20} + \left(\alpha_\lambda - \sigma_\lambda\right)\right\}\cos\left(\frac{2\pi}{\lambda}\alpha_\lambda\right) \qquad (2)$$

Both α_λ and σ_λ vary non-linearly with λ for most of the birefringent crystals. Consequently, the change of the shape of the axial irradiance curve and the shift of the central plane are not uniform for equal variation in wavelength throughout the spectrum. Thus, the polychromatic axial irradiance distribution, which is an incoherent superposition of individual monochromatic components, becomes asymmetric for the said system. For a polychromatic input illumination with mean wavelength λ_0 and a bandwidth of $2\delta\lambda$, the expression for axial irradiance distribution for the proposed system (under diffraction-limited condition) with parallel- and crossed-polarizers configuration becomes (Mandal, Sanyal & Ghosh, 2007)

$$I_{poly-}^{\ +}(W_{20}) =$$

$$\frac{1}{2\delta\lambda} \times \frac{1}{4}\int_{\lambda_0 - \delta\lambda}^{\lambda_0 + \delta\lambda} \left\{ \left[\mathrm{sinc}^2\left\{W_{20} - \left(\alpha_\lambda + \sigma_\lambda\right)\right\} + \mathrm{sinc}^2\left\{W_{20} + \left(\alpha_\lambda - \sigma_\lambda\right)\right\}\right]\right.$$

$$\pm \frac{1}{2}\text{sinc}\left\{W_{20} - \left(\alpha_\lambda + \sigma_\lambda\right)\right\} \text{sinc}\left\{W_{20} + \left(\alpha_\lambda - \sigma_\lambda\right)\right\} \cos\left(\frac{2\pi}{\lambda}\alpha_\lambda\right) \Big\} d\lambda \tag{3}$$

The monochromatic IPSF of the proposed system under focused and defocused conditions with both parallel- and crossed-polarizers configurations is given by (Mandal & Ghosh, 2012).

$$IPSF_{\lambda\,-}^{\;+}(W_{20}) = 4\left\{\left[\int_0^1 \begin{matrix}\cos\\ \sin\end{matrix}\left\{k_\lambda\, \alpha_\lambda \left(1-r^2\right)\right\} \cos\left[k_\lambda \left(W_{20} - \sigma_\lambda\right) r^2\right] J_0\left(\rho_\lambda r\right) r\, dr\right]^2 + \left[\int_0^1 \begin{matrix}\cos\\ \sin\end{matrix}\left\{k_\lambda\, \alpha_\lambda \left(1-r^2\right)\right\} \sin\left[k_\lambda \left(W_{20} - \sigma_\lambda\right) r^2\right] J_0\left(\rho r\right) r\, dr\right]^2\right\} \tag{4}$$

where $J_0(\xi)$ stands for the zero order Bessel function of first kind with argument ξ and ρ is the radial coordinate of the image plane. For a polychromatic illumination having flat-top spectral profile with mean wavelength λ_0 and a spread of $2\delta\lambda$, the above expression becomes

$$IPSF_{poly\,-}^{\;\;+}(W_{20}) = \frac{4}{2\delta\lambda}\int_{\lambda_0-\delta\lambda}^{\lambda_0+\delta\lambda} \left\{\left[\int_0^1 \begin{matrix}\cos\\ \sin\end{matrix}\left\{k_\lambda\, \alpha_\lambda \left(1-r^2\right)\right\} \cos\left[k_\lambda \left(W_{20} - \sigma_\lambda\right) r^2\right] J_0\left(\rho_\lambda r\right) r\, dr\right]^2 + \left[\int_0^1 \begin{matrix}\cos\\ \sin\end{matrix}\left\{k_\lambda\, \alpha_\lambda \left(1-r^2\right)\right\} \sin\left[k_\lambda \left(W_{20} - \sigma_\lambda\right) r^2\right] J_0\left(\rho r\right) r\, dr\right]^2\right\} d\lambda \tag{5}$$

The polychromatic OTF of the proposed system under both parallel-polarizers and crossed-polarizers configurations comes out as (Mandal, 2014)

$$OTF_{poly-}^{\;\;+}\left(W_{20}\right) = \frac{1}{\int_{\lambda_1}^{\lambda_2} N_-^+\left(\alpha_\lambda\right) d\lambda}\left[\int_{\lambda_1}^{\lambda_2}\int_{x=0}^{a_\lambda}\int_{y=0}^{b_\lambda} \cos\left[2k_\lambda \left\{\left(W_{20} - \sigma_\lambda\right) - \alpha_\lambda\right\} s_\lambda x\right] dx\,dy\,d\lambda\right.$$

$$+\int_{\lambda_1}^{\lambda_2}\int_{x=0}^{a_\lambda}\int_{y=0}^{b_\lambda}\cos\left[2k_\lambda\left\{\left(W_{20}-\sigma_\lambda\right)+\alpha_\lambda\right\}s_\lambda x\right]dxdyd\lambda$$

$$\pm\int_{\lambda_1}^{\lambda_2}\int_{x=0}^{a_\lambda}\int_{y=0}^{b_\lambda}\cos\left\{2k_\lambda\alpha_\lambda\left(x^2+y^2+\frac{s_\lambda^{\,2}}{4}-1\right)+2k_\lambda\left(W_{20}-\sigma_\lambda\right)s_\lambda x\right\}dxdyd\lambda \tag{6}$$

where $a_\lambda = 1-\frac{s_\lambda}{2}$ and $b_\lambda = \left[1-\left(x+\frac{s_\lambda}{2}\right)^2\right]^{\frac{1}{2}}$.

All the above image assessment parameters under polychromatic input illumination are calculated numerically using 96-point Legendre-Gauss quadrature technique of integration.

RESULTS AND DISCUSSIONS

The proposed system provides the maximum depth-of-focus for $\alpha = 0.3142\lambda$ with parallel-polarizers configuration when illuminated with a monochromatic source of light (Sanyal & Ghosh, 2000). We intend to study the effect of polychromatic illumination on the focusing characteristics of the system for this particular case. α is a combination of the birefringence value and the central thickness of the lens. Since α is small in this case and very small thickness of the lens is not feasible, so we need to select a material with a very small birefringence for the birefringent lens. We, therefore, consider quartz crystal as the lens material. The variation of both δn and $\frac{2\left(\bar{n}-1\right)}{\delta n}$ with λ is non-linear for quartz in general (Mandal, Sanyal & Ghosh, 2007). In addition, σ_λ is a non-linear function of α_λ. Non-linear characteristics of these parameters introduce asymmetry in the axial irradiance distribution curve. As a result the proposed system tends to lose its inherent property of high depth-of-focus. Thus the variation of these parameters with λ should not be very large if the symmetry of the axial irradiance curve and the high focal depth are to be maintained. For this reason, the operating wavelength range should be considered around and above 600 nm for quartz crystal where the wavelength dependence of these parameters is not very significant. The transmittance of crystal quartz range is also very high (about 90%) in this wavelength region and almost uniform.

The axial irradiance distribution for the said system (diffraction-limited condition) obtained with $\alpha_{\lambda 0} = 0.3142\,\lambda_0$ under parallel-polarizers configuration at 800 nm is shown in Figure 2 with spectral spreads ($2\delta\lambda$) of 0 nm, 50 nm, 100 nm, 150 nm, 200 nm, 300 nm and 400 nm in each case. Though not shown in the figure, it is found that the allowable wavelength spread is approximately 50 nm at the 550 nm operating wavelength (yellow light) and 60 nm at 632 nm (red light). As shown in Fig. 2 the allowable spectral spread is 100 nm at 800 nm (infra-red radiation). Here, it may be mentioned that 632.8 nm is the operating wavelength of a standard low-power He-Ne laser. Hence it may be concluded that red region of the visible light may be suitable for illuminating the system and even a white light source can be used in conjunction with a wavelength-selective filter. Similar situations may be observed with crossed-polarizers configuration. It is also found that for lower wavelength operation, the usable wavelength band is found to be smaller in both parallel-polarizers and crossed-polarizers configurations. On the contrary, much wavelength spread may be allowed if the system is designed to work under IR

Figure 2. Axial irradiance distribution curve of a quartz crystal lens

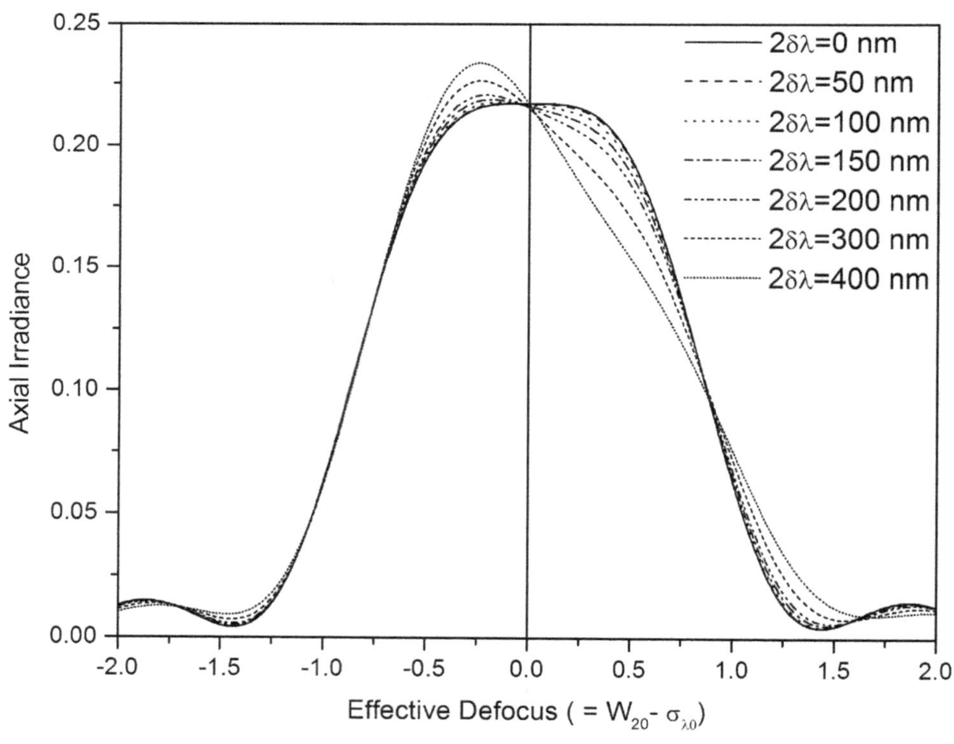

illumination. Moreover, the variation of the refractive indices as well as the birefringence with wavelength at the IR range is nearly linear, thus the asymmetry for the axial irradiance distribution also becomes less (Mandal, Sanyal & Ghosh, 2007).

For $\alpha \leq 0.375\,\lambda$, the proposed system shows enhanced resolution under parallel-polarizers configuration while with crossed-polarizers configuration the spread of the point spread function is more than normal indicating a lesser resolution under monochromatic illumination (Mandal & Ghosh, 2012). $\alpha_{\lambda 0} = 0.25\,\lambda_0$ is considered for further investigation. The variations of the IPSF for the quartz crystal lens at the Gaussian image plane under parallel- and crossed-polarizers configurations are shown in Figures 3 and 4 respectively. The operating wavelength (λ_0) is considered to be 632 nm and spectral spreads ($2\delta\lambda$) are selected as 0 nm, 200 nm and 400 nm. The IPSF for an equivalent isotropic lens at the Gaussian image plane has also been drawn for comparison. As evident from the figures, such a low-power birefringent

Figure 3. Polychromatic IPSF of the proposed system at the Gaussian image with $\alpha = 0.25\,\lambda_0$ *under parallel-polarizers configuration*

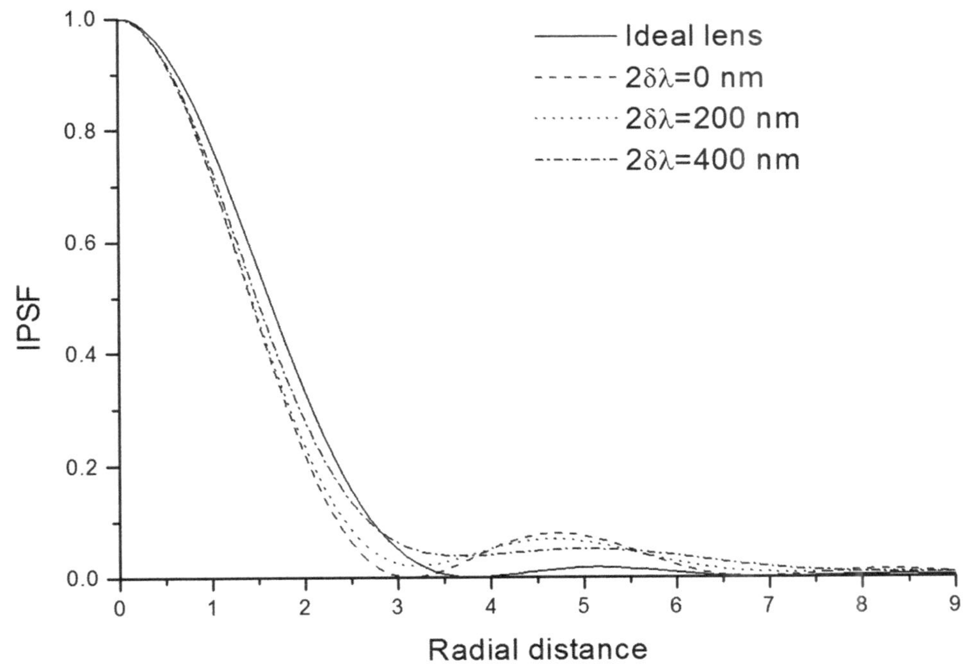

Figure 4. Polychromatic IPSF of the proposed system at the Gaussian image with $\alpha = 0.25\lambda_0$ under crossed-polarizers configuration

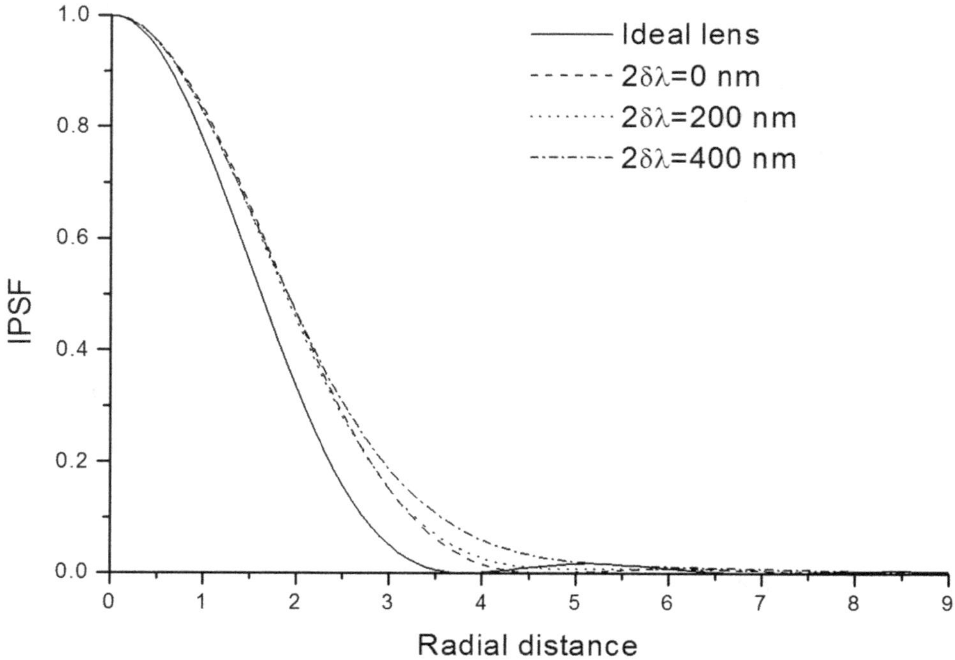

lens (small α) retains the inherent characteristics of enhanced resolution under parallel-polarizers configuration and apodization under crossed-polarizers configuration even when the spectral spread is 400 nm (Mandal & Ghosh, 2012).

Figures 5 and 6 show the variation of the polychromatic IPSF for the quartz lens with defocus under parallel-polarizers and crossed-polarizers configurations respectively. The operating wavelength is considered as 632 nm with a spectral spread 100 nm. The same has also been drawn for an isotropic lens for comparative study. The proposed system still retains all of its inherent property even when the effective defocus changes to $\lambda_0 / 2$. It was also found that for $0.375\lambda < \alpha \leq 0.625\lambda$ with crossed-polarizers configuration, the said system shows enhanced resolution when monochromatic source of light is used as input. Figure 7 shows the polychromatic IPSF curve at the Gaussian image plane for $\alpha_{\lambda 0} = 0.6044\lambda_0$ under crossed-polarizers configuration (Mandal & Ghosh, 2012). Here, it may be mentioned that the birefringent lens shows the maximum depth-of-focus under diffraction-limited condition for

Figure 5. Polychromatic IPSF of the quartz lens when $\alpha = 0.25\,\lambda_0$ *under parallel-polarisers configuration*

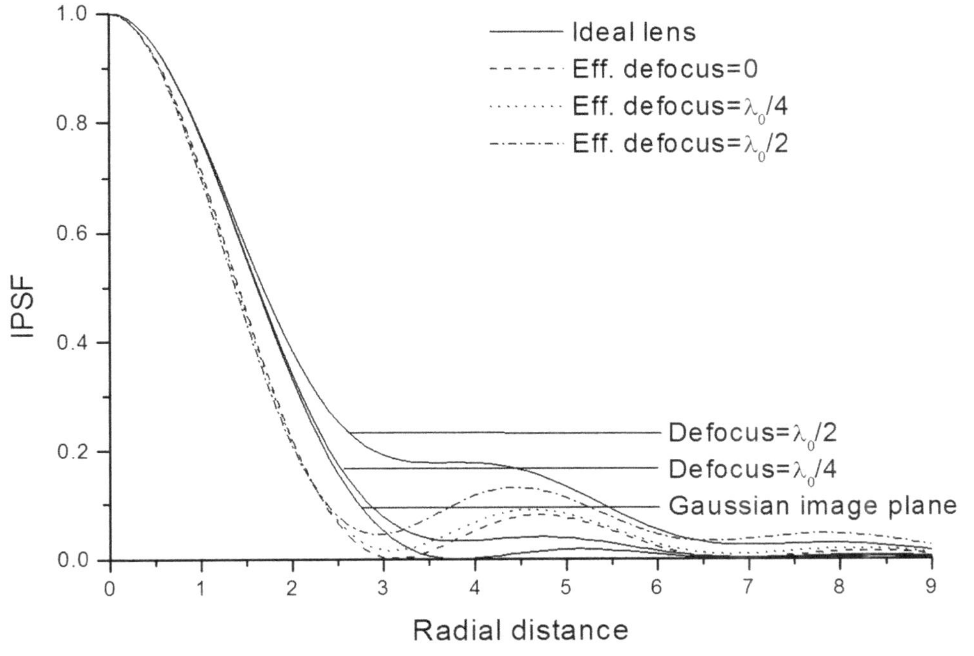

$\alpha = 0.6044\,\lambda$ under crossed-polarizers configuration when illuminated with a monochromatic source of light. A significant change in the IPSF curve is found to take place in this case. If α increases further, the said system starts behaving as a double focus lens. Figures 8 and 9 depict the polychromatic IPSF curve for $\alpha_{\lambda 0} = 0.75\lambda_0$ under crossed-polarizers configuration at a mean wavelength of 632 nm and spectral spreads 100 nm and 400 nm respectively. The said system retains its bifocal nature when the spectral spreads are small. However, the quartz lens loses this property at higher spectral spread due to incoherent superposition of resultant monochromatic vectors and unequal phase retardation suffered by different wavelengths constituting the polychromatic beam along the system axis (Mandal & Ghosh, 2012).

In order to study the polychromatic OTF we again consider $\alpha = 0.25\,\lambda$ because with this value of α the same system offers enhanced resolution under parallel-polarizers configuration and apodization with crossed-polarizers configuration. λ_1

Figure 6. Polychromatic IPSF of the quartz lens when $\alpha = 0.25\,\lambda_0$ under crossed-polarisers configuration

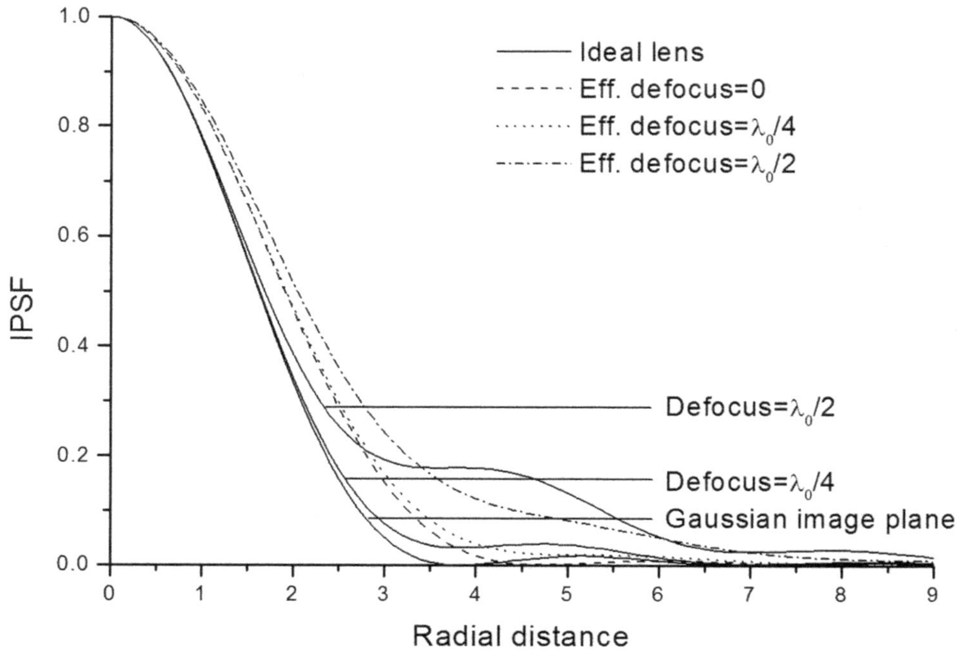

is taken as 400 nm as this is approximately the lower cut-off value for the visible light. The polychromatic OTF curves at the Gaussian image plane for $\alpha_{\lambda 1} = 0.25\lambda_1$ under both parallel- and crossed-polarizers configurations have been shown in Figs. 10 and 11 respectively with spectral spreads ($\Delta\lambda = \lambda_2 - \lambda_1$) of 0 nm, 100 nm and 300 nm. The same for an identical conventional lens has been drawn for comparison. As evident from the figures, a suppression of low frequencies and an enhancement of high frequencies occur in the parallel-polarizers configuration under monochromatic illumination. On the contrary, a suppression of high frequencies with an enhancement of low frequencies occurs under crossed-polarizers configuration with the same source. These amply justify the use of a single system for enhanced resolution under parallel-polarizers configuration as well as apodization for crossed-polarizers configuration, also revealed by the respective IPSF curves. The said system maintains the property of enhanced resolution under parallel-polarizers configuration even when the spectral spread is as high as 100 nm. However, at a very high value of

Figure 7. Polychromatic IPSF of the proposed system under crossed-polarizers configuration with $\alpha = 0.6044\,\lambda_0$

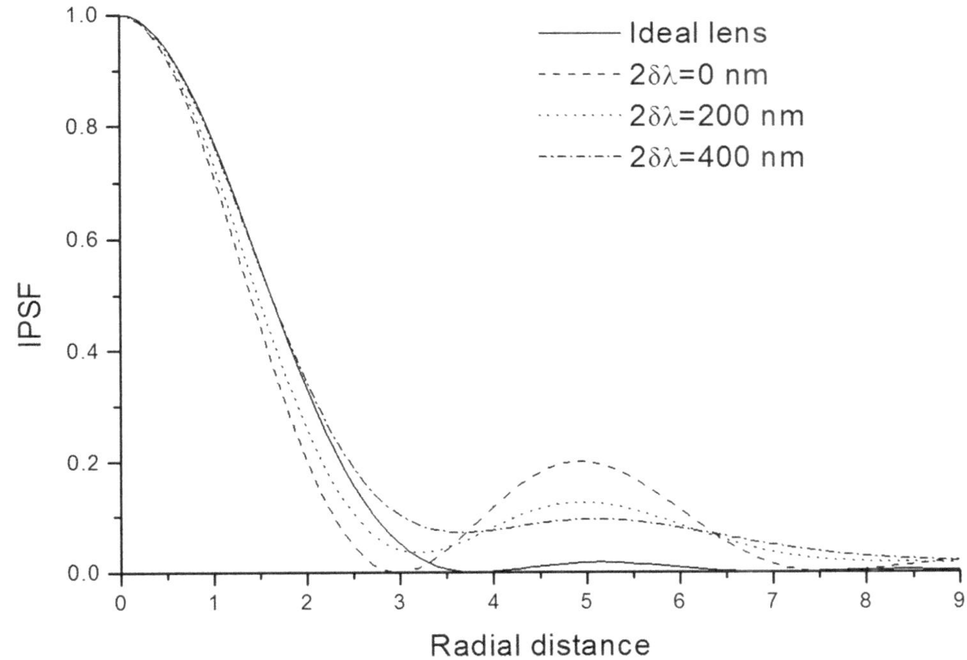

spectral spread (300 nm, say) the OTF for the said system decreases almost uniformly with spatial frequency under parallel polarizers configuration resulting in deviation from its imaging characteristics under strictly monochromatic illumination. The corresponding polychromatic IPSF curve obtained in this case supports the fact.

The proposed system can also be used for edge enhancement of a degraded image. Instead of using two different filters, two orthogonal settings of the analyzer can be selected to get two different OTF curves, which are eventually subtracted from each other to enhance the high frequency contents of the image. Fig. 12 shows the polychromatic OTF curves of the birefringent lens with $\alpha_{\lambda 1} = 0.625\lambda_1$ at the Gaussian image plane for both parallel- and crossed-polarizers configurations taking $\Delta\lambda$ as 0 nm, 50 nm, 100 nm and 200 nm. The lesser variation of the OTF curve with spectral spread for both configurations ensures that the said system may be used for edge enhancement of the degraded image with up to 100 nm spectral spread (Mandal, 2014).

Figure 8. Polychromatic IPSF of the proposed system under crossed-polarizers configuration with $\alpha = 0.75\lambda_0$, λ_0 *= 632 nm and* $2\delta\lambda$ *= 100 nm*

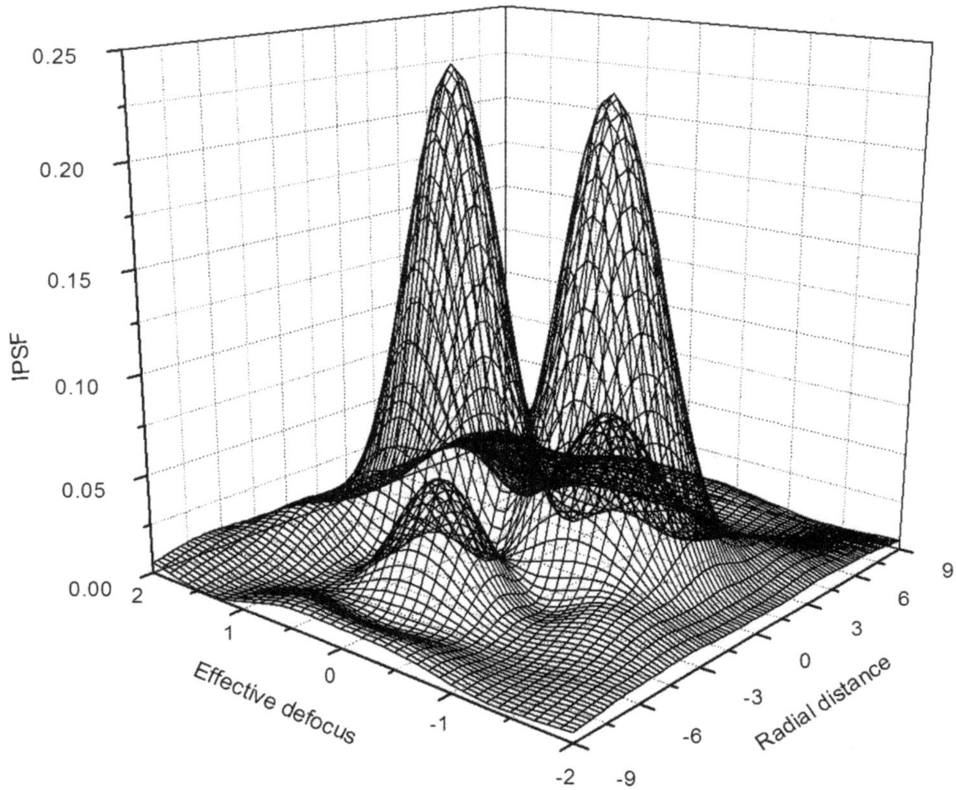

CONCLUSION

Earlier investigations revealed that if the relative orientation of the two polarizers is altered, the imaging behavior of the proposed system is also modified. As a result both apodization and enhanced resolution may be achieved with the same system even when a monochromatic source of light is used. These properties can effectively be utilized in both spectroscopy and astronomy. Under monochromatic illumination, the said system has two foci in general. The separation of these two foci can be modified by adjusting the lens parameters during manufacturing process. It is also possible to achieve a large focal depth for the system and the focal depth may be even larger than an identical isotropic lens for certain specific shape of the lens.

Figure 9. Polychromatic IPSF of the proposed system under crossed-polarizers configuration with $\alpha = 0.75\lambda_0$, λ_0 *= 632 nm and* $2\delta\lambda$ *= 400 nm*

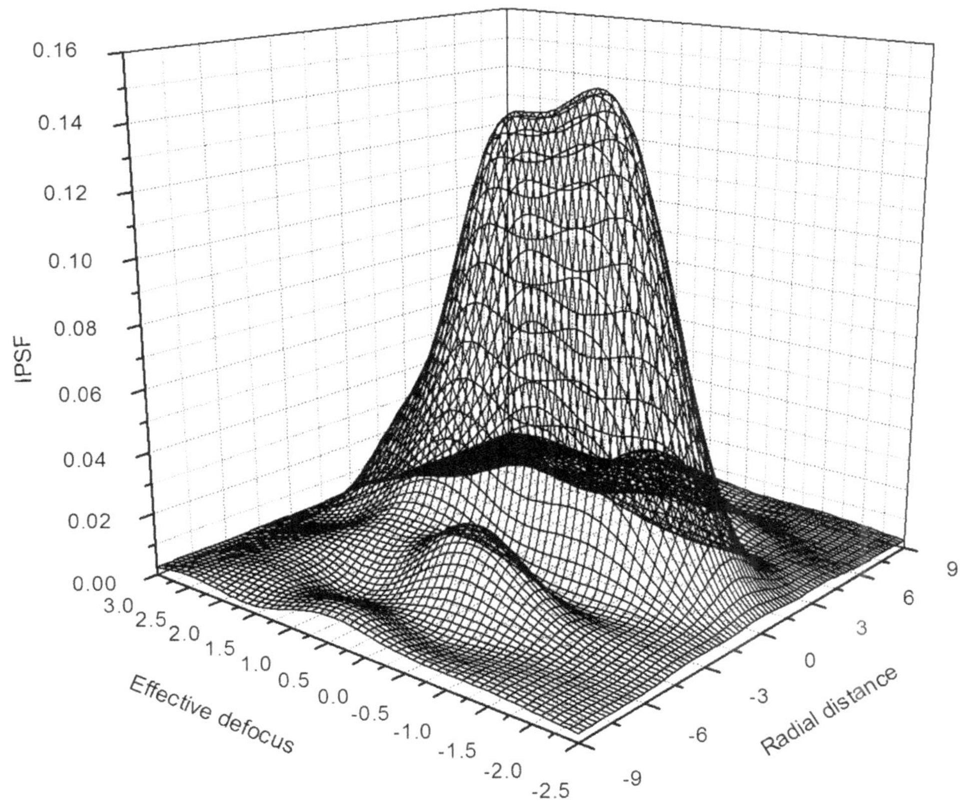

In some applications such as medical imaging, imaging with microscopes and in microelectronics industry, an optical system with high focal depth is required. The present investigation reveals that the said system retains all the above-mentioned good imaging qualities under diffraction-limited condition under polychromatic light illumination.

The focusing characteristics of the said system under broadband illumination are investigated using the polychromatic axial irradiance distribution function as the assessment parameter. Studies reveal that it maintains the property of large depth-of-focus in the visible and near IR regions for large bandwidth. However, the allowed spectral spread is larger in the near IR region. The polychromatic IPSF is introduced to study whether the same system exhibits the properties of enhanced resolution and

Figure 10. Polychromatic OTF of the proposed system with $\alpha_{\lambda 1} = 0.25\lambda_1$ *under parallel-polarizers configuration*

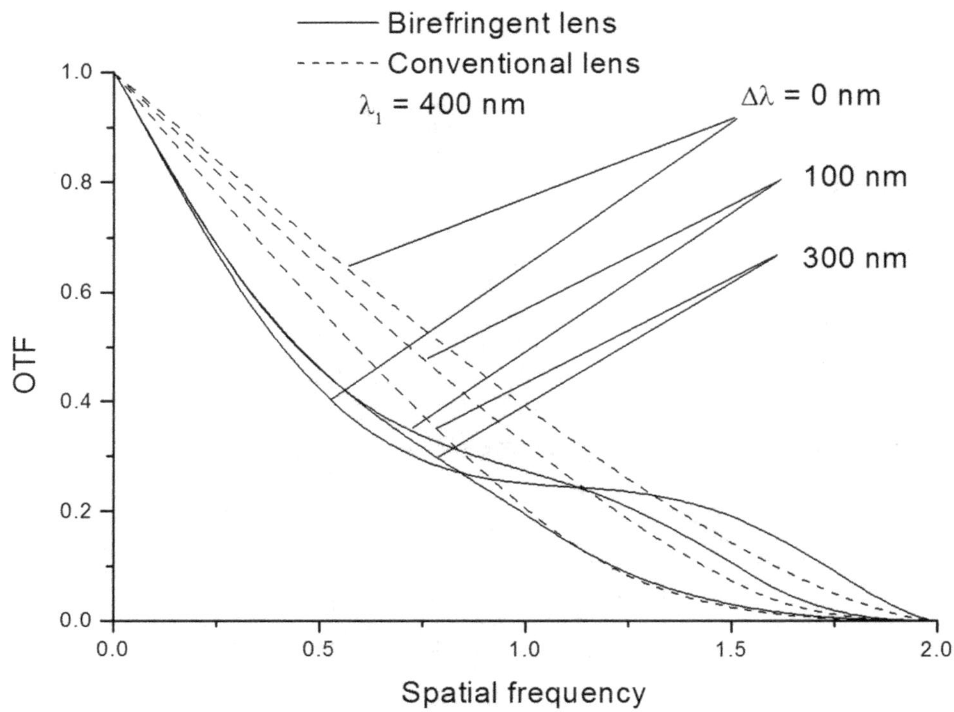

apodization for different directions of the transmission axis of the second polarizer when white light is used as the input. From the results it may be concluded that the said system retains both properties for certain specific design parameters even with a large spectral bandwidth around the mean wavelength. However, the system loses its bifocal nature at a large spectral spread.

Birefringent lenses with small values of α, are found to offer most of the advantages mentioned above. A lens material with low value of birefringence should be chosen for small values of α. Crystal quartz is definitely the best material available for such a lens. The birefringence value, n_o and n_e change very slowly with the change of wavelength for this material and almost linear at the higher end of the visible light (red light), near IR and SWIR regions. Hence, the birefringence parameters for the birefringent lens change very slowly and almost linearly with λ in these regions. As a result, when a polychromatic source with larger bandwidth is used for illuminating

Figure 11. Polychromatic OTF of the proposed system with $\alpha_{\lambda 1} = 0.25\lambda_1$ *under crossed-polarizers configuration*

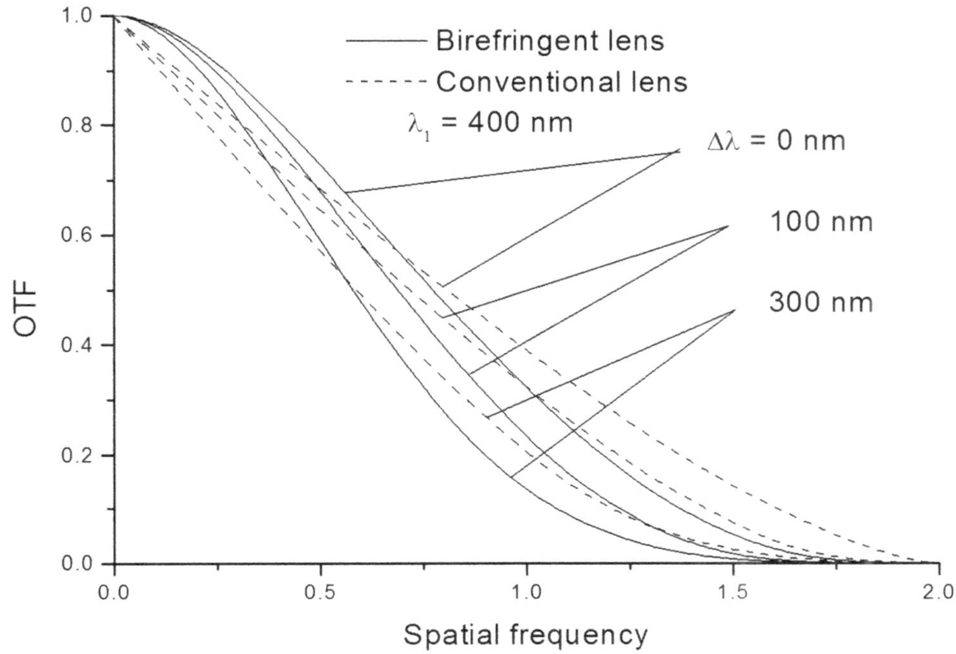

the said lens, its imaging properties does not change significantly from those when illuminated with a monochromatic source of light. Non-linear variation of birefringence, n_o and n_e with wavelength would result in asymmetric axial irradiance distribution curve and the said system would lose its inherent property of high focal depth. The transmittance of the quartz crystal is also very high in the wavelength range of 190-2900 nm.

The imaging characteristic of the proposed system under white light illumination is also investigated using the polychromatic OTF as the image assessment parameter. The results obtained in this case support the results of the polychromatic IPSF. Since the orientation of the two polarizers as well as the lens parameters (α and σ) also modifies the OTF of the system, the frequency response characteristics of the lens may be modified as per users' choice, a prerequisite for image processing applications. In fact, a birefringent lens with $\alpha = 0.625\lambda$ can be utilized for high frequency enhancement of a degraded image when illuminated with a polychromatic source of light. In this case, the images of an object are obtained under both parallel- and

Figure 12. Polychromatic OTF of the proposed with $\alpha_{\lambda 1} = 0.625\lambda_1$ *under both parallel- and crossed-Polarizers configurations*

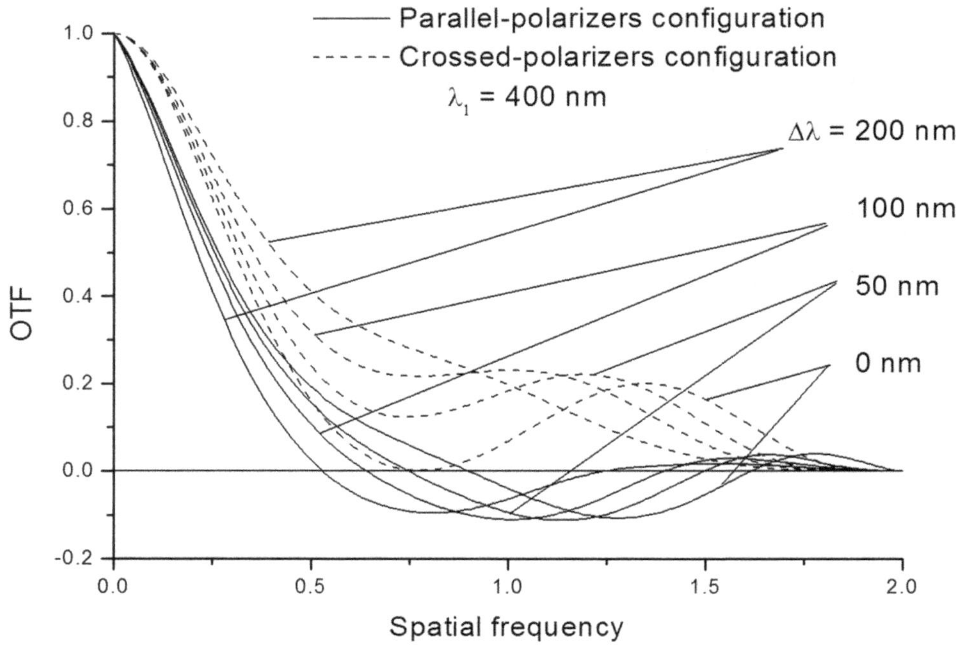

crossed-polarizers configurations and are subtracted from one another resulting in a significant improvement of contrast.

However, the proposed system suffers from some serious disadvantages. It may not be suitable to operate under low light illumination as an appreciable amount of light energy is absorbed in the polarizers. One solution to this problem may be to illuminate the system with a high-power source. Secondly, the thickness of the lens should be kept at a small value to get the desired effects. Fabrication of such a thin lens is definitely not an easy task. The problem may be alleviated using a doublet with the crystal optic axis of the individual elements orthogonal to each other, which in turn reduces the effective birefringence of the combination. Care must be taken for proper alignment of the optic axis of the birefringent crystal during the fabrication of such lenses. Finally, it may be concluded that a single birefringent lens may not be very useful as an imaging device; however, when a birefringent lens is employed as one of the components in a properly designed lens system it may provide an additional parameter to control the characteristics of that optical imaging system.

REFERENCES

Chou, C., Shyu, J. C., Huang, Y. C., & Yuan, C. K. (1998). Common-path optical heterodyne profilometer: A configuration. *Applied Optics*, *37*(19), 4137–4142. doi:10.1364/AO.37.004137 PMID:18285853

Davis, J. A., Evans, G. H., Crabtree, K., & Moreno, I. (2004). Programmable birefringent lenses with a liquid-crystal display. *Applied Optics*, *43*(34), 6235–6241. doi:10.1364/AO.43.006235 PMID:15619832

Dyson, L. (1957). Common-path interferometer for testing purposes. *Journal of the Optical Society of America*, *47*(5), 386–390. doi:10.1364/JOSA.47.000386

Fiala, W. (1990). Birefringent multi-focal lenses: Theory and application to the correction of refractive error. *Optometry and Vision Science*, *67*(10), 787–791. doi:10.1097/00006324-199010000-00006 PMID:2247302

Fiala, W. (1992). Multi-focal intra-ocular lenses fabricated from media exhibiting tuned birefringence. *Optometry and Vision Science*, *69*(4), 329–333. doi:10.1097/00006324-199204000-00013 PMID:1565436

Fiala, W., Finkelmann, H., & Gfeller, W.F. (1993). *Bifocals manufactured from birefringent materials*. Prague Meeting on Macromolecules. Special Lecture SL4.

Hazra, L. N. (2006). Walsh filters for tailoring of resolution in microscopic imaging. *Micron (Oxford, England)*, *38*(2), 129–135. doi:10.1016/j.micron.2006.07.003 PMID:16908176

Jacquinot, P., & Roizen-Dossier, B. (1964). *Apodization in Progress in Optics* (Vol. 3). Amsterdam: North-Holland.

Kubota, H., & Inoue, S. (1959). Diffraction images in the polarizing microscope. *Journal of the Optical Society of America*, *49*(2), 191–198. doi:10.1364/JOSA.49.000191 PMID:13621312

Kubota, H., & Saito, H. (1960). Further study on diffraction Images in the polarizing microscope. *Journal of the Optical Society of America*, *50*(10), 1020–1024. doi:10.1364/JOSA.50.001020

Lit, J. W. Y. (1971). Effects of a half wave filter in a portion of an aperture of a perfect lens. *Journal of the Optical Society of America*, *61*(3), 297–302. doi:10.1364/JOSA.61.000297

Luneberg, R. K. (1944). *Mathematical Theory of Optics*. Brown University Press.

Mandal, S. (2014). Evaluation of polychromatic image quality of a birefringent lens by means of optical transfer function. *Optik-International Journal for Light and Electron Optics*, *125*(1), 121–125. doi:10.1016/j.ijleo.2013.06.028

Mandal, S., & Ghosh, A. (2012). Polychromatic intensity point spread function of a birefringent lens. *Optik-International Journal for Light and Electron Optics*, *123*(18), 1623–1626. doi:10.1016/j.ijleo.2011.08.045

Mandal, S., Sanyal, S., & Ghosh, A. (2007). Imaging characteristics of a birefringent lens under broadband illumination. *Optik-International Journal for Light and Electron Optics*, *118*(7), 335–339. doi:10.1016/j.ijleo.2006.04.005

Marathay, A. S. (1969). Realization of complex filters with polarized light. *Journal of the Optical Society of America*, *59*(6), 748–752. doi:10.1364/JOSA.59.000748

Mezouari, S., & Harvey, A. R. (2003). Phase pupil functions for reduction of defocus and spherical aberrations. *Optics Letters*, *28*(10), 771–773. doi:10.1364/OL.28.000771 PMID:12779141

Ojeda-Castaneda, J., Andres, P., & Diaz, A. (1986). Annular apodizers for low sensitivity to defocus and to spherical aberration. *Optics Letters*, *11*(8), 487–489. doi:10.1364/OL.11.000487 PMID:19738664

Ojeda-Castaneda, J., & Berriel-Valdos, L. R. (1988). Arbitrarily high focal depth with finite apertures. *Optics Letters*, *13*(3), 183–185. doi:10.1364/OL.13.000183 PMID:19742021

Ojeda-Castaneda, J., & Berriel-Valdos, L. R. (1990). Zone plate for arbitrarily high focal depth. *Applied Optics*, *29*(7), 994–997. doi:10.1364/AO.29.000994 PMID:20562947

Ojeda-Castaneda, J., Berriel-Valdos, L. R., & Montes, E. (1985). Spatial filter for increasing the depth of focus. *Optics Letters*, *10*(11), 520–522. doi:10.1364/OL.10.000520 PMID:19730471

Ojeda-Castaneda, J., & Gomez-Sarabia, C. M. (1989). Focal depth: Optimum annular apodizer. *Applied Optics*, *28*(20), 4263–4264. doi:10.1364/AO.28.004263 PMID:20555858

Osterberg, H., & Wilkins, J. E. Jr. (1949). The resolving power of a coated objective. *Journal of the Optical Society of America*, *39*(7), 553–557. doi:10.1364/JOSA.39.000553

Ray, M., Sarkar, S. K., Chakraborty, R. N., & Basuray, A. (2008). Birefringence measurement using a confocal setup with a pair of birefringent lenses. *Optical Engineering (Redondo Beach, Calif.)*, *47*(3), 033604–033608. doi:10.1117/1.2899095

Sanyal, S., Bandyopadhyay, P., & Ghosh, A. (1998). Vector wave imagery using a birefringent lens. *Optical Engineering (Redondo Beach, Calif.)*, *37*(2), 592–599. doi:10.1117/1.601859

Sanyal, S., & Ghosh, A. (1999). Imaging characteristics of birefringent lenses under focused and defocused condition. *Optik-International Journal for Light and Electron Optics.*, *110*(11), 513–520.

Sanyal, S., & Ghosh, A. (2000). High focal depth with a quasi-bifocus birefringent lens. *Applied Optics*, *39*(14), 2321–2325. doi:10.1364/AO.39.002321 PMID:18345140

Sanyal, S., & Ghosh, A. (2002). Frequency response characteristics of a birefringent lens. *Optical Engineering (Redondo Beach, Calif.)*, *41*(3), 592–597. doi:10.1117/1.1430420

Sanyal, S., & Ghosh, A. (2003). Image assessment of a birefringent lens based on the factor of encircled energy. *Optical Engineering (Redondo Beach, Calif.)*, *42*(4), 1058–1064. doi:10.1117/1.1555736

Sanyal, S., Kawata, Y., Ghosh, A., & Mandal, S. (2004). Frequency response characteristics of a birefringent lens with off-axis aberrations. *Applied Optics*, *43*(19), 3838–3847. doi:10.1364/AO.43.003838 PMID:15250550

Whiting, A. I., Abouraddy, A. F., Saleh, B. E. A., Teich, M. C., & Fourkas, J. T. (2003). Polarization-assisted transverse and axial optical superresolution. *Optics Express*, *11*(15), 1714–1723. doi:10.1364/OE.11.001714 PMID:19466051

Wilkins, J. E. Jr. (1950). The resolving power of a coated objective II. *Journal of the Optical Society of America*, *40*(4), 222–224. doi:10.1364/JOSA.40.000222

Yang, X., Xiang, C., & Zhang, G. (1993). Birefringent common-path interferometer for testing large convex spherical surfaces. *Optical Engineering (Redondo Beach, Calif.)*, *32*(5), 1080–1083. doi:10.1117/12.133348

Chapter 10
Investigating Opto-Electronic Properties of Surface Plasmon Structure for Spectroscopic Applications

Pratibha Verma
National Institute of Technology Andhra Pradesh, India

Arpan Deyasi
RCC Institute of Information Technology, India

ABSTRACT

This chapter is proposed with an approach to analyze reflectance as a function of negative index material thickness for different parameters under the surface plasmon condition and extended approach towards the field enhancement of electric field as function of incidence angle and transmittance as function of incidence angle has been analyzed. This chapter can reflect the good comparison between 3 layer medium and n layer medium model. Characteristic impedance of MIM surface plasmon structure is analytically calculated considering the effect of both Faraday inductance and kinetic inductance. Effect of metal layer thickness, insulator thickness, and electron density are tailored to observe the impedance variation with frequency. Wavelength dependence of characteristic impedance and quality factor of MIM (metal-insulator-metal) surface plasmon structure is analyzed. Structural parameters and damping ratio of the structure is tuned within allowable limit to analyze the variation after detailed analytical computation.

DOI: 10.4018/978-1-5225-8531-2.ch010

Copyright © 2019, IGI Global. Copying or distributing in print or electronic forms without written permission of IGI Global is prohibited.

INTRODUCTION

Plasmonic

Nowadays in communication system, larger amount of data with high speed are transferred based on either electronics or photonics. Advanced technology named as 'plasmonics' is the study of the interaction between electromagnetic field of light with free electrons in metal under specific conditions (Born & Wolf 1999; Kittel 1996). Plasmonic is helpful to transfer both optical and electronic data at faster rate. Free electrons in the metal surface does collective oscillation excited by electric field of the incident light (Wood 1902; Ritchie 1957). Moreover, due to ohmic losses and electron-core interactions, losses are unavoidable for the plasmon oscillation, which is usually deleterious to most plasmonic devices (Knoll 2000). By proper designing metal patterns, the oscillation of free electrons in the metal surface can be controlled and at the same time, the absorption of light at any specific wavelength can be enhanced greatly in the metal for surface plasmon (SP) excitation (Raether & Kretschmann 1968; Veselago 1968). Optimized design of Metal-Insulator-Metal (MIM) device can make it work in a wide-spectrum from infrared to visible.

Surface Plasmon

Plasma, which is one of the four fundamental states of matter, has equal number of unbound positive and negative particles (Yamamoto 2008). When the electronic charges on the surface of metal or any conducting medium, has rapid oscillation of electronic charges, then this oscillation is called surface plasma oscillation (Pendry *et. al.,* 1999). Light is known as form of electromagnetic energy. When light falls on the surface of metal or any conducting medium, then it creates electromagnetic field. At critical incidence angle, the magnitude of electromagnetic field is maximum, because at critical angle, light will propagate along the surface of metal. More than critical incidence angle, total internal reflection will occur and hence electromagnetic field will be totally into the metal (Pechprasarn & Somekh 2011). In direction of electromagnetic field along the surface, the electrons will be displaced opposite to the applied electric field (since light which is establishing electromagnetic field has two components, electric field and magnetic field (Garcia-Vidal *et. al.,* 1991; Kolb 1982). This electric field is treated as applied electric field.) . Due to presence of magnetic field as 'B' and electron is also moving opposite to the direction of electric field with velocity 'V' will create force on moving electron, i.e. F=q(V*B), so electron will have tendency to move either upward or downward perpendicular to the surface of metal. Now, when light will be reflected back from surface, the direction of electric field will be opposite to the previous, but direction of magnetic field will be same (Abbasi & Bahrampour 2008; Otto 2000). So now, F=q(-V*B) will be in opposite direction and will affect the movement of electron opposite to the previous case (Kuo & Moghaddam 2006). The electron will move upward and then downward or vice-versa and will continue in same way. Hence, electron will oscillate perpendicular to the surface of metal but this is confined at the boundary and will vanishat both sides of the metal surface. So, this oscillation is called surface plasmon (Caloz *et. al.,* 2003). Because plasmon is nothing but the quantum of plasma oscillation, oscillation of electron is signifying plasma oscillation. Plasma oscillation is created by heating a gas or applying a strong magnetic field in any conducting medium (Ziolkowski & Heyman 2001). This phenomenon is happening along the surface or boundary between two medium (such as surface between metal and air), so that why it is called surface plasmon (Pendry 2000).

If the materials one has having real part of permittivity or dielectric constant is positive and other is negative at the given frequency of incident light then surface plasmon is occurred at the interface between two materials. The examples of material having value of real part of their relative permittivity as positive are glass, air, vacuum etc (Huang *et. al.,* 2009; Satija *et. al.,* 2011; Wang *et. al.,* 2011). On the other hand some material such as metal (e.g. gold) and heavily doped semiconductors have negative value of their relative permittivity. The condition to make light bound to the surface is to consider the materials whose magnitude of the real part of the permittivity in the negative permittivity region greater than the magnitude of the permittivity in the positive permittivity region. In the results discussed below the Surface Plasmon is formed by using the interface between gold and air (Khare & Nema 2012).

Reflectance or reflection coefficient depends on the radiated power (or energy) from the surface (Usha *et. al.,* 2016). Since electron oscillation (plasma oscillation) is absorbing some part of radiated energy which is affecting reflectance (Sreekanth *et. al.,* 2016; Nikitin *et. al.,* 2005). By this way surface plasmon has effect on reflectance.

Surface Plasmon Polariton

The electromagnetic excitation which propagates evanescently in the perpendicular direction at the interface between a conductor and dielectric is known as Surface plasmon polariton (Jetté-Charbonneau *et. al.,* 2005; Zavats *et. al.,* 2005; Ruppin 2000). The coupling of the electromagnetic fields of the incident light to oscillations of the conductor's electron plasma is responsible for rise of electromagnetic surface waves (infrared or visible-frequency electromagnetic waves). In the metal, surface plasmon is due to motion of electrons or charges considered in form of wave whereas polariton is the electromagnetic wave in the dielectric (Popov *et. al.,* 2007). The wavelength of Polariton is lower than the incident light (photons) and due to this there is compactness in spatial confinement and higher local field intensity of polariton. The propagation of surface plasmon polaritons along the interface is until absorption of its energy into the metal or scatter of energy into other directions (Ding *et. al.,* 2016; Scheumann *et. al.,* 1999). Surface plasmon polariton is excited by both electrons and photons. Firing electrons into the bulk of metal generates some excitation by electrons. The transfer of energy into the bulk plasma is due to scattering of the electrons which establish a surface plasmon polariton for the component of the scattering vector parallel to the surface. Other than this frequency and momentum can be kept same to excite an SPP through photons (Demetriadou 2015).

Surface Plasmon Resonance

The rapid oscillation of free electrons at the interface of metal-dielectric interface is Surface Plasmon for the complex value of dielectric constant of the metal surface (Homola *et. al.,* 1999; Li *et. al.,* 2011) where there is delocalization and coherence of electrons (Jiang *et. al.,* 2016). If the frequency of incident photon or light is matched with the oscillation frequency of conduction electrons at the interface between the positive permittivity material and negative permittivity material (Liu *et. al.,* 2016; Herminghaus *et. al.,* 1988) then there is existence of Surface Plasmon Resonance Condition (Nylander *et. al.,* 1983; Frey *et. al.,* 1988). Measurement of the energy of the incident light absorbed by the material is maximum at resonance condition. From Rough surface part of energy re-emitted in the form of light for the interaction of surface plasmon wave with a rough surface or any other kind of irregularity (Ishimaru *et. al.,* 2005; Sharma *et. al.,* 2007). From the galaxy, the larger number of low and high frequency signals cannot be noticed easily, there may be the possibility of identification of those frequencies through surface plasmon resonance condition (Jorgenson & Yee 1993). Other than this, surface plasmons are used to study second harmonic generation, Raman Scattering, the surface sensitivity of several spectroscopic measurements including fluorescence, to detect molecules such as DNA or proteins etc.

Choice of MIM Structure

Metal-insulator-metal (MIM)-based sensor is used to provide higher performance than conventional metal-insulator (MI) by rising the condition of conventional metal-insulator (MI). In MIM two interfaces of Metal-insulator give more rise of surface plasmon condition. The effective signal to noise ratio (SNR) of signal is increased in MIM structure and so there is better resolution of signal. The field is more confined or localized in MIM compared to the other structure. In MIM there is higher propagation loss of the incident wave which give better field localization. High quality factor can easily achieved in MIM structure.

Mathematical Formulation

Mathematical formulation is the significant key to explain the effectiveness of the physics behind the research. Mathematical modelling of reflectance for 3 layer and n layer of IMI structure, reflectance for 3 layer of MIM structure, characteristic impedance and quality factor of MIM structure are demonstrated with respect to the different parameters.

Formulation of Reflectance

In this section, mathematical formulation of reflectance versus incident angle, thickness of metal layer and other parameters are derived for 3 and n layer of IMI structure.

Formulation of Reflectance for 3 Layer for IMI Structure

For P-wave:

$$(E_P - R_P)COS\theta = T_P COS\theta'' \tag{1}$$

$$n_1(E_P + R_P) = n_2 T_P \tag{2}$$

For S-wave:-

$$(E_s + R_s) = T_s \tag{3}$$

$$n_1(E_P + R_P)COS\theta = n_2 COS\theta'' T_S \tag{4}$$

$$r_P = \frac{R_P}{E_P} \tag{5}$$

$$E_P - R_P = T_P \frac{COS\theta''}{COS\theta} \tag{6}$$

$$E_P + R_P = \frac{n_2}{n_1} T_P \tag{7}$$

Adding eq(6) and eq(7), we get:

$$2E_P = T_P \frac{COS\theta''}{COS\theta} + \frac{n_2}{n_1} T_P \tag{8}$$

Subtracting eq.(6) from eq.(7), we get:

$$2R_P = -T_P \frac{COS\theta''}{COS\theta} + \frac{n_2}{n_1} T_P \tag{9}$$

Using eq.(5), eq.(8) and eq.(9), we get:

$$r_P = \frac{-T_P \frac{COS\theta''}{COS\theta} + \frac{n_2}{n_1} T_P}{T_P \frac{COS\theta''}{COS\theta} + \frac{n_2}{n_1} T_P} \tag{10}$$

$$r_P = \frac{n_2 COS\theta - n_1 COS\theta''}{n_2 COS\theta + n_1 COS\theta''} \tag{11}$$

Similarly we can say:

$$r_{ik}^{p} = \frac{\tilde{n}_k COS\theta_i - \tilde{n}_i COS\theta_k}{\tilde{n}_k COS\theta_i + \tilde{n}_i COS\theta_k} \tag{12}$$

$$k = \frac{2\pi}{\lambda} = \frac{\omega}{v} = \frac{\tilde{n}\,\omega}{c} \tag{13}$$

$$\tilde{n}^2 = \frac{k^2c^2}{\omega^2} = \mu\varepsilon + \frac{i\sigma\mu}{\varepsilon_0\omega} \tag{14}$$

$$\tilde{n} = n + ik \tag{15}$$

K is the wave vector in the direction of wave propagation.

$$k_{zk} = k_k COS\theta_k \tag{16}$$

$$k_{zi} = k_i COS\theta_i \tag{17}$$

$$r_{ik}^{p} = \frac{\tilde{n}_k \frac{k_{zi}}{k_i} - \tilde{n}_i \frac{k_{zk}}{k_k}}{\tilde{n}_k \frac{k_{zi}}{k_i} + \tilde{n}_i \frac{k_{zk}}{k_k}} \tag{18}$$

$$r_{ik}^{p} = \frac{\tilde{n}_k \frac{k_{zi}c}{n_i\omega} - \tilde{n}_i \frac{k_{zk}c}{\tilde{n}_k\,\omega}}{\tilde{n}_k \frac{k_{zi}c}{n_i\omega} + \tilde{n}_i \frac{k_{zk}c}{\tilde{n}_k\,\omega}} \tag{19}$$

$$r_{ik}^{p} = \frac{\tilde{n}_k^2 k_{zi} - \tilde{n}_i^2 k_{zk}}{\tilde{n}_k^2 k_{zi} + \tilde{n}_i^2 k_{zk}} \tag{20}$$

$$r_{ik}^{p} = \frac{\frac{k_{zi}}{\tilde{n}_i} - \frac{k_{zk}}{\tilde{n}_k^2}}{\frac{k_{zi}}{\tilde{n}_i} + \frac{k_{zk}}{\tilde{n}_k^2}} \tag{21}$$

$$r_{ik}^{p} = \frac{\frac{k_{zi}}{\tilde{\varepsilon}_i} - \frac{k_{zk}}{\tilde{\varepsilon}_k^2}}{\frac{k_{zi}}{\tilde{\varepsilon}_i} + \frac{k_{zk}}{\tilde{\varepsilon}_k^2}} \tag{22}$$

$$t_P = \frac{T_P}{E_P} \tag{23}$$

$$t_P = \frac{T_P}{\frac{T_p}{2}\left(\frac{COS\theta''}{COS\theta} + \frac{n_2}{n_1}\right)} \tag{24}$$

$$t_p = \frac{2n_1 COS\theta}{n_1 COS\theta'' + n_2 COS\theta} \tag{25}$$

Now,

$$r_P = \frac{n_2 COS\theta - n_1 COS\theta''}{n_2 COS\theta + n_1 COS\theta''} \tag{26}$$

$$1 + r_P = 1 + \frac{n_2 COS\theta - n_1 COS\theta''}{n_2 COS\theta + n_1 COS\theta''} \tag{27}$$

$$1 + r_P = \frac{2n_2 COS\theta}{n_2 COS\theta + n_1 COS\theta''} \tag{28}$$

$$1 + r_P = \frac{\frac{n_2}{n_1}(2n_1 COS\theta)}{n_2 COS\theta + n_1 COS\theta''} \tag{29}$$

$$1 + r_P = \frac{n_2}{n_1} t_p \tag{30}$$

$$t_P = \frac{n_2}{n_1}(1 + r_p) \tag{31}$$

$$t_{ik}^P = \frac{\tilde{n}_i}{\tilde{n}_k}(1 + r_{ik}^p) \tag{32}$$

$$t_{ki}^P = \frac{\tilde{n}_k}{\tilde{n}_i}(1 + r_{ki}^p) \tag{33}$$

$$t_{ki}^{P} = \frac{\tilde{n}_k}{\tilde{n}_i}(1 - r_{ik}^{p}) \tag{34}$$

$$t_{ik}^{P} t_{ki}^{P} = \frac{\tilde{n}_k}{\tilde{n}_i}(1 + r_{ki}^{p})\frac{\tilde{n}_k}{\tilde{n}_i}(1 - r_{ik}^{p}) \tag{35}$$

$$t_{ik}^{P} t_{ki}^{P} = (1 + r_{ki}^{p})(1 - r_{ik}^{p}) \tag{36}$$

The total reflection of the three-layer model becomes:

$$R = | r_{pr12}^{P} |^2 \tag{37}$$

It is the ratio of the reflected power(or flux) to the incident power of the light:

$$R = \frac{I' A COS\theta'}{IA COS\theta} \tag{38}$$

$$R = \frac{I'}{I} = \frac{n_1 R^2 / 2c\mu_0}{n_1 E^2 / 2c\mu_0} = r^2 \tag{39}$$

$$r_{ik}^{p} = \frac{\tilde{n}_k COS\theta_i - \tilde{n}_i COS\theta_k}{\tilde{n}_k COS\theta_i + \tilde{n}_i COS\theta_k} \tag{40}$$

$$r_{ik}^{p} = \frac{\frac{COS\theta_i}{\tilde{n}_i} - \frac{COS\theta_k}{\tilde{n}_k}}{\frac{COS\theta_i}{\tilde{n}_i} + \frac{COS\theta_k}{\tilde{n}_k}} \tag{41}$$

According to snell's law:

$$\tilde{n}_1 \sin\theta_1 = \tilde{n}_2 \sin\theta_2 = \tilde{n}_{pr} \sin\theta_{pr} \tag{42}$$

$$\tilde{n}_i \sin\theta_i = \tilde{n}_k \sin\theta_k = \tilde{n}_{pr} \sin\theta_{pr} \tag{43}$$

Now,

$$COS\theta_k = \sqrt{(1 - \sin\theta_k^2)} \tag{44}$$

Putting equations of snell's law, we get:-

$$COS\theta_k = \sqrt{(1 - \frac{n_{pr}^2}{\tilde{n}_k^2} \sin\theta_{pr}^2)} \tag{45}$$

$$COS\theta_k = \sqrt{(\frac{\tilde{n}_k^2 - \tilde{n}_{pr}^2 \sin\theta_{pr}^2}{\tilde{n}_k^2})} \tag{46}$$

So,

$$\tilde{n}_k COS\theta_k = \sqrt{(n_k^2 - \tilde{n}_{pr}^2 \sin\theta_{pr}^2)} \tag{47}$$

$$\tilde{n}_i COS\theta_i = \sqrt{(n_i^2 - \tilde{n}_{pr}^2 \sin\theta_{pr}^2)} \tag{49}$$

We know:

$$r_{ik}^P = \frac{\frac{COS\theta_i}{\tilde{n}_i} - \frac{COS\theta_k}{\tilde{n}_k}}{\frac{COS\theta_i}{\tilde{n}_i} + \frac{COS\theta_k}{\tilde{n}_k}} \tag{50}$$

$$r_{ik}^{P} = \frac{\dfrac{\tilde{n}_i COS\theta_i}{\tilde{n}_i^2} - \dfrac{\tilde{n}_k COS\theta_k}{\tilde{n}_k^2}}{\dfrac{\tilde{n}_i COS\theta_i}{\tilde{n}_i^2} + \dfrac{\tilde{n}_k COS\theta_k}{\tilde{n}_k^2}} \tag{51}$$

$$r_{ik}^{P} = \frac{\dfrac{(\tilde{n}_i^2 - \tilde{n}_{pr}^2 \sin^2\theta_{pr})^{1/2}}{\tilde{n}_i^2} - \dfrac{(\tilde{n}_k^2 - \tilde{n}_{pr}^2 \sin^2\theta_{pr})^{1/2}}{\tilde{n}_k^2}}{\dfrac{(\tilde{n}_i^2 - \tilde{n}_{pr}^2 \sin^2\theta_{pr})^{1/2}}{\tilde{n}_i^2} + \dfrac{(\tilde{n}_k^2 - \tilde{n}_{pr}^2 \sin^2\theta_{pr})^{1/2}}{\tilde{n}_k^2}} \tag{52}$$

$$r_{ik}^{P} = \frac{\dfrac{(\varepsilon_i - \tilde{n}_{pr}^2 \sin^2\theta_{pr})^{1/2}}{\varepsilon_i} - \dfrac{(\varepsilon_k - \tilde{n}_{pr}^2 \sin^2\theta_{pr})^{1/2}}{\varepsilon_k}}{\dfrac{(\varepsilon_i - \tilde{n}_{pr}^2 \sin^2\theta_{pr})^{1/2}}{\varepsilon_i} + \dfrac{(\varepsilon_k - \tilde{n}_{pr}^2 \sin^2\theta_{pr})^{1/2}}{\varepsilon_k}} \tag{53}$$

$$r_{pr1}^{P} = \frac{\dfrac{(\tilde{n}_{pr}^2 - \tilde{n}_{pr}^2 \sin^2\theta_{pr})^{1/2}}{\tilde{n}_{pr}^2} - \dfrac{(\tilde{n}_k^2 - \tilde{n}_{pr}^2 \sin^2\theta_{pr})^{1/2}}{\tilde{n}_1^2}}{\dfrac{(\tilde{n}_{pr}^2 - \tilde{n}_{pr}^2 \sin^2\theta_{pr})^{1/2}}{\tilde{n}_{pr}^2} + \dfrac{(\tilde{n}_1^2 - \tilde{n}_{pr}^2 \sin^2\theta_{pr})^{1/2}}{\tilde{n}_1^2}} \tag{54}$$

$$r_{pr1}^{P} = \frac{\dfrac{\tilde{n}_{pr}(1 - \sin^2\theta_{pr})^{1/2}}{\tilde{n}_{pr}^2} - \dfrac{(\tilde{n}_k^2 - \tilde{n}_{pr}^2 \sin^2\theta_{pr})^{1/2}}{\tilde{n}_1^2}}{\dfrac{\tilde{n}_{pr}(1 - \sin^2\theta_{pr})^{1/2}}{\tilde{n}_{pr}^2} + \dfrac{(\tilde{n}_1^2 - \tilde{n}_{pr}^2 \sin^2\theta_{pr})^{1/2}}{\tilde{n}_1^2}} \tag{55}$$

$$r^{P}_{pr1} = \frac{\dfrac{COS\theta}{\tilde{n}_{pr}} - \dfrac{(\tilde{\varepsilon}_1 - \tilde{n}^2_{pr}\sin^2\theta_{pr})^{1/2}}{\tilde{\varepsilon}_1}}{\dfrac{COS\theta_{pr}}{\tilde{n}_{pr}} + \dfrac{(\tilde{\varepsilon}_1 - \tilde{n}^2_{pr}\sin^2\theta_{pr})^{1/2}}{\varepsilon_1}} \quad (56)$$

Now,

$$r^{P}_{12} = \frac{\dfrac{(\tilde{n}^2_1 - \tilde{n}^2_{pr}\sin^2\theta_{pr})^{1/2}}{\tilde{n}^2_1} - \dfrac{(\tilde{n}^2_2 - \tilde{n}^2_{pr}\sin^2\theta_{pr})^{1/2}}{\tilde{n}^2_2}}{\dfrac{(\tilde{n}^2_1 - \tilde{n}^2_{pr}\sin^2\theta_{pr})^{1/2}}{\tilde{n}^2_1} + \dfrac{(\tilde{n}^2_2 - \tilde{n}^2_{pr}\sin^2\theta_{pr})^{1/2}}{\tilde{n}^2_2}} \quad (57)$$

$$R = | r^{p}_{pr12} |^2 = | \frac{r^{P}_{pr1} + r^{P}_{12} e^{2ikz_1 d_1}}{1 + r^{P}_{pr1} r^{P}_{12} e^{2ikz_1 d_1}} |^2 = | r^{P}_{p12} |^2 \quad (58)$$

$$r^{p}_{pr12} = \frac{r^{P}_{pr1} + r^{P}_{12} e^{2ikz_1 d_1}}{1 + r^{P}_{pr1} r^{P}_{12} e^{2ikz_1 d_1}} \quad (59)$$

$$r^{p}_{pr12} = \frac{r^{P}_{pr1} + r^{P}_{12} e^{2ikz_1 d_1}}{1 + r^{P}_{pr1} r^{P}_{12} e^{2ikz_1 d_1}} \quad (60)$$

$$r^{p}_{pr12} = (r^{P}_{pr1} + r^{P}_{12} e^{2ikz_1 d_1})(1 + r^{P}_{pr1} r^{P}_{12} e^{2ikz_1 d_1})^{-1} \quad (61)$$

$$r^{p}_{pr12} = (r^{P}_{pr1} + r^{P}_{12} e^{2ikz_1 d_1})(1 - r^{P}_{pr1} r^{P}_{12} e^{2ikz_1 d_1} + (r^{P}_{pr1})^2 (r^{P}_{12})^2 e^{41kz_1 d_1} - \ldots\ldots\ldots\ldots\ldots\ldots\ldots\ldots) \quad (62)$$

$$r^{p}_{pr12} = r^{P}_{pr1} - r^{2P}_{pr1} r^{P}_{12} e^{2ikz_1 d_1} + (r^{P}_{pr1})^3 (r^{P}_{12})^2 e^{4ikz_1 d_1} + r^{P}_{12} e^{2ikz_1 d_1} - r^{P}_{pr1} (r^{P}_{12})^2 e^{4ikz_1 d_1} + (r^{P}_{pr1})^2 (r^{P}_{12})^3 e^{4ikz_1 d_1} \tag{63}$$

$$r^{p}_{pr12} = r^{P}_{pr1} + r^{P}_{12} e^{2ikz_1 d_1} (1 - r^{2}_{pr1}) + (r^{2p}_{pr1} - 1) r^{P}_{pr1} (r^{P}_{12})^2 e^{4ikz_1 d_1} + (r^{P}_{pr1})^2 (r^{P}_{12})^3 e^{41kz_1 d_1} \tag{64}$$

$$r^{p}_{pr12} = r^{P}_{pr1} + r^{P}_{12} e^{2ikz_1 d_1} (1 + r_{pr1})(1 - r_{pr1}) + (r_{pr1} + 1)(r_{pr1} - 1) r^{P}_{pr1} (r^{P}_{12})^2 e^{41kz_1 d_1} + (r^{P}_{pr1})^2 (r^{P}_{12})^3 e^{4ikz_1 d_1} \tag{65}$$

$$r^{p}_{pr12} = r^{P}_{pr1} + r^{P}_{12} e^{2ikz_1 d_1} (1 + r_{pr1})(1 - r_{pr1}) + (r_{pr1} + 1)(r_{pr1} - 1) r^{P}_{pr1} (r^{P}_{12})^2 e^{41kz_1 d_1} + (r^{P}_{pr1})^2 (r^{P}_{12})^3 e^{4ikz_1 d_1} \tag{66}$$

$$r^{p}_{pr12} = r^{P}_{pr1} + (1 + r_{pr1}) r^{P}_{12} (1 - r_{pr1}) e^{2ikz_1 d_1} + (1 + r_{pr1}) r^{P}_{12} (1 - r_{pr1}) e^{2ikz_1 d_1} + (1 + r_{pr1}) r^{P}_{12} (-r_{pr1}) r^{P}_{12} (1 - r_{pr1})^2 e^{41kz_1 d_1} \tag{67}$$

$$r^{p}_{pr12} = r^{P}_{pr1} + t^{P}_{pr1} r^{P}_{12} t^{P}_{1pr} + t^{P}_{pr1} r^{P}_{12} r^{P}_{1pr} r^{P}_{12} t^{P}_{1pr} e^{41kz_1 d_1} \tag{68}$$

Where phase factor $kz_1 d_1 = k_1 (d_1 COS\theta_1)$ is optical path length.

$$kz_1 d_1 = k_1 d_1 COS\theta_1 \tag{69}$$

$$kz_1 d_1 = \frac{\tilde{n}_1 \omega}{c} d_1 (1 - \sin^2 \theta)^{1/2} \tag{70}$$

$$kz_1d_1 = \frac{\tilde{n}_1\omega}{c}d_1(1 - \frac{n_{pr}^2}{\tilde{n}_1^2}\sin^2\theta_{pr})^{1/2} \tag{71}$$

$$kz_1d_1 = \frac{\tilde{n}_1\omega}{c}d_1(\frac{\tilde{n}_1^2 - n_{pr}^2\sin^2\theta_{pr}}{\tilde{n}_1})^{1/2} \tag{72}$$

$$kz_1d_1 = \frac{\omega}{c}d_1(\tilde{\varepsilon}_1 - n_{pr}^2\sin^2\theta_{pr})^{1/2} \because \tilde{n}_1^2 = \tilde{\varepsilon}_1 \tag{73}$$

Formulation of Reflectance for n Layer

Reflectance of perpendicular and parallel polarized waves for complex index of refraction:

$$r_\perp = \frac{n_1\cos(\theta_{incident}) - n_2\cos(\theta_{refracted})}{n_1\cos(\theta_{incident}) + n_2\cos(\theta_{refracted})} \tag{74}$$

$$r_\parallel = \frac{n_2\cos(\theta_{incident}) - n_1\cos(\theta_{refracted})}{n_2\cos(\theta_{incident}) + n_1\cos(\theta_{refracted})} \tag{75}$$

Perpendicular polarized or transverse electric, light cannot excite surface Plasmon resonance because of the oscillations of the electric field do not couple with the longitudinal modes of vibration of the sample media.

Fresnel co-efficient for a parallel polarized incident waves is:

$$r_\parallel = \frac{\left(\frac{n_2}{n_1}\right)^2\cos(\theta) - i\sqrt{\sin^2\theta - \left(\frac{n_2}{n_1}\right)^2}}{\left(\frac{n_2}{n_1}\right)^2\cos(\theta) + i\sqrt{\sin^2\theta - \left(\frac{n_2}{n_1}\right)^2}} \tag{76}$$

For physical systems:

$$R_{\|} = \left[\frac{n_1 \sqrt{1 - \left[\frac{n_1}{n_2} \sin(\theta_{incidence}) \right]^2} - n_2 \cos(\theta_{incidence})}{n_1 \sqrt{1 - \left[\frac{n_1}{n_2} \sin(\theta_{incidence}) \right]^2} + n_2 \cos(\theta_{incidence})} \right]^2 \tag{77}$$

When $n_1 < n_2$, external reflection results and when $n_1 > n_2$, internal reflection occurs. If $n_1 = n_2$, then no reflection occurs because there is no optical boundary to interact with.

Reflectance for two layer system as:

$$R = \frac{\left(n_2 - n_1\right)^2 + n_2^2 k_2^2}{\left(n_2 + n_1\right)^2 + n_2^2 k_2^2} \tag{78}$$

For multilayer, reflectance co-efficient between layer I and j is:

$$r_{ij} = \frac{n_j^2 \sqrt{n_i^2 - n_{ambient}^2 \sin^2 \theta} - n_i^2 \sqrt{n_j^2 - n_{ambient}^2 \sin^2 \theta}}{n_j^2 \sqrt{n_i^2 - n_{ambient}^2 \sin^2 \theta} + n_i^2 \sqrt{n_j^2 - n_{ambient}^2 \sin^2 \theta}} \tag{79}$$

The absorptive term also contributes to a phase shift at each layer by a phase factor φ_i of i.

$$\varphi_i = \frac{2\pi d}{\lambda} \sqrt{n_i^2 - n_{ambient}^2 \sin^2 \theta}$$

where d_i is the thickness of layer i.

For parallel polarized systems, the characteristic matrix describing each layer is:-

$$M_i = \begin{bmatrix} \cos\left(\frac{2\pi}{\lambda} n_i h_i \cos\theta_i\right) & -i\sin\left(\frac{2\pi}{\lambda} n_i h_i \cos\theta_i\right) \\ -i\sqrt{\frac{\mu_i}{n_i^2}}\cos\left(\theta_i\right)\sin\left(\frac{2\pi}{\lambda} n_i h_i \cos\theta_i\right) & \cos\left(\frac{2\pi}{\lambda} n_i h_i \cos\theta_i\right) \end{bmatrix} \tag{80}$$

The field strength in an n-layer system and their Fresnel behavior is:

$$Q_i = M_i^{-1} \prod_{i=1}^{n-1} M_i Q_{n-1} \tag{81}$$

Where Q_i is defined as the matrix of the field strength at each layer is:

$$Q_i = \begin{bmatrix} H_Y^0 \\ E_Z^0 \end{bmatrix} \tag{82}$$

So,

$$Q_1 = M_1^{-1} \prod_{i=1}^{n-1} M_i Q_{n-1} \tag{83}$$

$$Q_1 = M_1^{-1}\left[M_1 M_2 M_3 ....M_{n-1}\right] Q_{n-1}$$

$$Q_1 = M_2 M_3 ....M_{n-1} Q_{n-1} \tag{84}$$

For P-wave at boundary:

$$U_i = H_Y^T + H_Y^R$$

$$V_i = E_x^T + E_x^R$$

So,

$$Q_1 = \begin{bmatrix} U_1 \\ V_1 \end{bmatrix}$$

So, elaborating eq(11):

$$\begin{bmatrix} U_1 \\ V_1 \end{bmatrix} = M_2 M_3 ....M_{n-1} \begin{bmatrix} U_{n-1} \\ V_{n-1} \end{bmatrix} \tag{85}$$

$$\begin{bmatrix} U_1 \\ V_1 \end{bmatrix} = M \begin{bmatrix} U_{n-1} \\ V_{n-1} \end{bmatrix} \tag{86}$$

$$M_{ij} = \left(\prod_{k=2}^{n-1} M_k \right)_{ij} , \ i,j = 1,2 \tag{87}$$

So, For i=j=1,n=1

$$M_{11} = \left(\prod_{k=2}^{1-1} M_k \right) \tag{88}$$

$$M_{11} = \left(\prod_{k=2}^{0} M_k \right) \tag{89}$$

$$M_{11} = M_0 M_1 M_2 \tag{90}$$

Similarly, For i=1, j=2,n=2

$$M_{12} = \left(\prod_{k=2}^{1} M_k \right) \tag{91}$$

$$M_{12} = M_1 M_2 \tag{92}$$

$$\because q_i = \left(\frac{\mu_i}{\tilde{\varepsilon}_i}\right)^{1/2} \cos\theta_i \tag{92}$$

$$q_i = \left(\frac{1}{\tilde{\varepsilon}_i}\right)^{1/2} \cos\theta_i \tag{93}$$

$$\because q_i = \frac{1}{\tilde{n}_i}\cos\theta_i \because \varepsilon_i = n_i^2 \tag{94}$$

We know, for two layer:

$$r_{||} = \frac{n_2\cos(\theta_{incident}) - n_1\cos(\theta_{refracted})}{n_2\cos(\theta_{incident}) + n_1\cos(\theta_{refracted})} \tag{95}$$

$$r_{||} = \frac{\dfrac{\cos(\theta_{incident})}{n_1} - \dfrac{\cos(\theta_{refracted})}{n_2}}{\dfrac{\cos(\theta_{incident})}{n_1} + \dfrac{\cos(\theta_{refracted})}{n_2}} \tag{96}$$

$$r_{||} = \frac{q_1 - q_2}{q_1 + q_2} \tag{97}$$

The reflection and transmission coefficient for P-wave™ is

$$r^p = \frac{\left(M_{11} + M_{12}q_N\right)q_1 - \left(M_{21} + M_{22}q_N\right)}{\left(M_{11} + M_{12}q_N\right)q_1 + \left(M_{21} + M_{22}q_N\right)} \tag{98}$$

$$R_p = \left| r_p \right|^2 \tag{99}$$

$$r_p = R_p^{1/2} e^{i\varphi r_p} \tag{100}$$

$$\varphi_p^r = \arg\left(r^p\right) \tag{101}$$

$$t_H^P = \frac{2q_1}{\left(M_{11} + M_{12}q_N\right)q_1 + \left(M_{21} + M_{22}q_N\right)} \tag{103}$$

$$t_E^P = \frac{\mu_N n_1}{\mu_1 \tilde{n}_N} t_H^P \tag{104}$$

$$T_P = \frac{\mu_N \operatorname{Re}\left(\tilde{n}_N \cos\theta_N / \tilde{n}_N^2\right)}{\mu_1 n_1 \cos\theta_1 / n_1^2} \left| t_H^P \right|^2 \tag{105}$$

$$\varphi_P^t = \arg\left(t_E^P\right) \tag{106}$$

For s wave(TE) the above equation hold except

$$q_k = \sqrt{\frac{\tilde{\epsilon}_k}{\mu_k}} \cos\theta_k \tag{107}$$

Formulation of Reflectance for MIM Structure

In this section, mathematical formulation of reflectance versus incident angle, thickness of metal layer and other parameters are derived for 3 layer of MIM structure.

Let, θ_{m1} = Angle of incidence on first metal layer of MIM (Metal-Insulator-Metal) structure between metal-insulator interface.

Let, θ_i = Angle of refraction on first metal layer of MIM (Metal-Insulator-Metal) structure between metal-insulator interface.

Let, θ_{m2} = Angle of refraction on second metal layer of MIM (Metal-Insulator-Metal) structure between insulator-metal interface.

$$\theta_i = \sin^{-1}((\frac{n_m}{n_i}) \times \sin\theta_{m1})$$

$$\theta_{m2} = \sin^{-1}((\frac{n_i}{n_m}) \times \sin\theta_i)$$

$$\omega = \frac{2 \times \pi \times c}{\lambda} \text{ [} \because \omega \text{ =angular frequency]}$$

$$k_i = n_i \times \frac{\omega}{c}$$

$$k_i = \frac{2n_i\pi}{\lambda} \text{ [} \because k_i \text{ =Wave number]}$$

$k_{zi} = k_i \cos\theta_i$ [$\because$ c =velocity of light in free space and k_{zi} =Propagation vector in z direction]

Let, r_{mi} = Reflectivity between metal and insulator

Let, $\because$ r_{im} = Reflectivity between insulator and metal

$$r_{mi} = \frac{n_i \cos\theta_{m1} - n_m \cos\theta_i}{n_i \cos\theta_{m1} + n_m \cos\theta_i}$$

$$r_{im} = \frac{n_m \cos\theta_i - n_i \cos\theta_{m2}}{n_m \cos\theta_i + n_i \cos\theta_{m2}}$$

$$\because r_{mim} = \left| \frac{r_{mi} + r_{im} \exp(i2k_{zi}d)}{1 + r_{mi}r_{im} \exp(i2k_{zi}d)} \right|$$ [$\because d$ =distance between two metal layer or, thickness of insulator]

$$\because r_{mim} = \frac{[\{r_{mi} + r_{im} \cos(2k_{zi}d)\}^2 + r_{im}^2 \times \sin^2(2k_{zi}d)]}{[\{1 + r_{mi}r_{im} \cos(2k_{zi}d)\}^2 + r_{im}^2 r_{mi}^2 \sin^2(2k_{zi}d)]}$$

$$\because r_{mim} = \frac{\left[\left\{\left(\frac{n_i \cos\theta_{m1} - n_m \cos\theta_i}{n_i \cos\theta_{m1} + n_m \cos\theta_i}\right) + \left(\frac{n_m \cos\theta_i - n_i \cos\theta_{m2}}{n_m \cos\theta_i + n_i \cos\theta_{m2}}\right)\cos(2k_{zi}d)\right\}^2 + r_{im}^2 \times \sin^2(2k_{zi}d)\right]}{\left[\left\{1 + \left(\frac{n_i \cos\theta_{m1} - n_m \cos\theta_i}{n_i \cos\theta_{m1} + n_m \cos\theta_i}\right)\left(\frac{n_m \cos\theta_i - n_i \cos\theta_{m2}}{n_m \cos\theta_i + n_i \cos\theta_{m2}}\right)\cos(2k_{zi}d)\}^2 + \left(\frac{n_i \cos\theta_{m1} - n_m \cos\theta_i}{n_i \cos\theta_{m1} + n_m \cos\theta_i}\right)^2\left(\frac{n_m \cos\theta_i - n_i \cos\theta_{m2}}{n_m \cos\theta_i + n_i \cos\theta_{m2}}\right)^2 \sin^2(2k_{zi}d)\right\}\right]}$$

If both two plates are made of same metal then,

$$\theta_{m1} = \theta_{m2}$$

Hence,

$$r_{mim} = \frac{\left\{n_i^2 \cos^2\theta_m - n_m^2 \cos^2\theta_i + (n_m^2 \cos^2\theta_i^2 - n_i^2 \cos^2\theta_m)\cos(2k_{zi}d)\right\}^2 + r_{im}^2 \sin^2(2k_{zi}d)(n_m \cos\theta_i + n_i \cos\theta_m)^4}{\left\{\left(n_i \cos\theta_m + n_m \cos\theta_i\right)^2 - \left(n_m \cos\theta_i - n_i \cos\theta_m\right)\cos(2k_{zi}d)\right\}^2 + (n_i \cos\theta_m - n_m \cos\theta_i)^4 \sin^2(2k_{zi}d)}$$

Formulation of Impedance

In this section, mathematical formulation of characteristic impedance versus frequency with respect to the faraday inductance, kinetic inductance and resistance are derived for 3 layer of MIM structure.

Formulation of Characteristic Impedance for MIM Structure

Since we know,

$$L_k = \frac{1}{\omega^2 \varepsilon_o (1 - \varepsilon_r^{'}(\omega))} \frac{(Length)}{(Area)} \text{ [} L_k \text{ =kinetic inductance]}$$

$$R = \frac{\varepsilon_r^{''}(\omega)}{\omega \varepsilon_o \left|1 - \varepsilon_r^{'}(\omega)\right|^2} \frac{(Length)}{(Area)} \text{ [} R \text{ =resistance]}$$

$$L_f = \frac{\mu_o d}{W}(Length) \text{ [} L_f \text{ =faraday inductance, } W = \text{Width of the metal structure]}$$

$$Area = W\delta \text{ [} \delta = \text{skin depth of the penetration of the electromagnetic fields]}$$

where,

$$\varepsilon_r = 1 - \left(\frac{\omega_p}{\omega}\right)^2$$ [ω_p =angular Plasmon frequency, ω =angular frequency of the incident light]

$$\varepsilon_r^{'} = 1 + 2\frac{\omega_p^2}{\omega^4}$$

$$\varepsilon_r^{''} = 1 - 6\frac{\omega_p^2}{\omega^4}$$

Then,

$$L_k = \frac{1}{\omega^2 \varepsilon_o \left(1 - 1 - 2\frac{\omega_p^2}{\omega^4}\right)} = -\frac{\omega^2}{2\varepsilon_o \omega_p^2}$$

$$R = \frac{1-6\frac{\omega_p^2}{\omega^4}}{\omega\varepsilon_o\left|1-1-2\frac{\omega_p^2}{\omega^4}\right|^2} = \frac{\omega^3\left(\omega^4-6\omega_p^2\right)}{4\omega_p^4}$$

$$Z = R + j\omega L_k + j\omega L_f$$

$$|Z| = \sqrt{\left(R^2 + \omega^2 L_k^2 + \omega^2 L_f^2\right)}$$

Putting the value of R, L_k, L_f we get

$$|Z| = \sqrt{\left(\left(\frac{\omega^3\left(\omega^4-6\omega_p^2\right)}{4\omega_p^4}\right)^2 + \omega^2\left(-\frac{\omega^2}{2\varepsilon_o\omega_p^2}\right)^2 + \omega^2\left(\frac{\mu_o d}{W}(Length)\right)^2\right)}$$

Formulation of Quality Factor

In this section, quality factor with respect to wavelength is deduced for 3 layer of MIM structure.

Formulation of Quality Factor for MIM Structure

Quality factor of the MIM (metal-insulator-metal) structure is:

$$Q_{mat} = \frac{d(\omega\varepsilon')/d\omega}{2\varepsilon''} \tag{1}$$

[Since ‘ ω ’is the angular frequency of incident wave]

Since relative permittivity of the metal is:-

$$\varepsilon_r = 1 - \frac{\omega_{pe}^2}{\omega^2 + \xi^2} \tag{2}$$

[Since 'ξ' is the damping ratio; 'ω_{pe}' is the angular plasma frequency]

Differentiating relative permittivity with respect to angular frequency, we get:

$$\varepsilon_r^{'}(\omega) = 1 - \frac{\omega_{pe}^2}{\omega^2 + \xi^2} \tag{3}$$

$$\varepsilon_r^{''}(\omega) = \frac{\xi\omega_{pe}^2}{\omega(\omega^2 + \xi^2)} \tag{4}$$

Differentiating equation (3 with respect to, we get:

$$\frac{d(\omega\varepsilon')}{d\omega} = \frac{d}{d\omega}\{\omega(1 - \frac{\omega_{pe}^2}{\omega^2 + \xi^2})\}$$

$$\frac{d(\omega\varepsilon')}{d\omega} = 1 - \frac{\omega_{pe}^2}{\omega^2 + \xi^2} + \omega(\frac{2\omega\omega_{pe}^2}{(\omega^2 + \xi^2)^2})$$

$$\frac{d(\omega\varepsilon')}{d\omega} = 1 - \frac{\omega_{pe}^2}{\omega^2 + \xi^2} + \frac{2\omega\omega_{pe}^2}{(\omega^2 + \xi^2)^2} \tag{5}$$

Putting the value of equation (4) and (5) in equation (1), we get:

$$Q_{mat} = \frac{1 - \frac{\omega_{pe}^2}{\omega^2 + \xi^2} + \frac{2\omega^2\omega_{pe}^2}{(\omega^2 + \xi^2)^2}}{\frac{\xi\omega_{pe}^2}{\omega(\omega^2 + \xi^2)}}$$

$$Q_{mat} = \frac{\omega[(\omega^2 + \xi^2)^2 - \omega_{pe}^2(\omega^2 + \xi^2) + 2\omega^2\omega_{pe}^2]}{\xi\omega_{pe}^2}$$

$$Q_{mat} = \frac{\omega(\omega^4 + 2\omega^2\xi^2 + \xi^4 + \omega_{pe}^2\omega^2 - \omega_{pe}^2\xi^2)}{\xi\omega_{pe}^2}$$

$$Q_{mat} = \frac{(\frac{2\pi c}{\lambda})^4((\frac{2\pi c}{\lambda})^4 + 2(\frac{2\pi c}{\lambda})^2\xi^2 + \xi^4 + \omega_{pe}^2(\frac{2\pi c}{\lambda})^2 - \omega_{pe}^2\xi^2)}{\xi\omega_{pe}^2}$$

[Since 'λ' is wavelength of incident light and 'c' is the speed of light]

RESULTS OF REFLECTANCE

Observation for 3 Layer Medium for IMI Structure

From depicted results, reflected radiant energy and transmittance is decreased quickly due to total internal reflection for condition such asfall of distinct values of thickness of gold layer at the constant critical angle of incidence (Gordon- II & Swalen 1977). Other than this, it is shown that electric field intensity at peak value is higher in thin layer compared to thick layer. According to matching condition of frequency generated by the plasmonic structure at resonance and frequency generated by surface plasmon due to incident light, then medium's resistance is at utmost and thus the power of the incident light is absorbed in the metal, sothe reflectance is reduced as radiated energy from the surface is attenuated.

Result and Discussion for 3-Layer Medium

In Figure 1, the thickness of gold layer profile with respect to reflectance is depicted for different values of incidence angle as mentioned the work (Verma *et. al.,* 2018). The magnitude of reflectance is increased with increase of incidence angle. Higher the strength of the reflected radiant energy is observed by the rate of increment of reflectance with increase of thickness of gold layer. After certain limit of increasing the thickness of gold layer, the magnitude of reflectance is saturated.

In Figure 2, the thickness of gold layer profile with reflectance is shown for different values of wavelength of He-Ne laser light. Magnitude of reflectance is decreased with increase of wavelength of He-Ne laser light as mentioned in the work (Verma *et. al.,* 2018). But on the other hand side, the magnitude of reflectance is increased with decrease of the thickness of gold layer. Distance travelled by the light deeper into the material is due to higher values of wavelength and because of

Figure 1. Thickness vs. Reflectance for different values of incidence angle

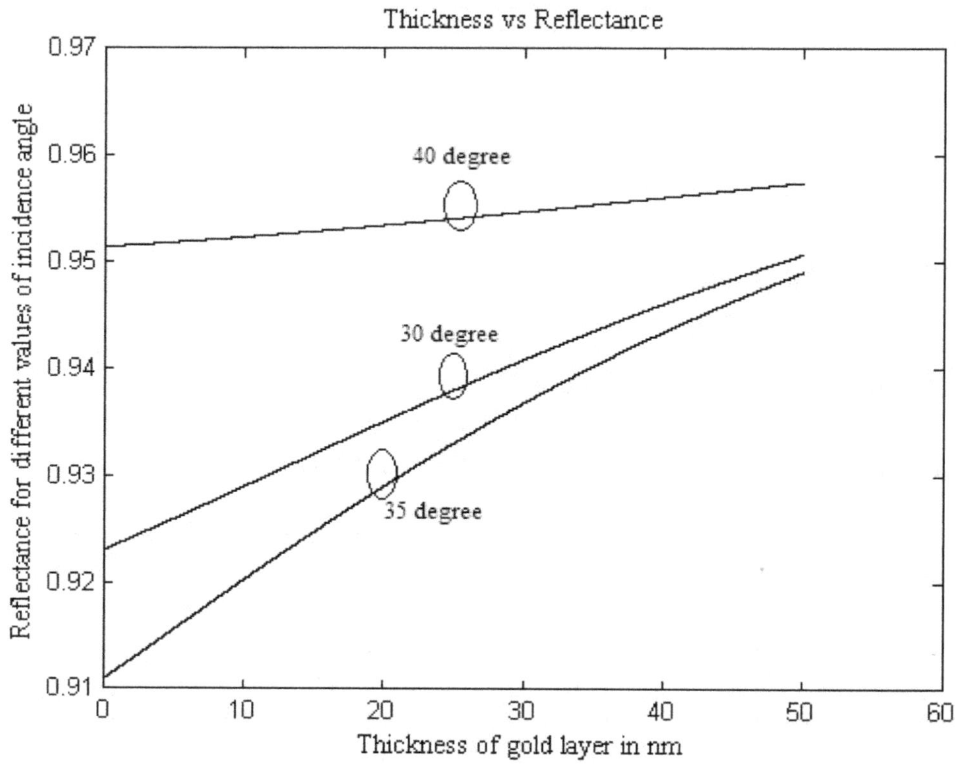

impact of it the energy of the incident light is less radiated from the surface after reflection and causes reduction in the magnitude of reflectance.

The incident angle profile with reflectance for different values of thickness of gold layer is shown in figure 3 as mentioned in the work (Verma *et. al.,* 2018). An abrupt dip of reflectance is observed around critical angle of 37.2^0 for the condition of the propagation constant of the incident wave along the surface is almost equal to propagation constant of the surface plasmon between glass medium and gold. The reflectance value is converged independent of the thickness of the gold layer after certain angle. It is also observed that reflectance is decreased before critical angle in small amount. The power of the surface plasmon is absorbed by internal absorption in the metal for maximum value of resistance of medium when resonance condition is fulfilled, and thereby magnitude of reflectance is reduced because of reduction in the radiative energy from the surface.

The incident angle profile with reflectance is shown in figure 4 for different values of wavelength of He-Ne laser light as mentioned in the work (Verma *et.*

Figure 2. Thickness vs. Reflectance for different values of wavelength of He-Ne laser light

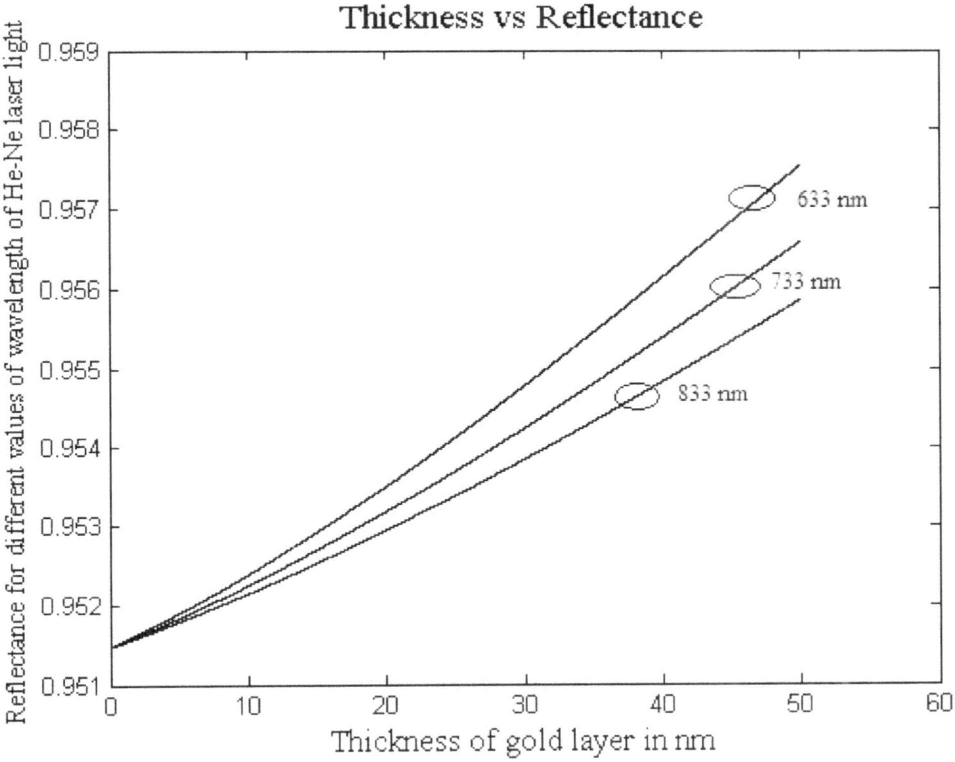

al., 2018). Keeping the value of wavelength constant as 633 nm, the reflectance is becoming zero at particular incident angle. At this wavelength, frequency of the light is matched with the natural oscillation frequency of the electronic charges on the metal boundary. This matching condition establishes increment in the fluctuations of charges. Much more part of power of incident signal is lost in oscillation and fluctuation. Hence reflectance is reduced. Now higher is the wavelength lower is the frequency which agitate the charges in small amount thus the reflectance is increased with respect to the 633nm wavelength.

Observation for n Layer Medium of IMI Structure

For n-layer, result based on incident angle profile with respect to different parameters such as field enhancement, transmittance and reflectance are described. It depicts

Figure 3. Incident angle vs. Reflectance for different values of thickness of gold layer

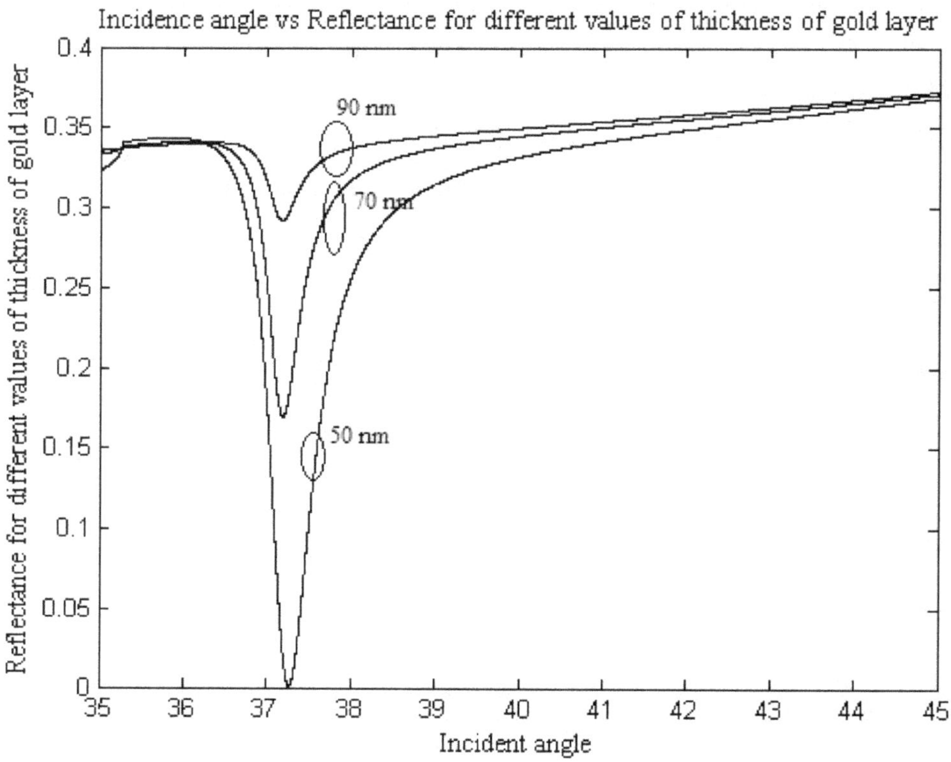

that rise in value of incident angle is one of the reason to heighten up the value of transmittance. But it is also observed that transmittance suddenly fall down at the critical angle of incident light. On the other hand, maximum value of Field enhancement is noted at the critical angle of incident light having the symmetry nature at both sides of maximum value point. The same nature is followed by the 4-layer as well as 3-layer IMI structure for the profile of incident angle w.r.t reflectance. The advantage to use 4-layer structure is the magnitude of reflectance is enhanced.

Result and Discussion for n-Layer Medium of IMI Structure

For different values of incidence angle, the transmittance of n layer profile is shown in Figure 5. The transmittance is increased with an increase of incidence angle (Deyasi *et. al.*, 2018). When incident angle is equal to the critical angle, then the transmittance has highest value just before or at the critical angle due to propagation

Figure 4. Incidence angle vs. Reflectance for different values of wavelength of He-Ne laser light

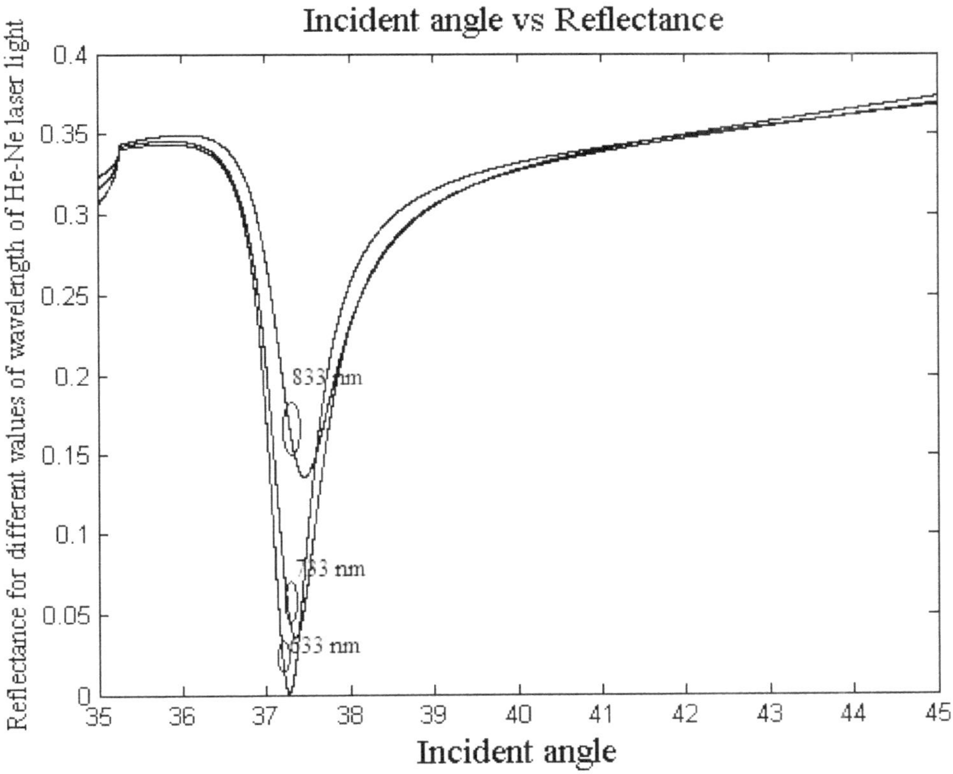

of most part of the energy along the surface. But beyond critical angle, transmittance is decreased sharply due to the total internal reflection.

The transmittance for different values of thickness of gold layer profile with incidence angle is shown in Figure 6. It is showed that the transmittance of gold layer is increased with an increase of incident angle for given value of thickness of gold layer ((Deyasi *et. al.,* 2018). But at certain incidence angle, transmittance falls rapidly to zero value for any value of thickness of gold layer. It is known that critical angle is dependent upon the refractive index of the material. As refractive index of a material is fixed, the sudden falling of the magnitude of transmittance is fixed at certain critical angle of incidence independent of the different thickness of gold layer.

Figure 5. Incidence angle vs. transmittance for 4 layers

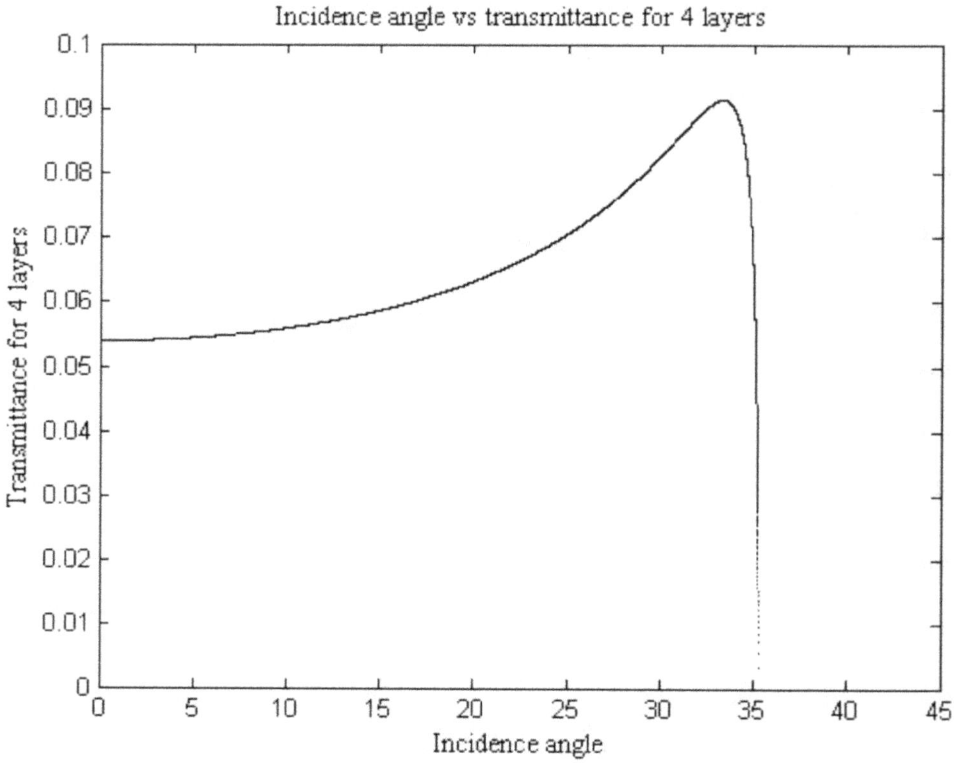

For different values of incidence angle, the transmittance for different values of wavelength of He-Ne laser light profile is depicted in Figure 7 ((Deyasi *et. al.*, 2018). The transmittance is increased with an increment in incidence angle as well as increase in the wavelength of He-Ne laser light. Deeper distance inside the material is travelled by the light with an increase in wavelength of light. But the transmittance is decreased rapidly as total internal reflection is increased beyond certain critical angle.

For different values of thickness of self-assembled monolayer, the transmittance profile for different values of incidence angle is shown in Figure 8 ((Deyasi *et. al.*, 2018). The critical angle is highly dependent on the refractive index of the material. If refractive index of a material is fixed then critical angle is also fixed. At the same critical angle of incidence, transmittance drop-off for different thickness of SAM layer. The variation of transmittance is minor for different values of thickness of the SAM layer than the thickness gold layer.

Figure 6. Incidence angle vs. Transmittance for different values of thickness of gold layer

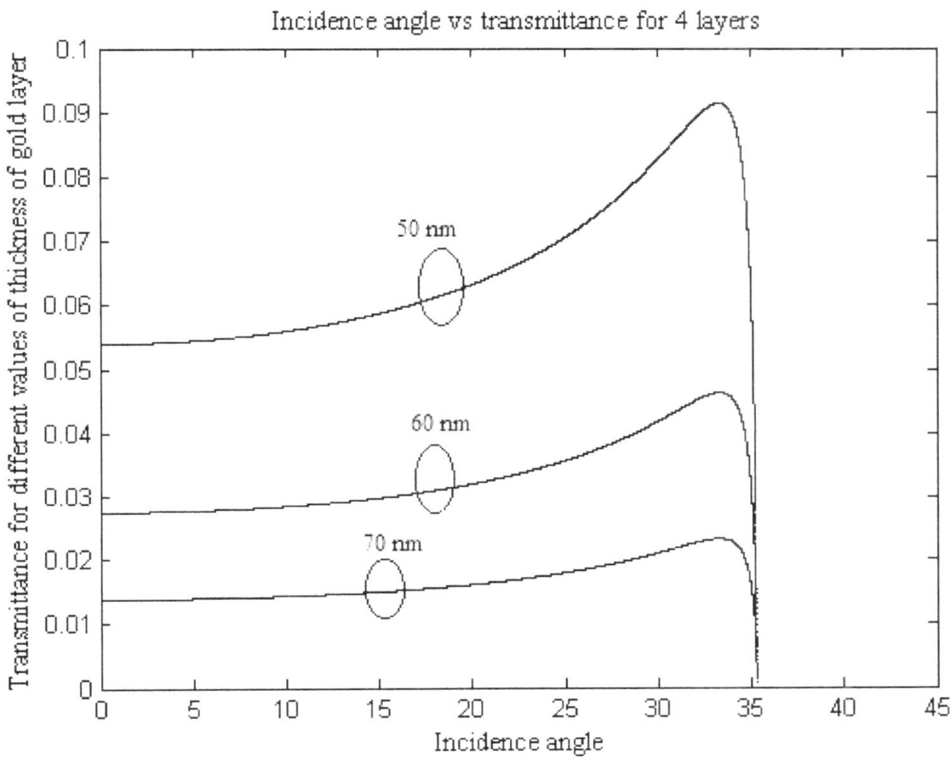

The field enhancement for n-layer profile with incidence angle is depicted in Figure 9. As when the incidence angle is increased the field intensity at the metal surface is also increased (Deyasi *et. al.,* 2018). The field intensity is at maximum at the metal surface as most of the energy is passed along the surface at critical incidence angle. If again the incidence angle is increased then energy is started to go inside the metal. As reflectivity is reduced, the field intensity is enhanced inside the metal which reduce the field intensity at the surface. Thus, it is observed that electric field is at the peak at critical angle.

For different values of gold layer, the field enhancement w.r.t to increment of incidence angle is shown in Figure 10. It is shown that the peak value of electric field is decreased with an increase in thickness of gold layer ((Deyasi *et. al.,* 2018). As in thin layer, the field accumulation is more prominent and so the field density is more eminent in thin layer as compared to the thick layer.

Figure 7. Incidence angle vs. Transmittance for different values of wavelength of He-Ne laser light

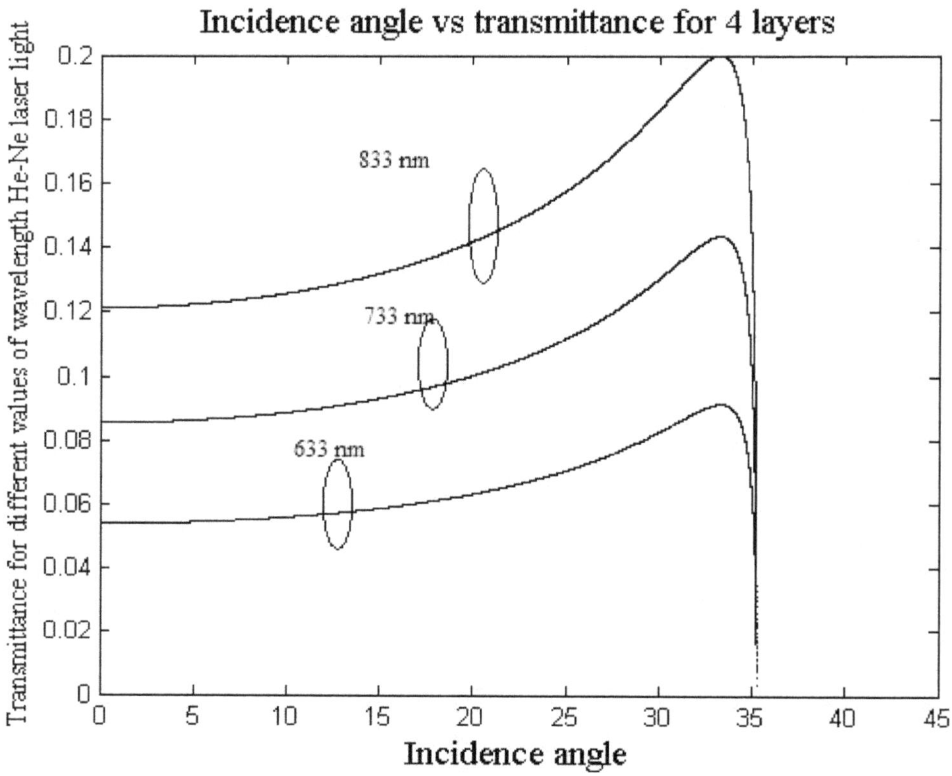

For different values of thickness of self-assembled, the field enhancement monolayer profile for w.r.t to increment of values of incidence angle is shown in Figure 11. At the layer surface, as the incidence angle is step-up, the field intensity is also raised (Deyasi *et. al.,* 2018). As most of the energy propagate along the surface at the critical angle, the field intensity is at peak at the surface. If again the incidence angle is incremented beyond the critical angle then most part of the energy is reflected back from the surface due to total internal reflection and smaller amount of energy passes along the surface, thus the field intensity at the surface is scaled down. Thus the maximum value of electric field at the surface is observed at critical angle. The difference between using gold layer and self-assembled layer is field intensity profile is changed when thickness of gold layer is varied whereas field intensity is same for variation in the thickness of self-assembled monolayer.

Figure 8. Incidence angle vs. Transmittance for different values of thickness of SAM layer

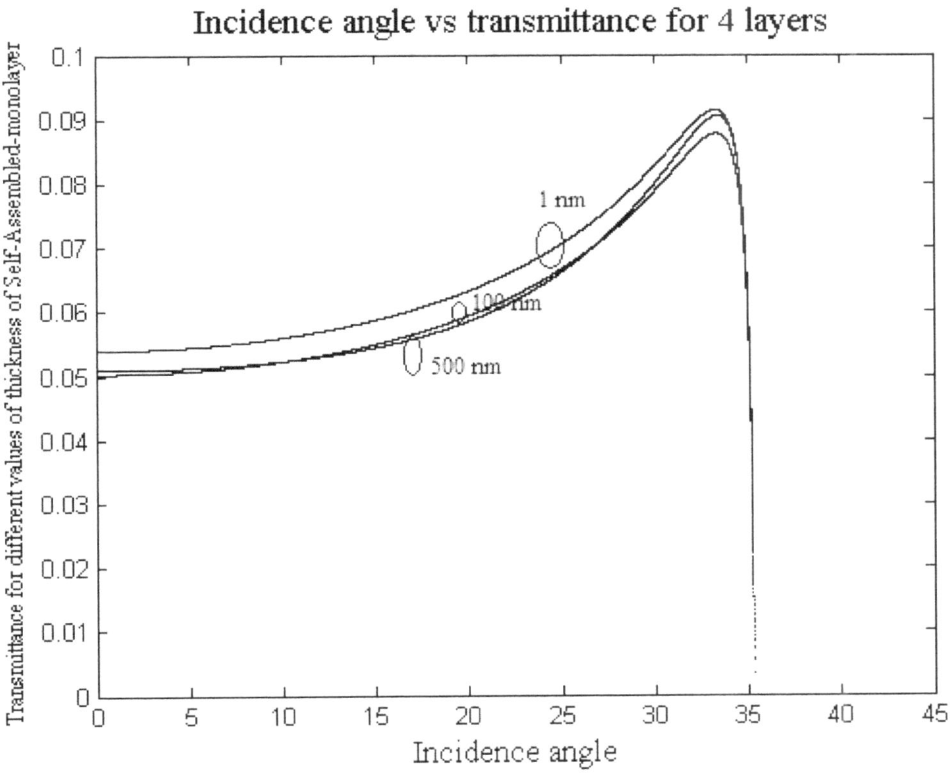

Therefore it is observed that field distribution is overlying one on the other for different values of the thickness of self-assembled monolayer.

The reflectance for n-layer profile with incidence angle is depicted in Figure 12. The sudden fall of reflectance is observed at critical angle of incidence when propagation constant of the incident wave is matched with propagation constant of the surface plasmon (Deyasi *et. al.*, 2018). The value of reflectance is converged after critical angle of incidence. It is also observed that reflectance is decreased with a small amount before critical angle. At critical angle total power or energy of the signal is propagated along the surface of an interface between two different medium and thereby no power or energy is reflected back into the first medium and causes to reduction of reflectance.

Figure 9. Incidence angle vs. Field enhancement for 4 layers

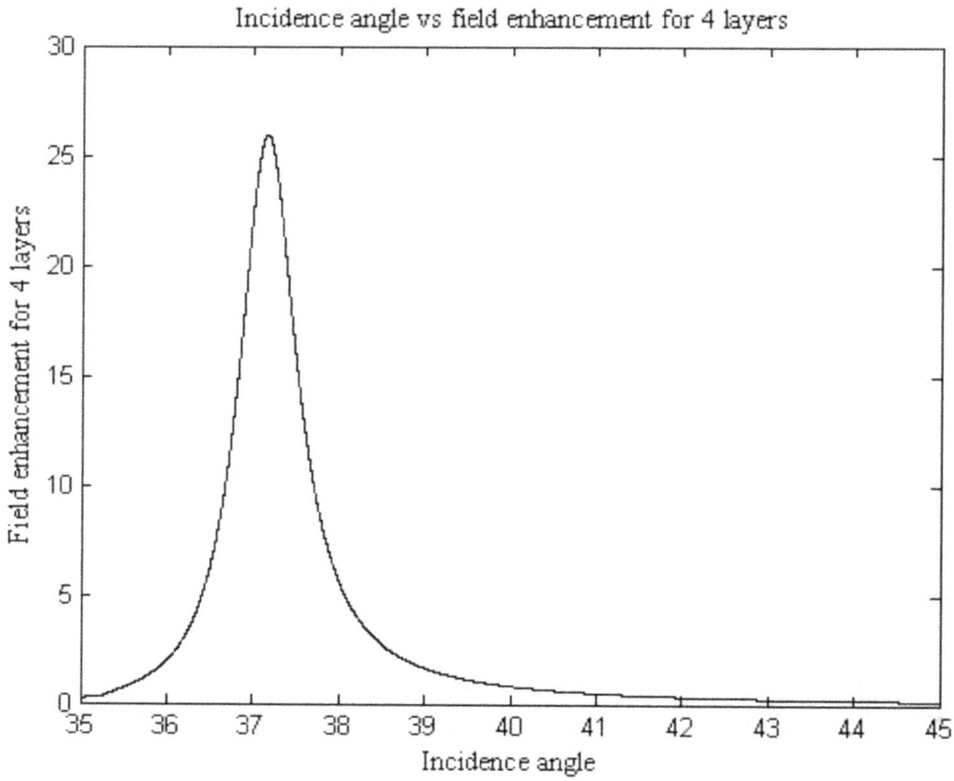

For different values of thickness of gold layer, the reflectance profile is shown in Figure 13. One of the reason behind the sudden fall of reflectance at critical angle is satisfaction of matching condition of propagation constant of the incident wave along the surface equal to propagation constant of the surface plasmon between glass medium and gold layer (Deyasi *et. al.*, 2018). Reflectance is converged irrespective of the thickness of the gold layer after certain angle of incidence. Other than this, power of the surface plasmon is absorbed by internal absorption in the metal at resonance condition occurred near to critical angle when resistance of the medium is a peak point. Hence there is reduction of energy from the surface which reduces the value of reflectance.

For different values of thickness of self-assembled monolayer, the reflectance profile with respect to change in value of incidence angle is shown in Figure 14. The sudden fall of reflectance is observed for the condition propagation constant of the

Figure 10. Incidence angle vs. Field Enhancement for different values of thickness of gold layer

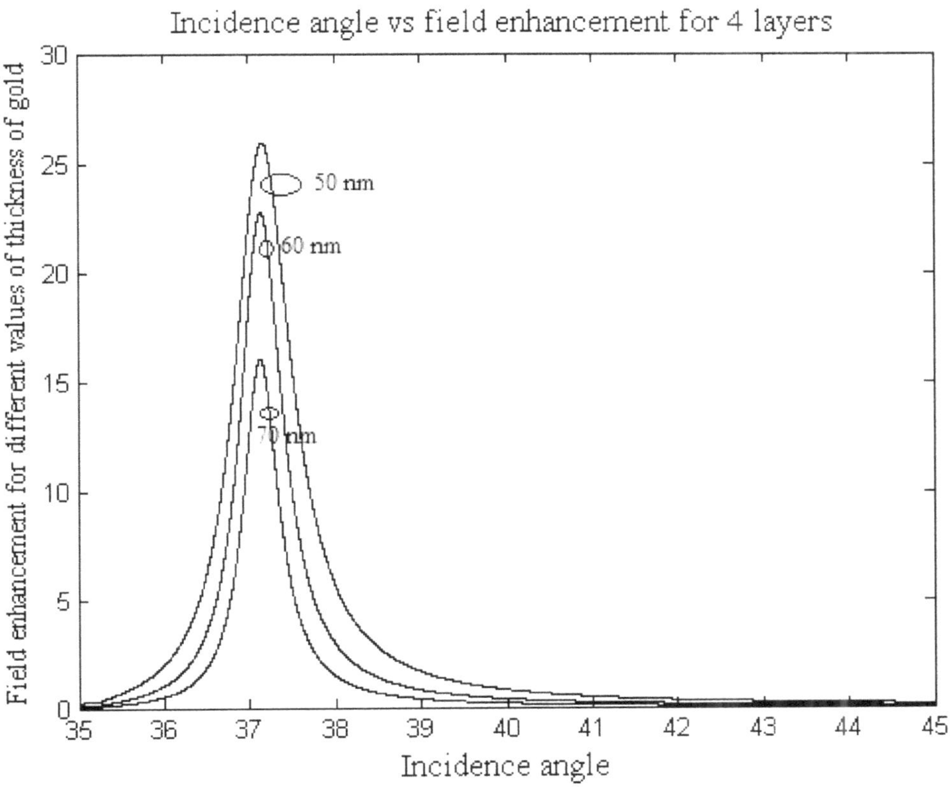

surface plasmon to be equal to the propagation constant of the incident wave in the self-assembled monolayer. Reflectance is reduced in small amount before critical angle. The resonance condition is generally occurred at critical angle. The power of the surface plasmon is absorbed internally in the metal at resonance when the resistance of medium is at peak. Hence radiated energy from the surface is reduced which decrease the reflectance. It can be seen that there is no variation for different values of thickness of self-assembled monolayer in the graph as all are overlapped one on the other. But on the other hand, there is variation in the graph by considering different values of thickness of gold layer.

For different values of wavelength of He-Ne laser light, the incident angle profile with reflectance is depicted in Figure 15. At 633 nm wavelength, natural frequency of oscillation of the electronic charges on the metal boundary is equal to

Figure 11. Incidence angle vs. Field Enhancement for different values of SAM layer

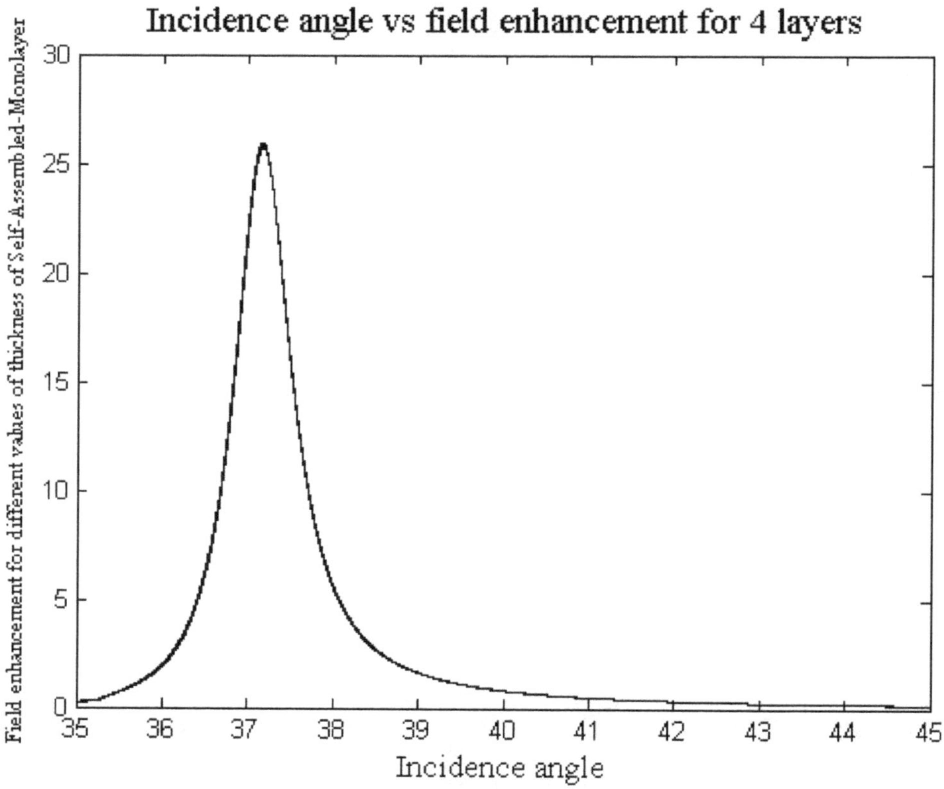

frequency of the incident light. Due to effect of this matching condition, fluctuation of charges is heightened up and so maximum power of incident light is absorbed in this fluctuation phenomena and because of this condition reflectance is decreased. As wavelength and frequency are inversely proportional to each other. Lower value of frequency is responsible for agitation of the charges in small amount in the metal and so reflectance is increased with respect to the wavelength of 633 nm.

Observation for 3 Layer Medium for 3 Layer Medium of MIM Structure

In this part, results about thickness vs. Reflectance of MIM with variation of different parameters are shown where distance between two plates is raised then transmission is stepped down and reflection is stepped up which provides the effect

Figure 12. Incidence angle vs. Reflectance for 4 layers

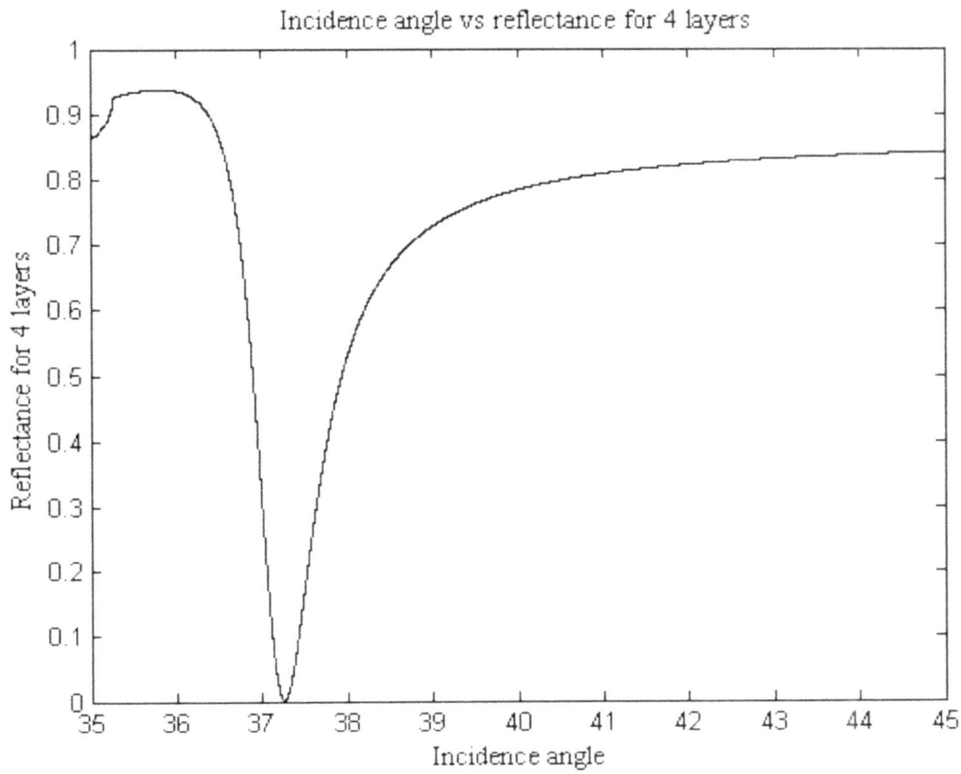

of peak in the Reflectance after certain increment of thickness but again Reflectance is decreased after this maximum value of reflectance when thickness is increased more as transmittance is dominant at this point.

Result and Discussion for 3-Layer Medium for MIM Structure

Reflectance vs. thickness of metal layer is shown in Figure 16. Resonance condition of Surface Plasmon phenomenon of both Metal-Insulator and Insulator-Metal is coupled only when distance between two plates (d) is slight. It causes more transmission of incident light into the MIM (Metal-Insulator-Metal) structure than reflection. Combined effect of resonance condition is less dominated with a rise of distance between plates. This results less transmission and more reflection. A peak in Reflectance is observed after certain increment of thickness or distance between

Figure 13. Incidence angle vs. Reflectance for 4 layers for different values of thickness of gold layer

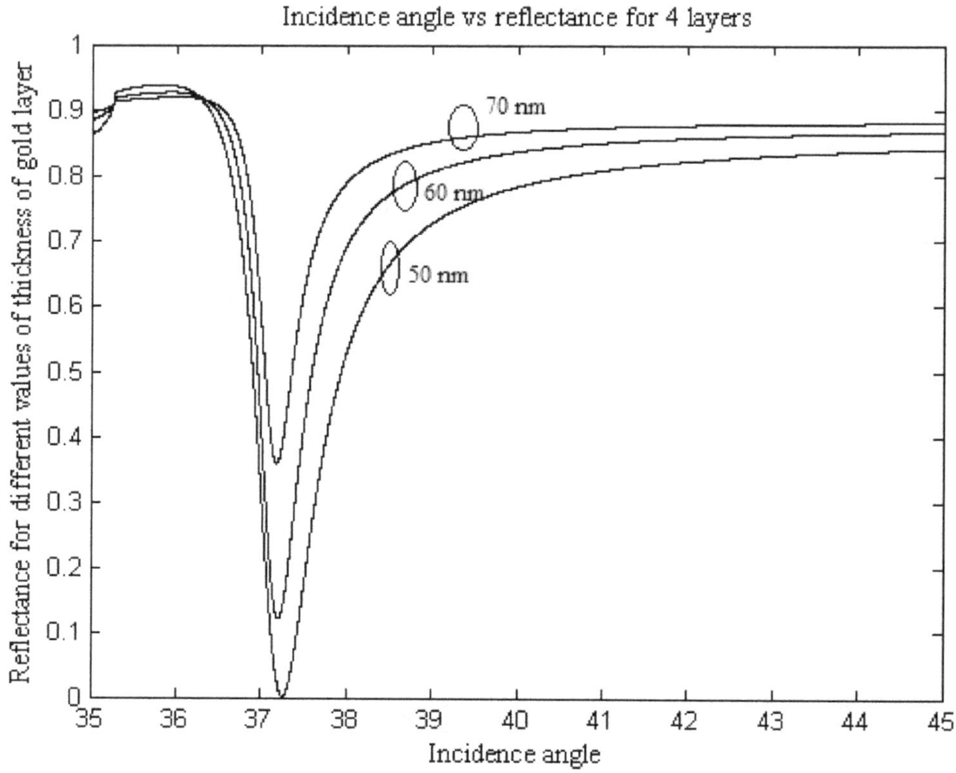

plates. When 'd' is 3-5 times greater than the decay length of the Plasmon field, then surface Plasmon modes on the two interfaces are again dependent to each other. Hence, high transmission of incident light and negative peak value of reflectance is created. Transmission of incident light is dominant after this thickness.

Reflectance vs. thickness of medium between two metal plates for different values of wavelength of incident light is shown in Figure 17. Peak of reflectance is shifted towards right with growing value of wavelength of light as resonance condition of Double Surface Plasmon phenomenon is shifted. Less oscillation of electrons in surface Plasmon is created by more wavelength of incident light. Thus, resonance condition is shifted. More the wavelength more the incident light to propagate deeper into the MIM structure, thus transmission is more and reflectivity is less for same thickness at certain limit.

Figure 14. Incidence angle vs. Reflectance for 4 layers for different values of thickness of SAM layer

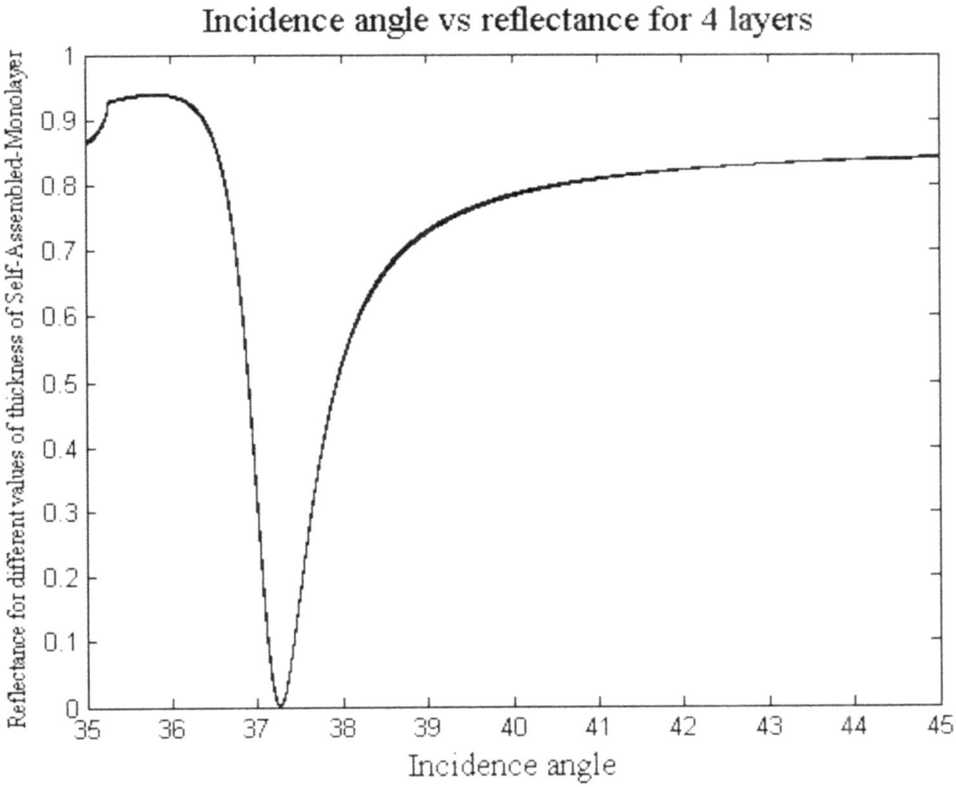

Reflectance vs. Thickness of medium between two metal plates for different values of incident angle is shown in Figure 18. As incident angle is increased the peak value of reflectance is shifted towards left. As incident angle is increased, reflectance of light is more than transmission as it depicts the effectiveness of the reflecting radiant energy. The peak value of reflectance is decreased with an decrease of incidence angle.

Reflectance vs. Thickness of medium between two metal plates for different values of refractive index is shown in Figure 19. More density of the medium is indicated by more value of refractive index. Hence lower value of refraction and higher reflection is caused by the lower velocity of light. Thus high peak value of reflectance is created among all three types of refractive index medium. For same

Figure 15. Incidence angle vs. Reflectance for 4 layers for different values of thickness of He-Ne light

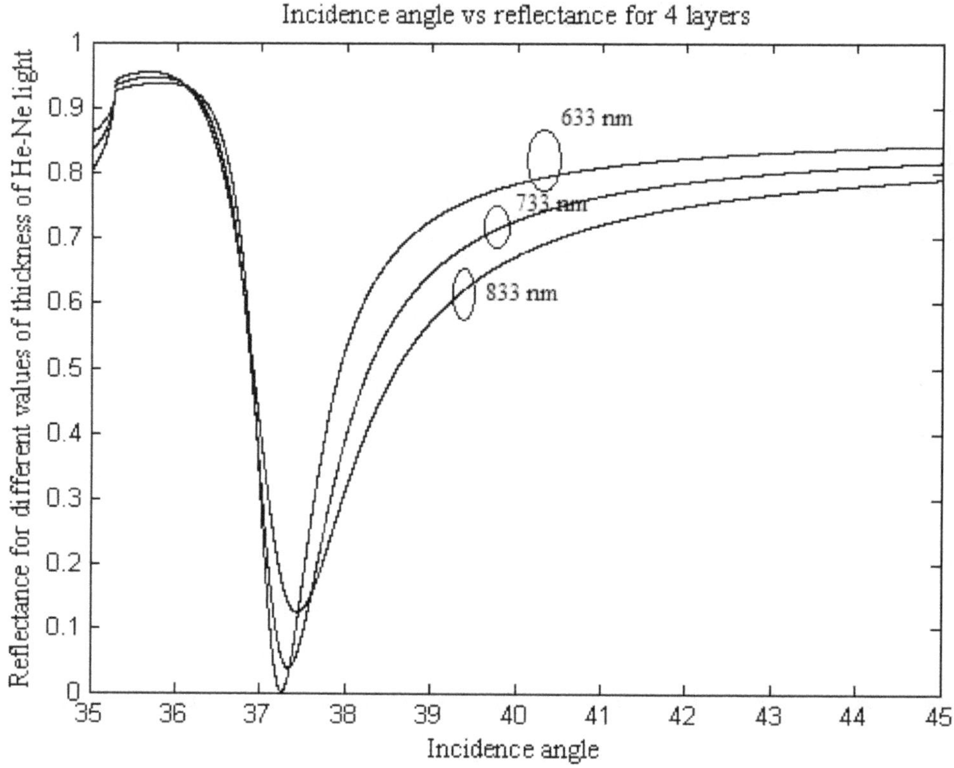

thickness occurrence of peak value of reflectance is near because of increasing of effectiveness of radiant energy and less propagation of light into the MIM structure.

RESULTS OF CHARACTERISTIC IMPEDANCE

Observation of Characteristic Impedance for 3 Layer Medium of MIM Structure

In this part, impedance is varied with frequency by considering variation in the thickness of metal layer, thickness of insulator layer and electron density (Deyasi *et. al.,* 2017). At very low frequency, antisymmetric and symmetric modes are

Figure 16. Thickness of metal layer vs. Reflectance for MIM structure

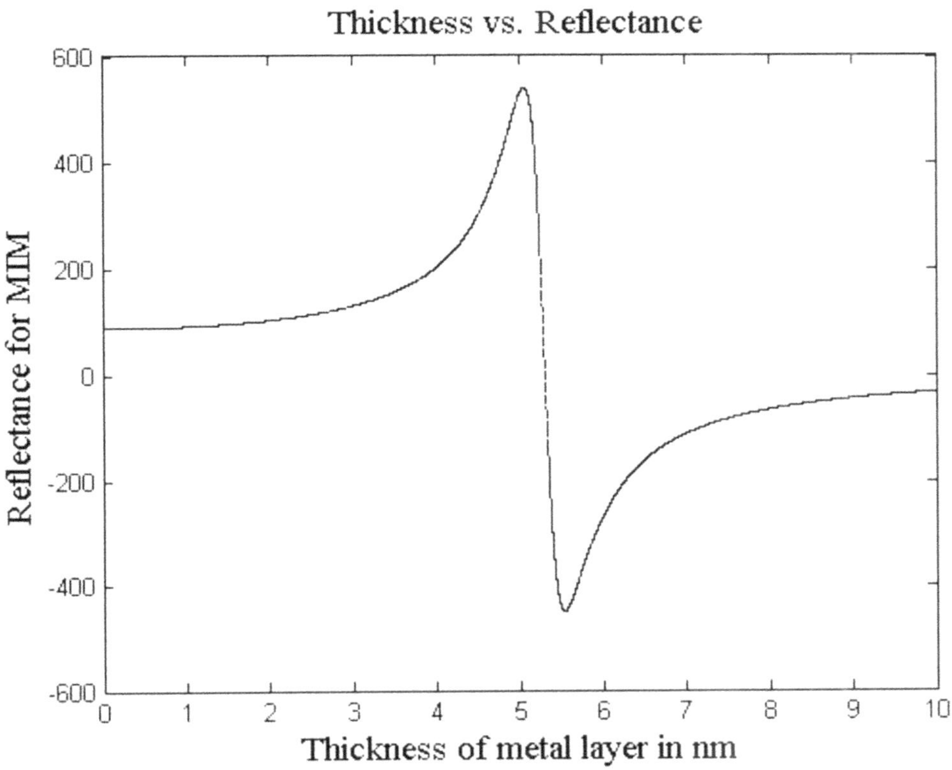

coupled by magnitude of optimum impedance which is one of the reason to give rise to double sided surface plasmon. Moreover, it is observed that resistive part is dominated than other type of impedance such as Faraday inductance and Kinetic inductance at lower frequency ranges. Without kinetic and faraday inductances the antisymmetric and symmetric mode cannot be present at lower frequency range.

Result and Discussion for 3-Layer Medium for MIM Structure

For Metal-Insulator-Metal parallel plate structure (MIM) Frequency vs. magnitude of impedance is depicted in Figure 20.

The graph is nonlinear due to presence of faraday inductance and kinetic inductance which exhibit in the MIM slab structure. At higher frequency, kinetic impedance is dominant than the other to step up the impedance magnitude. In metal, the origin

Figure 17. Thickness of medium between two metal plates vs. Reflectance for different values of wavelength of laser light

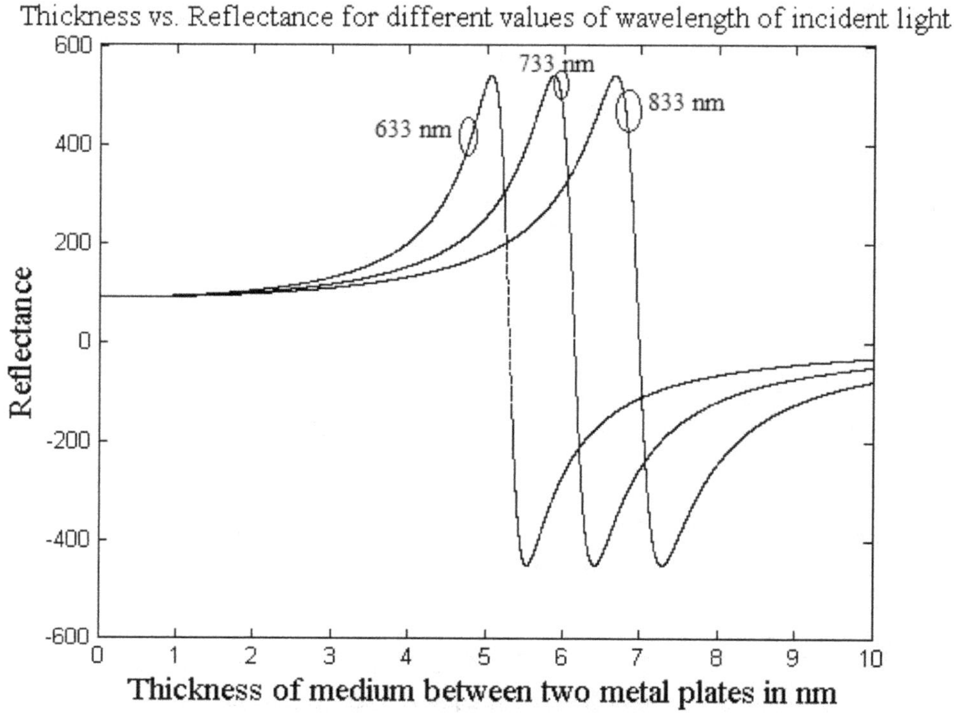

of kinetic inductance is the negative real part of complex relative dielectric which give rise to imaginary part of resistivity ((Deyasi *et. al.,* 2017). If imaginary part of dielectric constant is more prominent than its real part then resistance in the metal is considered and this condition happens at lower frequency (Deyasi & Verma 2018). At initial point of frequency, the cause of generation of Faraday inductance is the change in magnetic flux associated with the metal and hence effect of Faraday inductance and resistance rise the magnitude of impedance (|z|).At little higher frequency, Faraday inductance is accepted to be act as open circuit as changes in magnetic becomes very rapid with increase of frequency but at steady state Faraday inductance behaves as short circuit and thus |z| is limited to approximately zero value. But with higher increase of frequency, kinetic inductance comes into play for enhancement of |z|. At higher frequency whenResistivity is imaginary and Faraday is accepted to be as short circuit and for smaller physical dimensions then only kinetic inductance is significant.For negative range of frequency, since

Figure 18. Thickness of medium between two metal plates vs. Reflectance for different values of incident angle

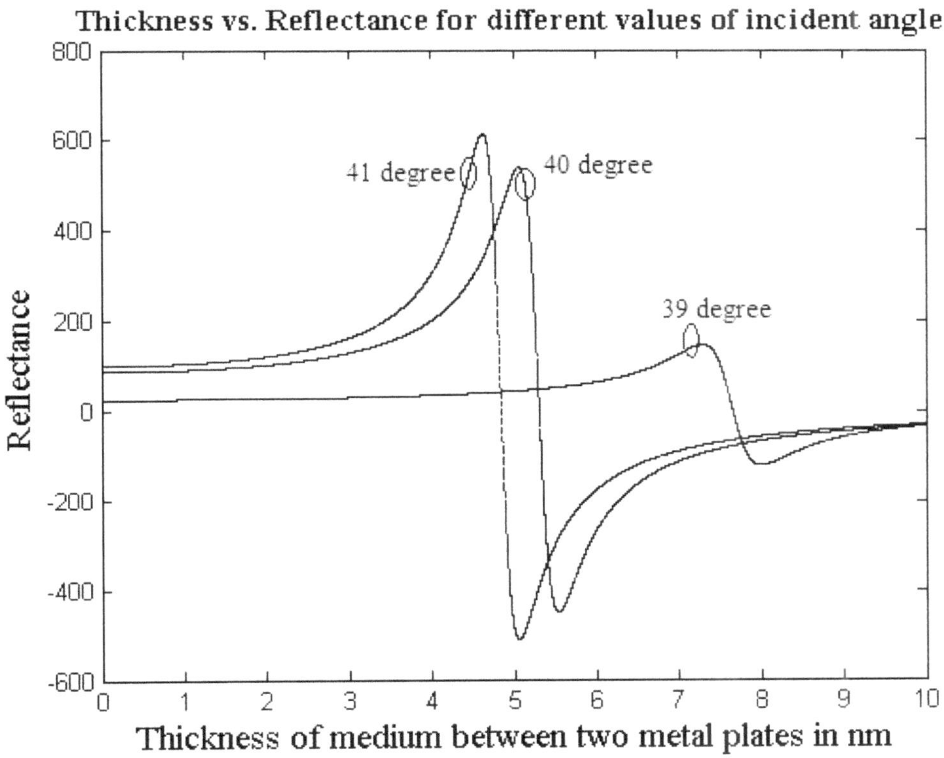

Surface plasmon effect is not established because of having only positive value of permeability and relative permittivity but still impedance is present for any Metal-Insulator-Metal structure in which resistance part is significant with smaller effect of Faraday inductance and kinetic inductance.

For different values of thickness of insulator layer Frequency vs. magnitude of impedance is shown in Figure 21. Higher value of metal layer thickness is responsible for minimization of impedance magnitude, |Z| due to presence of kinetic inductance. But, |Z| at positive lower value frequency is enhanced because of rising value of Faraday inductance comparatively ((Deyasi *et. al.*, 2017). As faraday inductance is changing proportionally w.r.t to change in thickness of insulator layer (d) and so |Z| is increased w.r.t to increase in d (Deyasi & Verma 2018). on the other side, as frequency Kinetic inductance(L_K) is highly dependent on increasing value of frequency

Figure 19. Thickness of medium between two metal plates vs. Reflectance for different values of refractive index

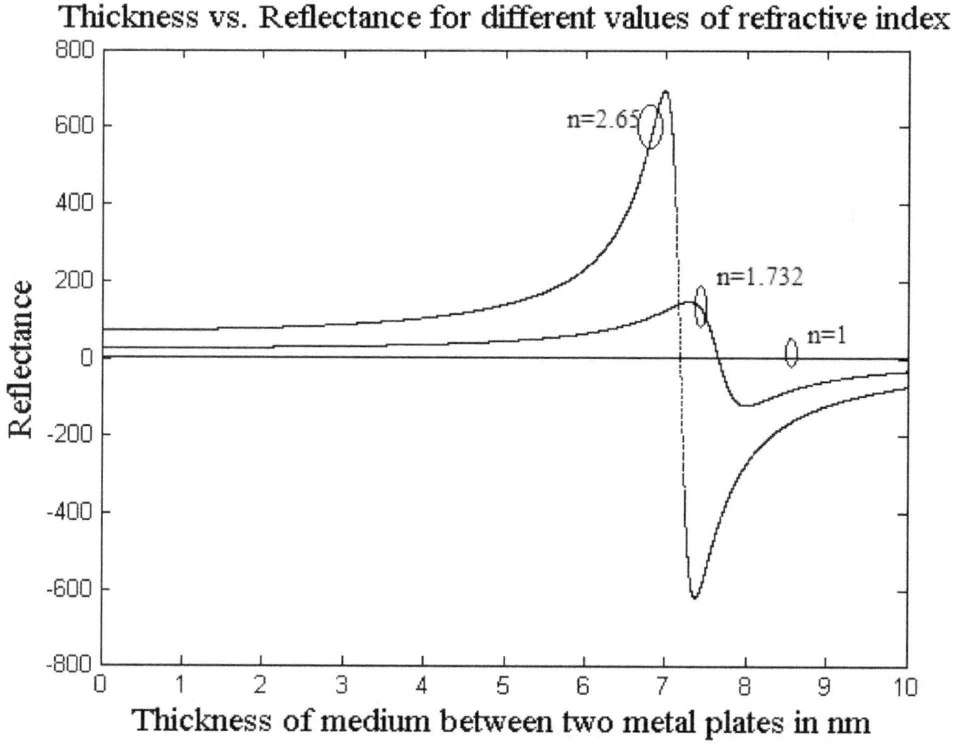

because of acceleration of free electrons for the conductors which are having the high carrier mobility. The generated electromotive force between two metal plates is decreased with a higher acceleration of free electrons in metal when d value is heightened up. When the frequency value in the negative range is considered, then increment of 'd' value causes to increase in resistance value.

Frequency vs. magnitude of impedance for different values of electron density is depicted in Figure 22. |z| due to kinetic inductance is reduced because of change in plasma frequency (w_p) with an increment of electron density (Deyasi *et. al.*, 2017). The value of relative permittivity of metal is controlled by Plasma frequency. Kinetic inductance and resistance value is controlled by relative permittivity of the metal. The value of resistance at some range of frequency is heightened as well as the value of kinetic inductance at higher range of frequency is minimized by the plasma frequency, and accordingly magnitude of impedance is varied.

Figure 20. Frequency vs. magnitude of impedance

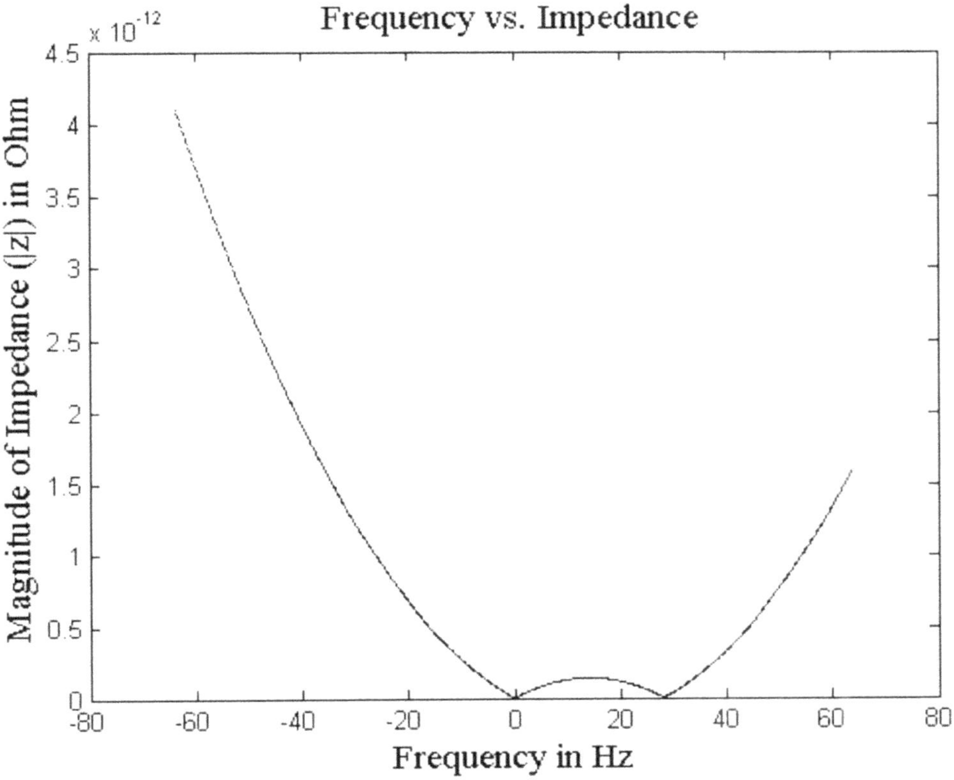

For different values of width of each layer frequency vs. magnitude of impedance is shown in Figure 23. Higher number of free conduction electrons on the surface are raised with rising of width (w) of the metal layer which enhance the value of inductance (Deyasi *et. al.*, 2017). Therefore the magnitude of impedance is step up with increment of 'w' for the frequencies having higher ranges. But, cross sectional area of the metal layer is increased as width is increased which causes decrement in resistance value of both metal plates and thereby |z| is decreased. When width of the metal layer is considered as 4m, then |z| is minimum for the frequency range of -100 Hz to 57 Hz in comparison to other two widths of the metal layer as 2m and 3m. Between this considered frequency -100 Hz to 57 Hz, at first the value of resistance is superior than the other and at second faraday inductance.

Figure 21. Frequency vs. magnitude of impedance for different values of thickness of insulator

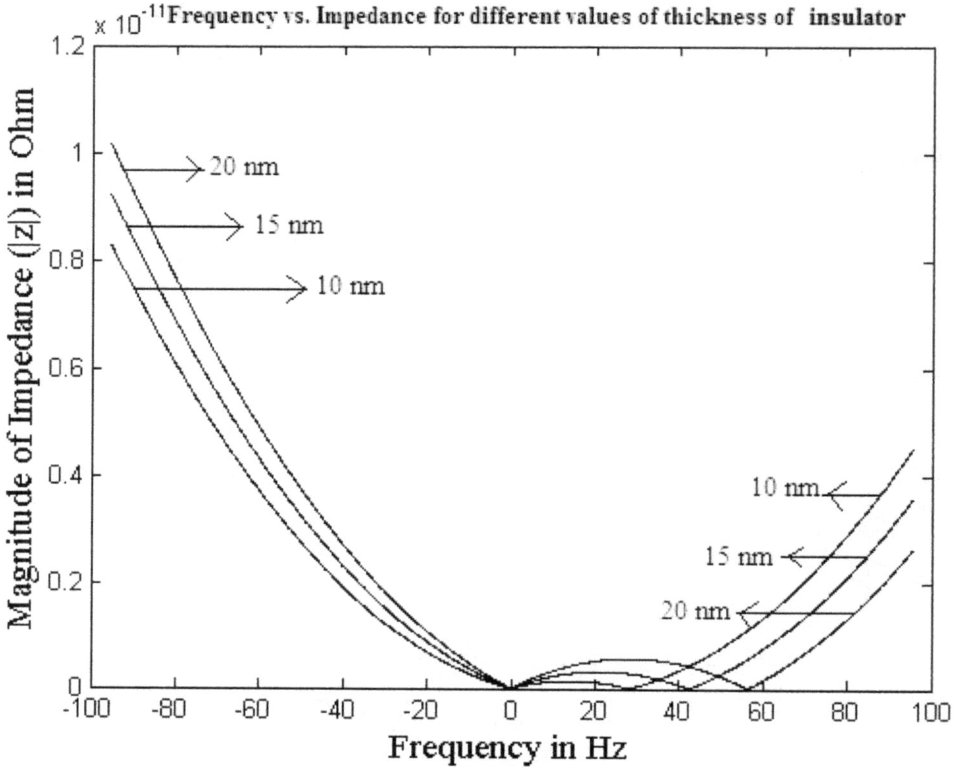

For Metal-Insulator-Metal parallel plate structure (MIM), wavelength vs. magnitude of impedance is shown in Figure 24. The effective value of resistance of the conductor is increased at higher frequencies due to skin effect. Since at lower values of wavelength, the skin effect is lower that causes reduction in the effective cross-section of the conductor. As metal conductor can be penetrated more by signal at lower wavelength and thereby energy of the signal is more absorbed and causes the structure to have step up in impedance at lower wavelength. It is also observed that the impedance is decreased to zero after certain range of wavelength.

It is already observed that Faraday impedance and Kinetic impedance share a part in total impedance. For different widths of the metal layer Faraday inductance has the same value w.r.t change in wavelength as shown in Figure 25. But on the other

Figure 22. Frequency vs. magnitude of impedance for different values of electron density

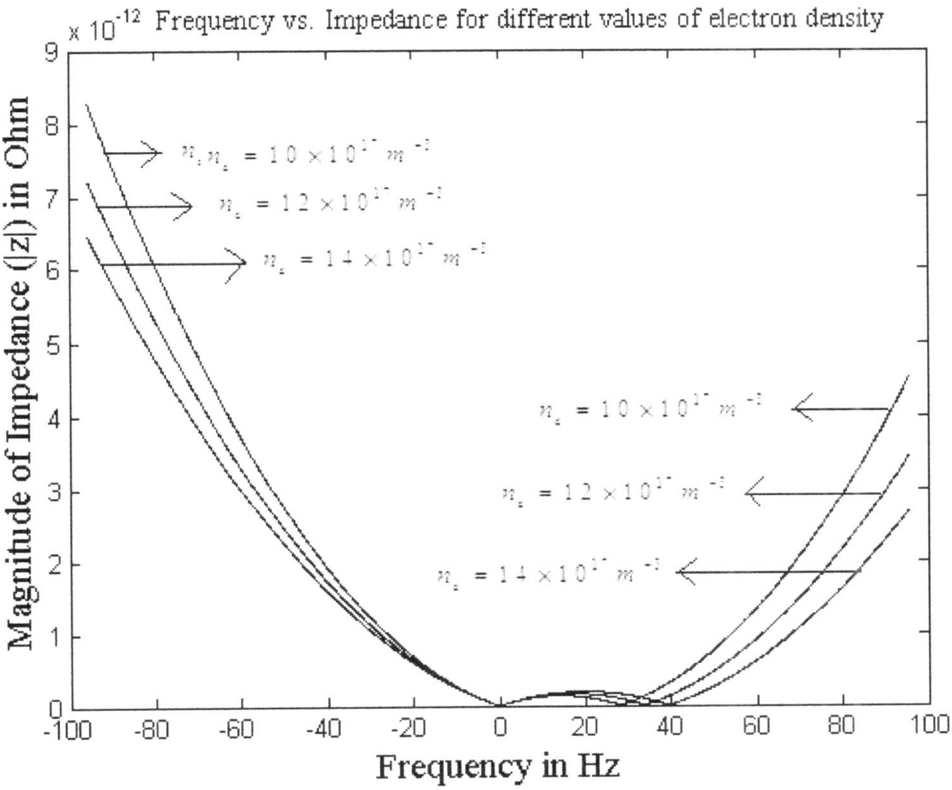

hand, Kinetic inductance is reduced exponentially w.r.t change in wavelength. It is observed from Figure 26 that in this context that we have considered the absolute values of kinetic inductance as overall impedance does not consider the reduction trend of it with operating frequency. For decreasing value of wavelength, inductance value is higher for lower wavelength and higher width of the metal layer whereas differences in inductance become insignificant for different value of width of metal layer for moderate value of wavelength. This is because of the fact that effect of surface plasmon frequency is undistinguished for higher wavelength.

For different values of width of metal plate, wavelength vs. magnitude of impedance is demonstrated in Figure 27. Magnitude of kinetic inductance is enhanced when

Figure 23. Frequency vs. magnitude of impedance for different values of width of each layer

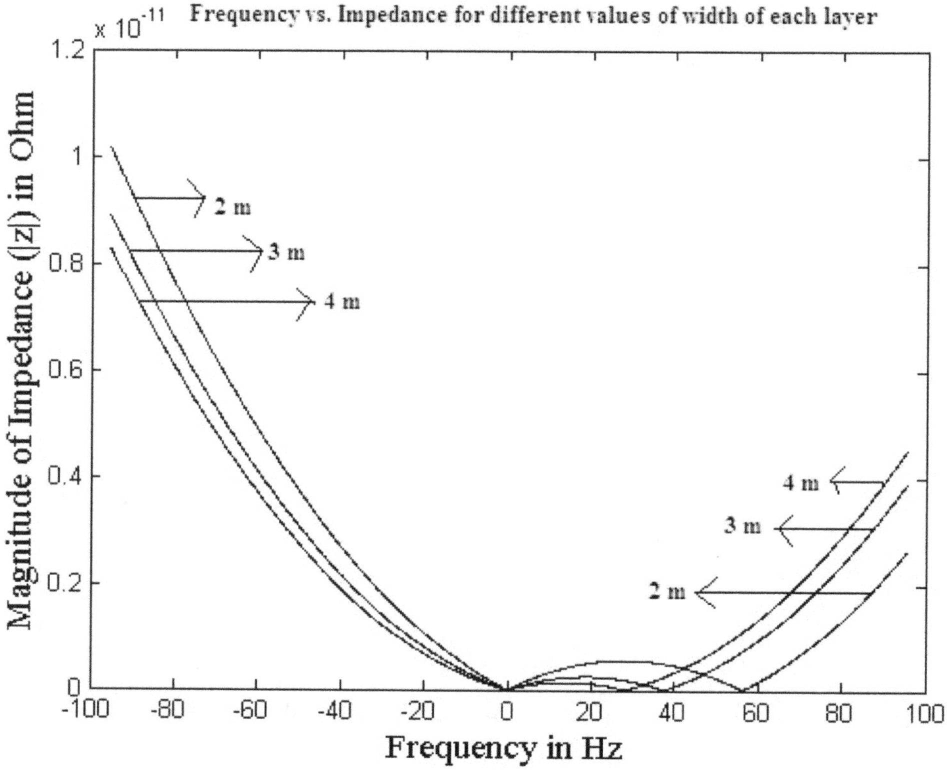

width (w) of the metal layer is increased as because free conduction electrons on the surface of the metal layer is raised. For frequencies having higher ranges or wavelength having lower ranges, the impedance magnitude is raised for increasing value of 'w'. But cross sectional area of the metal is increased with increasing value of 'w' and because of this resistance of both the metal plates is decreased which results to decrease the value of |z|. more free number of conduction electron is increased with increase in width of the metal layers. Thereby effective resistance is decreased having higher value of flowing of current because of higher number of electrons present in the metal layer. So higher value of width of the metal layer is inversely proportional to impedance. It is also observed that value of impedance is limited to zero after certain range of wavelength.

Figure 24. Impedance profile with wavelength for MIM surface plasmon structure

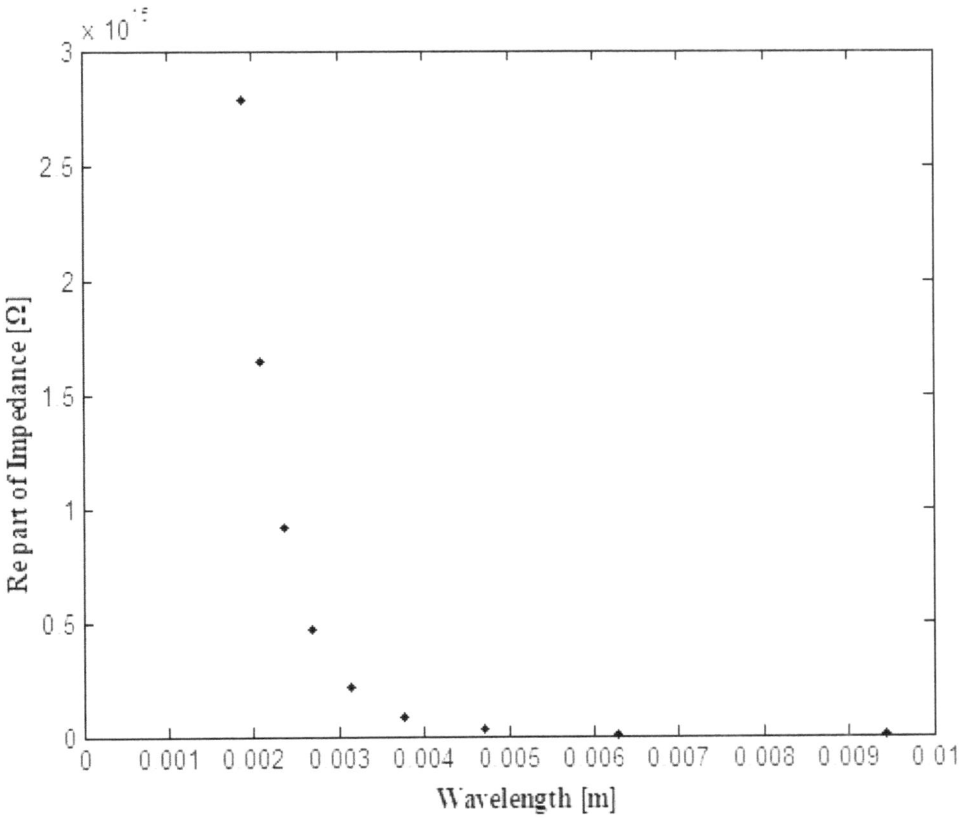

For different values of length of metal plate, wavelength vs. magnitude of impedance is depicted in Figure 28. Long distance is travelled by the free conduction electrons which are present on the surface of the metal layer if there is a flow of current through a series path which enhance the kinetic inductance and thereby causes to increase in the effective impedance ((Verma *et. al.*, 2017). When length is increased keeping the cross sectional area of the metal plates as constant longer distance is travelled by free electrons along the metal plates and so resistance of the metal layer is increased. Therefore impedance magnitude is enhanced because of dominance of kinetic inductance and increment of resistance at higher length of the metal plates at higher frequency or lower wavelength. If the length is lower, then the impedance is also lower in the lower ranges of wavelength.

Figure 25. Faraday impedance profile for different widths of the metal plate

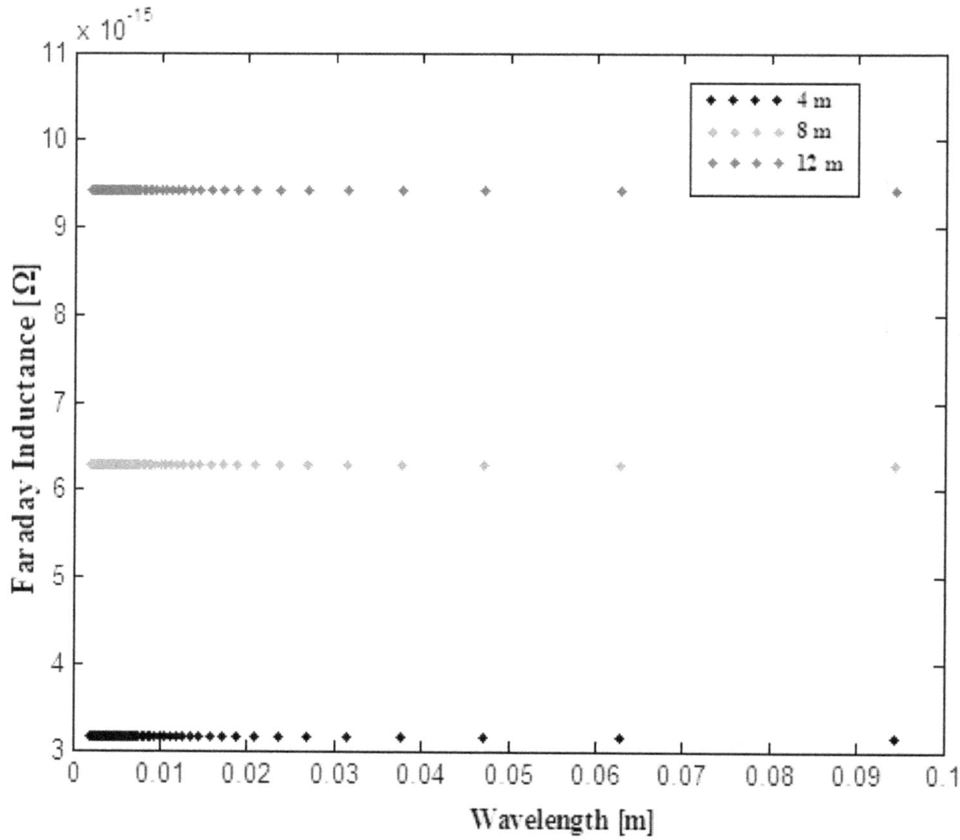

RESULTS OF QUALITY FACTOR

Observation for Quality Factor for MIM Structure

In this section, wavelength dependence of quality factor of MIM surface plasmon structure is analyzed. Structural parameters and damping ratio of the structure is tuned within allowable limit to analyze the variation after detailed analytical computation.

Result and Discussion for Quality Factor for MIM Structure

Wavelength versus Quality factor is depicted in Figure 29. The ratio of energy stored by the medium per cycle to the energy dissipated by the medium per cycle is the

Figure 26. Kinetic profile for different widths of the metal plate

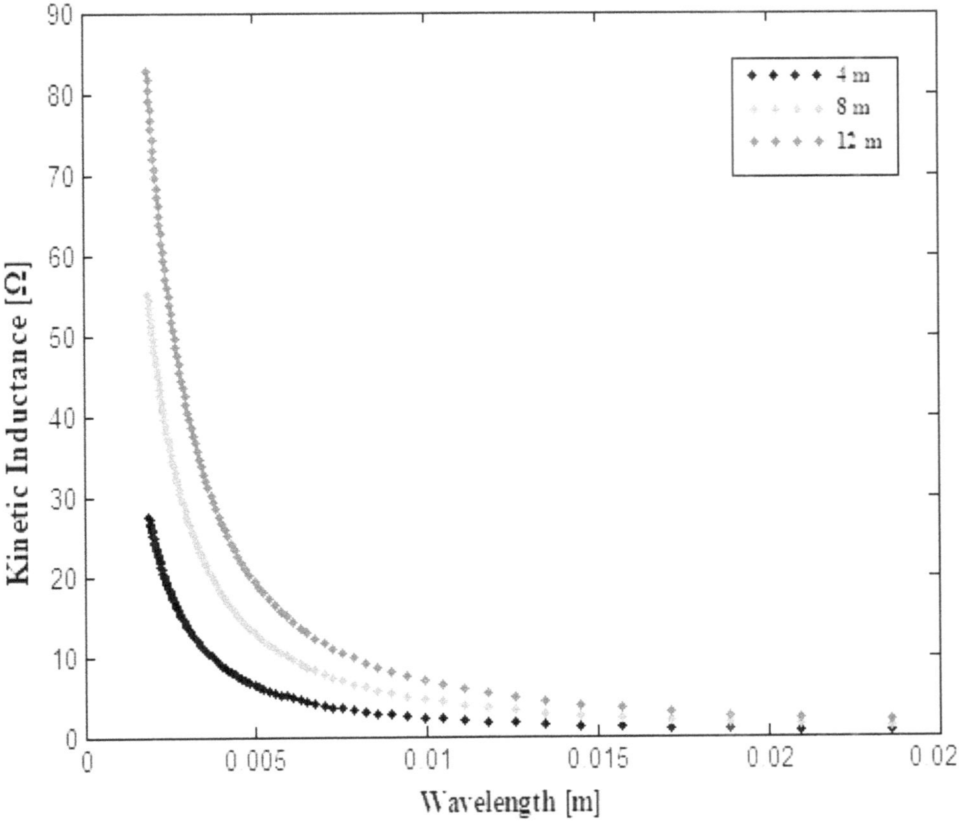

Quality factor. The metal is lossy in (metal-insulator-metal) MIM structure metal in comparison to dielectric or insulator. Quality factor is decreased exponentially w.r.t to rising in the value of wavelength (Verma *et. al.*, 2017). It is because of the reason, for higher value of wavelength the wave can propagate deeper into the Metal-insulator-metal (MIM) which cases the total power absorbed into the metal due to which quality factor is decreased. The frequency of incident light is decremented with an increase in wavelength and at certain wavelength, the frequency of the oscillation of the electron between the interface of metal and insulator is matched with frequency of incident light for the resonance condition of surface plasmon oscillation. When surface plasmon resonance condition is occurred then higher part

Figure 27. Impedance profile for different widths of the metal plate

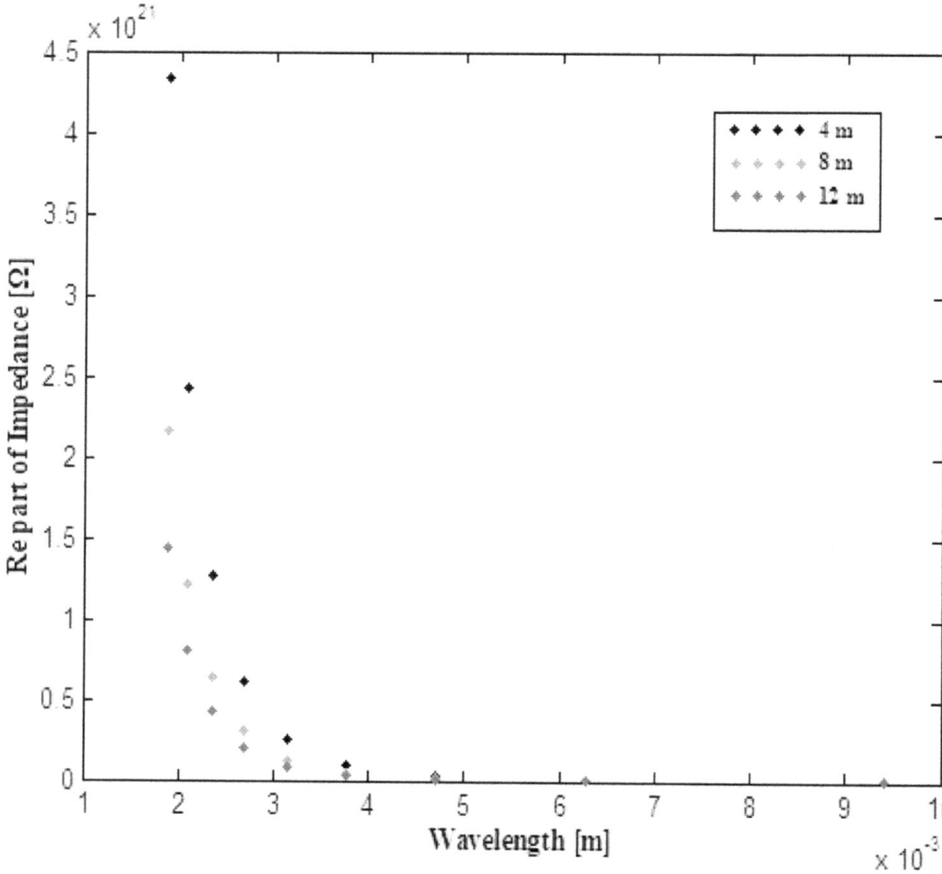

of the incident energy is transmitted into the medium. Resonant frequency is reduced with reduction in oscillation of surface plasmon when surface plasmon condition is mismatched. Hence quality factor is decreased when lower part of energy of light is transmitted into the MIM and most of the energy is reflected which causes the store of energy in little amount.

For different values of damping ratio, the wavelength versus Quality factor is shown in Figure 30. Magnitude of quality factor is decreased with increase of damping ratio ((Verma *et. al.*, 2017). After a certain disturbance, how an oscillation in a system is decayed is described by a dimensionless quantity named as damping ratio. Surface plasmon wave is decayed with higher value of damping ratio and due

Figure 28. Impedance profile for different lengths of the metal plate

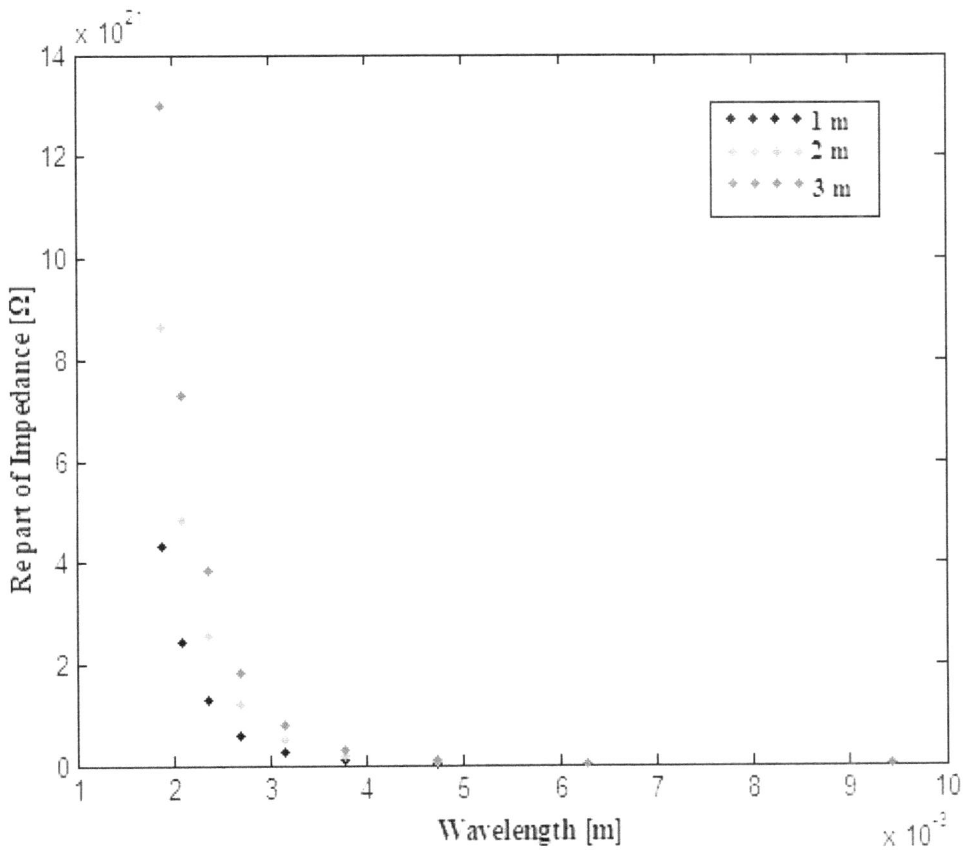

to this propagation of the wave is restricted deeper into the metal-insulator-metal (MIM) structure. Therefore most of the energy of is reflected than the transmitted energy into the metal. Hence quality factor is decreased as lower amount of energy is stored into the MIM structure.

CONCLUSION

In this work surface plasmon structure for spectroscopic applications and for optical sensors has been done by following ways:

At wider range of incidence angle having different thicknesses of gold layers, the reflectance of 3-layer surface plasmon based sensor is calculated.

Figure 29. Wavelength vs. quality factor

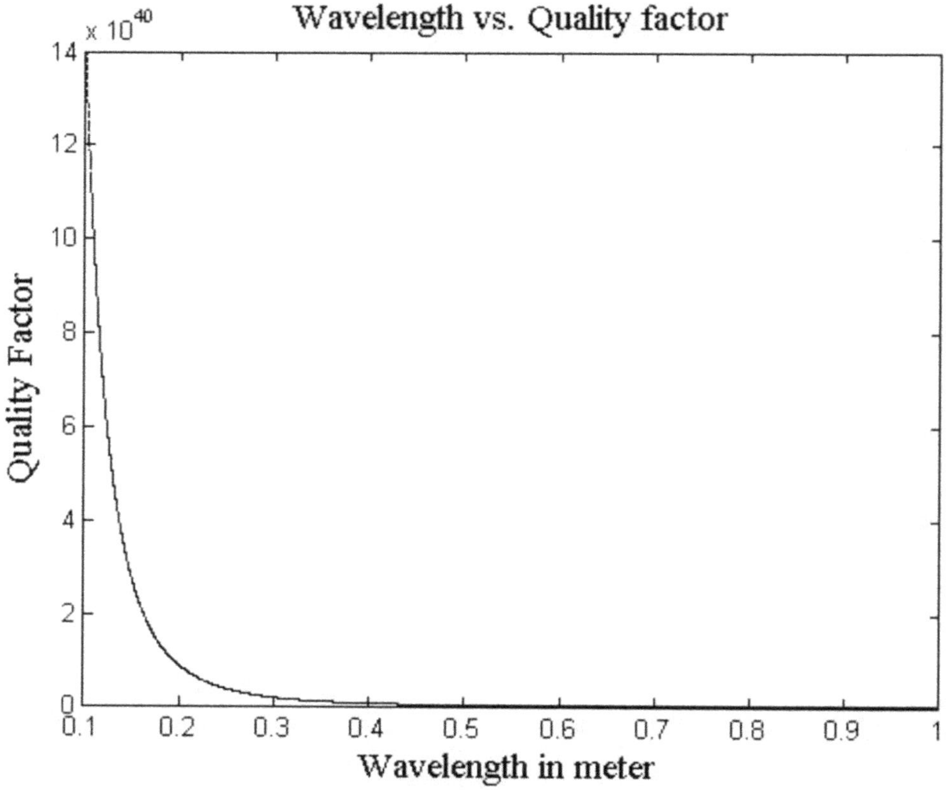

It is observed that when the propagation constant of surface plasmon wave is equal to the incident wave then there is sharp dip in the reflectance profile.

For 3 layer SPR based sensor, Wavelength of optical source, incident angle on reflectance and Impact of thickness of gold layer is analytically computed.

For different structural parameters, Field enhancement is also obtained.

- It is demonstrated in the result that resonance is occurred at the critical angle of 37-37.5 degree which maximize the value of field enhancement.
- Considering the effect of both Faraday inductance and kinetic inductance, characteristic impedance of MIM surface plasmon structure is analytically calculated.
- Variation in impedance w.r.t frequency is observed by changing the value of insulator thickness, metal layer thickness and electron density.

Figure 30. Wavelength vs. quality factor for different values of damping ratio

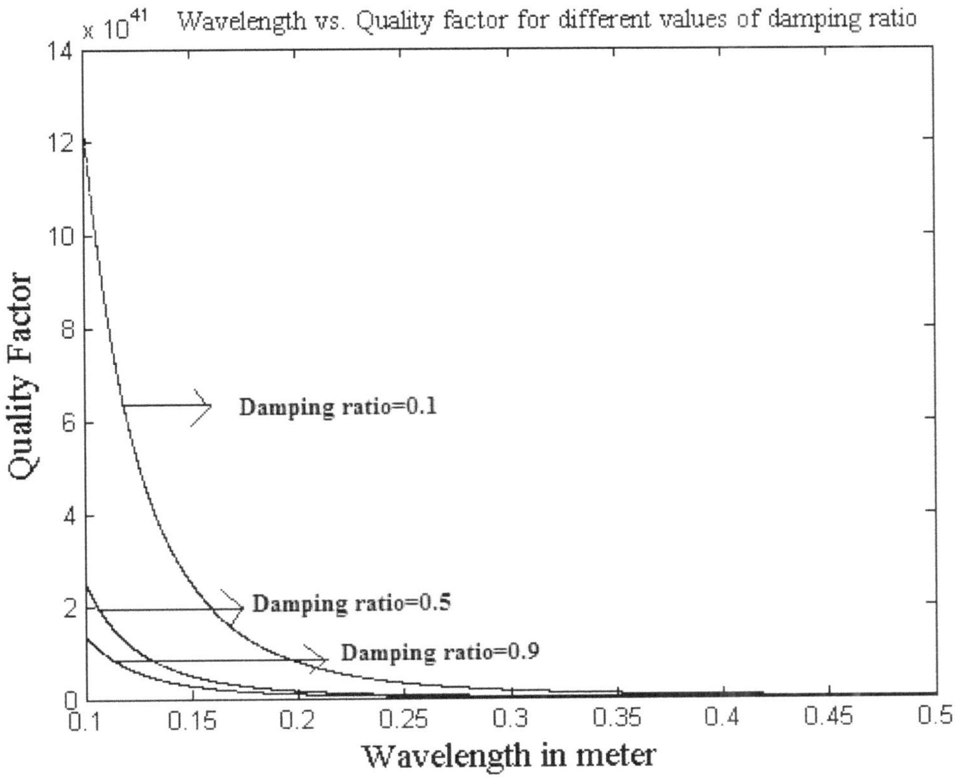

- At the very lower ranges of frequency, the Coupling between symmetric and anti-symmetric modes are shown by optimum magnitude of impedance which results for development in the effective double sided surface plasmon.
- It is analyzed that characteristic impedance and quality factor of MIM (Metal-Insulator-Metal) surface plasmon structure are highly dependent on wavelength.
- To analyze the variation after detailed analytical computation, Structural parameters and damping ratio of the structure is adjusted within allowable limit.
- It is observed that resistive part has higher value over the other impedances at higher frequency ranges, which signifies that there is no possibility of symmetric and anti-symmetric modes at the higher frequency ranges. Other than this, it also observed that Faraday inductance and kinetic inductance is highly effected by the wavelength.

REFERENCES

Abbasi, F., & Bahrampour, A. R. (2008). High harmonics generation in under dense plasma from Relativistic Thomson scattering. *IEEE 35th International Conference on Plasma Science*.

Born, M., & Wolf, E. (1999). *Principles of Optics* (7th ed.). Cambridge University Press.

Caloz, C., Sanada, A., & Itoh, T. (2003). Surface Plasmons at the Interface between Right-Handed and Left-Handed 2D Metamaterials. *IEEE Antennas and Propagation Society International Symposium*, *3*, 363 – 366.

Demetriadou, A. (2015). The impact of natural modes in plasmonic imaging. *Scientific Reports*, *5*, 18247.

Deyasi, A., Paul, P., & Verma, P. (2017). Calculating impedance variation with frequency of MIM surface plasmon structure for spectroscopic applications. *IEEE International Conference on Devices for Integrated Circuit*, 81-84.

Deyasi, A., & Verma, P. (2018). Computation of Skin Depth for MIM Surface Plasmon Structure at High Frequency. *Fifteenth International Conference on Wireless and Optical Communications Networks*.

Deyasi, A., Verma, P., & Sarkar, A. (2018). *Calculating Transmittance and Field Enhancement of n-Layer MIM Surface Plasmon Structure for Detection of Biological Nano-Objects*. Springer.

Ding, S. Y., Yi, J., Li, J. F., Ren, B., Wu, D. Y., Panneerselvam, R., & Tian, Z. Q. (2016). Nanostructure-based Plasmon-enhanced Raman spectroscopy for surface analysis of materials. *Nature Reviews Materials*, *1*, 16021.

Garcia-Vidal, F. J., Sanchez-Dehesa, J., Dechelette, A., Bustarret, E., Lopez-Rios, T., Fournier, T., & Pannetier, B. (1991). Localized surface plasmons in lamellar metallic gratings. *Journal of Lightwave Technology*, *17*(11), 2191–2195. doi:10.1109/50.803010

Gordon-II, J. G., & Swalen, J. D. (1977). The effect of thin organic films on the surface plasmaresonace on gold. *Optics Communications*, *22*(3), 374–376. doi:10.1016/S0030-4018(97)90032-8

Hecht, E. (2002). *Optics*. Addison-Wesley.

Herminghaus, S., Riedel, M., Leiderer, P., Weiland, U., Giebel, K. F., Bechinger, C., & Bastmeyer, M. (1999). Imaging of cell/substrate contacts of living cells with surface plasmon resonance microscopy. *Journal of Biophysics*, *76*(1), 509–516. doi:10.1016/S0006-3495(99)77219-X

Homola, J., Yee, S. S., & Gauglitz, G. (1999). Surface Plasmon resonance sensors [review]. *Sensors and Actuators. B, Chemical*, *54*(1-2), 3–15. doi:10.1016/S0925-4005(98)00321-9

Huang, H., Zhang, J., Ji, X., Zhou, J., Bao, M., & Huang, Y. (2009). Photonic Crystal Biosensors Based on Surface Plasmons. *IEEE 9th International Conference on Solid-State and Integrated-Circuit Technology*.

Ishimaru, A., Jaruwatanadilok, S., & Kuga, Y. (2005). Generalized Surface Plasmon Resonance Sensors Using Metamaterials And Negative Index Materials. *Progress in Electromagnetic Research, PIER*, *51*, 139–152. doi:10.2528/PIER04020603

Jetté-Charbonneau, S., Charbonneau, R., Lahoud, N., Mattiussi, G. A., & Berini, P. (2005). Bragg Gratings Based on Long-Range Surface Plasmon-Polariton Waveguides: Comparison of Theory and Experiment. *IEEE Journal of Quantum Electronics*, *41*(12), 1480–1491. doi:10.1109/JQE.2005.858466

Jiang, L., Zhao, X., Fei, Y., Yu, D., Qian, J., Tong, J., ... He, S. (2016). The effects of magnetic fields exposure on relative permittivity of saline solutions measured by a high resolution SPR system. *Scientific Reports*, *6*, 25111.

Jorgenson, R. C., & Yee, S. S. (1993). A fiber-optic chemical sensor based on surface plasmon resonance. *Sensors and Actuators. B, Chemical*, *12*(3), 213–220. doi:10.1016/0925-4005(93)80021-3

Khare, R., & Nema, R. (2012). Reflection Coefficient Analysis Of Chebyshev Impedance Matching Network Using Different Algorithms. *International Journal of Innovative Research in Science, Engineering and Technology*, *1*(2), 214–218.

Kittel, C. (1996). *Introduction to Solid State Physics* (7th ed.). New York: Wiley.

Knoll, W. (2000). Interfaces and thin films as seen by bound electromagnetic waves. *Annual Review of Physical Chemistry*, *49*(1), 569–638. doi:10.1146/annurev.physchem.49.1.569 PMID:15012436

Kolb, D. M. (1982). *The study of solid-liquid interfaces by surface plasmon polariton excitation*. Surface Polaritons. doi:10.1016/B978-0-444-86165-8.50014-1

Kuo, C. H., & Moghaddam, M. (2006). Backscattering enhancement of surface plasmons from multilayer rough surfaces. *IEEE Antennas and Propagation Society International Symposium*, 749 – 752.

Li, H., Peng, W., Wang, Y., & Hu, L. (2011). Theoretical analyses of localized surface Plasmon resonance spectrum with nanoparticles imprinted polymers. *IEEE 2011 Asia Communications and Photonics Conference and Exhibition*, 1-8.

Liu, Y., Liu, Q., Chen, S., Cheng, F., Wang, H., & Peng, W. (2016). Surface Plasmon Resonance Biosensor Based on Smart Phone Platforms. *Scientific Reports*, *5*, 12864.

Nikitin, P. V., Rao, K. V. S., Lam, S. F., Pillai, V., Martinez, R., & Heinrich, H. (2005). Power Reflection Coefficient Analysis for Complex Impedances in RFID Tag Design. *IEEE Transactions on Microwave Theory and Techniques*, *53*(9), 2721–2725. doi:10.1109/TMTT.2005.854191

Nylander, C., Liedberg, B., & Lundstrom, I. (1983). Surface plasmon resonance for gas and biosensing. *Sensors and Actuators*, *4*, 299–304. doi:10.1016/0250-6874(83)85036-7

Otto, A. (2000). Excitation of surface plasma waves in silver by the method of frustrated totalreflection. *Zeitschrift fur Physik*, *216*(4), 398–410. doi:10.1007/BF01391532

Pechprasarn, S., & Somekh, M. (2011). *Analyzing Surface Plasmon Microscopy with Rigorous Diffraction Theory. IEEE Functional Optical Imaging.*

Pendry, J. B. (2000). Negative refraction makes a perfect lens. *Physical Review Letters*, *85*(18), 3966–3969. doi:10.1103/PhysRevLett.85.3966 PMID:11041972

Pendry, J. B., Holden, A. J., Robbins, D. J., & Stewart, W. J. (1999). Magnetism from conductors and enhanced nonlinear phenomena. *IEEE Transactions on Microwave Theory and Techniques*, *47*(11), 2075–2084. doi:10.1109/22.798002

Popov, A. K., Myslivets, S. A., George, T. F., & Shalaev, V. M. (2007). *2007 Compensating losses in positive- and negative-index metamaterials through nonlinear-optical quantum switching, IEEE paper on Lasers and Electro-Optics - Pacific Rim, 2007*. CLEO/Pacific Rim.

Raether, H., & Kretschmann, E. (1968). Radaitive decay of non-radiative surface plasmons excited by light. *Zeitschrift fur Naturforschung. Teil B. Anorganische Chemie, Organische Chemie, Biochemie, Biophysik, Biologie*, *23A*, 2135–2136.

Ritchie, R. H. (1957). Plasma losses by fast electrons in thin films. *Physical Review*, *106*(5), 874–881. doi:10.1103/PhysRev.106.874

Ruppin, R. (2000). Surface polaritons of a left-handed medium. *Physics Letters. [Part A]*, *277*(1), 61–64. doi:10.1016/S0375-9601(00)00694-0

Satija, J., Shukla, G. M., & Mukherji, S. (2011). Potential of dendrimeric architecture in surface Plasmon resonance biosensor. *IEEE 2010 International Conference on Systems in Medicine and Biology*.

Scheumann, V., Zizlsperger, Z., Mack, J., Jung, G., Badia, A., Arnold, S., & Knoll, W. (1999). Probing the elecrochemical deposition and/or desorption of self-assembled and electropolymerizableorganic thin film by surface plasmon spectroscopy and atomic force microscopy. *Sensors and Actuators. B, Chemical*, *54*(1-2), 145–165. doi:10.1016/S0925-4005(98)00333-5

Sharma, A. K., Jha, R., & Gupta, B. D. (2007). Fiber-Optic Sensors Based on Surface Plasmon Resonance: A Comprehensive Review. *IEEE Sensors Journal*, *7*(8), 1118–1129. doi:10.1109/JSEN.2007.897946

Sreekanth, K. V., Alapan, Y., ElKabbash, M., Ilker, E., Hinczewski, M., Gurkan, U. A., ... Strangi, G. (2016). Extreme sensitivity biosensing platform based on hyperbolic metamaterials. *Nature Materials*, *15*(6), 621–627. doi:10.1038/nmat4609 PMID:27019384

Usha, S. P., Mishra, S. K., & Gupta, B. D. (2016). Fabrication and Characterization of a SPR Based Fiber Optic Sensor for the Detection of Chlorine Gas Using Silver and Zinc Oxide. *Materials (Basel)*, *8*(5), 2204–2216. doi:10.3390/ma8052204

Verma, P., Deyasi, A., & Paul, P. (2018). Computing Reflectance of Three-Layer Surface Plasmon-Based Sensor at Visible Spectra. In Lecture Notes in Networks and Systems: Industry Innovative Innovations in Science, Engineering and Technology. Springer.

Verma, P., Paul, P., & Deyasi, A. (2017). Wavelength dependence of impedance and quality factor for MIM surface plasmon structure. *1st IEEE International Conference on Electronics, Materials Engineering and Nano-Technology*. 10.1109/IEMENTECH.2017.8076946

Veselago, V. G. (1968). The electrodynamics of substances withsimultaneously negative values of ε and μ. *Soviet Physics - Uspekhi*, *10*(4), 509–514. doi:10.1070/PU1968v010n04ABEH003699

Wang, Y., Knoll, W., & Dostalek, J. (2011). Long range surface Plasmon resonance bacterial pathogen biosensor with magnetic nanoparticle assay. *IEEE 2011 International Workshop on Biophotonics*.

Wood, R. W. (1902). On a remarkable case of uneven distribution of light in a diffraction grating spectrum. *Phil. Mag. Ser. 6*, *4*(21), 396–402. doi:10.1080/14786440209462857

Yamamoto, M. (2008). Surface Plasmon (SPR) Theory, Tutorial. *Rev Polarogr.*, *48*. doi:10.5189/revpolarography.48.209

Zavats, A. V., Darmanyad, S. A., Gcrard, D., Salomnn, L., & De Fornel, F. (2005). Surface Plasmon Polaritons on Nanostructured Surfaces and Thin Films. *IEEE 10th International Conference on Mathematical Methods in Electromagnetic Theory*, 73 – 78.

Ziolkowski, R. W., & Heyman, E. (2001). Wave propagation in media having negative permittivity and permeability. *Physical Review. E*, *64*(5), 056625. doi:10.1103/PhysRevE.64.056625 PMID:11736134

Chapter 11
All-Optical Switching and Logic-Gates Design Using Mode (Polarization)-Conversion in Micro-Ring Resonator

Jayanta Kumar Rakshit
National Institute of Technology Agartala, India

Gaurav Kumar Bharti
National Institute of Technology Agartala, India

ABSTRACT

The realization of all-optical polarization switch and all-optical logic gates based on polarization-conversion on single silicon micro-ring resonator (MRR) is demonstrated. By adjusting the mode state of the input source as well as the pump light, the all-optical polarization switch, and hence, all-optical NOT, OR/NOR. AND-NAND logic gates are realized. The design is ultra-compact, ultrafast, and less optical power is required for all-optical polarization-conversion-based switch and logic gates, respectively. The MRR also shows outstanding performance as its Q (quality) factor is very high. The design is robust, simple, stable, easy-to-fabricate, and silicon-on-insulator (SOI) compatible. The structure is compatible for interconnects and capable for integrating in electronics as well as in plasmonics circuits.

DOI: 10.4018/978-1-5225-8531-2.ch011

Copyright © 2019, IGI Global. Copying or distributing in print or electronic forms without written permission of IGI Global is prohibited.

INTRODUCTION

Human life is enhanced to a large extent due to computers and its application break through into all areas of human civilization. The speed of computer systems is limited by the bandwidth of electronic circuits. Therefore, a remarkable way out to increase the speed and bandwidth is needed, and unless researchers' find out different mechanism towards different technology, it will not possible to further develop the computer performance. Signal processing circuits are also essential for optical communication and networking systems along with bio-sensing, space technology etc. Switching is basic building block in all form of signal processing units and networks. Though special types of optical switches are projected and designed, their uses are still in preliminary phase (Chattopadhyay et al, 2013).

To avoid bottleneck of optoelectronic conversions in quickly growing ultra-high speed optical system, switching, routing and signal processing desires to carry out in optical domain (Zoiros et al, 2004). All-optical switching is one of the most important technology into cutting-edge research, and might ultimately show the new path way for computing applications with faster processing speed and huge bandwidth as well as enhanced connectivity without electrical-optical-electrical conversions. Research in all optical domain can also be investigated new ideas.

All-optical switches (AOSs) attract researchers in the recent years due to its high compactness, fast response and low threshold. Due to huge potentials of optical computing, different all-optical circuits which include optical logic gates, different combinational and sequential logic circuits digital converter have been anticipated and designed by different researchers using various techniques such as semiconductor optical amplifier (SOA), Mach–Zehnder interferometer (MZI), non-linear material (NLM) and ring resonator etc (Nayar et al, 1991; Jinno et al, 1992). Some kinds of AOSs are based on intensity modulation which generally requires higher switching power. Polarization-dependent AOS can be realized by arranging waveguide in interferometric style. The fiber with Mach-Zehnder interferometer can be utilized to design polarization-dependent switch (Li et al, 2006). Some usual bulky polarization components such as wave-plates, beam splitter are used in the designing of interferometric based optical switch. These AOSs are not potential candidate in practice due to the bulky nature in structure, the large insertion losses, the strong instability caused by surroundings and so on. However, almost all of these designs are not compact and simple. Polarization rotation due to bend waveguides is shown in literature (Soref et al, 2006). In the recent years, silicon photonics has reached massive development in optical communication, information processing and designing of all-optical logic gates due to huge bandwidth and ultra-high speed of light (Soref et al, 2006). Silicon micro-ring resonators (MRRs) is suitable component for photonic integrated circuits due to ultra-compact size of MRRs and low power consumption. The MRRs are found application in signal processing, optical switches, logic gates, (Melloni et al, 2002; Rakshit et al, 2014). due to its all-optical behavior. MRRs are also used in designing adder/subtractor circuits, converters, multiplexer and demultiplexer circuits, etc (Khorasanindejdad et al, 2009) Therefore, study of the polarization-dependent AOS especially in MRR based architecture is very important.

Polarization-conveyor in micro-ring resonator and microsphere using birefringence effect is reported in (Bianucci et al, 2007; Lamouroux et al, 1982; Fietz et al, 2007) Nonlinear polarization conversion in ring resonator with square waveguide is discussed by C. Fietz and G. Shvets (2007) Polarizations in notch ring resonators using various parameters of the MRR are explained by various researchers (Bianucci, 2007; Lamouroux et al, 1982). Polarization rotators based on MRRs are presented in (Little et al, 2000; Melloni et al, 2004). Polarization-dependent switch can be designed by controlling the polarization properties of the optical pump signals on micro-ring resonator (Bharti et al, 2018)

The logical operations are very important for optical signal processing and computing. Until now, many traditional logic gates have been designed. Most of them based on nonlinear effects of silicon in which strong pulse signal needed. For polarization-based logic circuits and multiplexing/demultiplexing in an optical communication system, devices based on the mode-conversion technique, need to be developed and thus, some polarization (mode) based converters have been proposed in recent past (Kim et al, 2009). Mode-conversion based polarization switch can be controlled by an electrical pump which causes a refractive-index change in the quantum well MRR is proposed and realized. Despite low electrical power requirement for mode-conversion in the proposed ring, it is comparatively large in size (Suzuki et al, 2016).

THEORY OF MICRO-RING RESONATOR

In simple words, a micro-ring resonator (MRR) is an optical device used as an optical switch and an optical filter. There are basically two structures of MRRs-simple (conventional structure of) MRR, race-track MRR, each one may be employed as single-waveguide coupled MRR and double-waveguide coupled MRR, as depicted in Figure 1(a) and Figure 1(b) respectively.

Each structure possesses some advantages. The fundamental configuration of the MRR is shown in Figure 1(b). If the coupling length of MRR is large (or more than some μm, depending upon the radius of MRR), the structure is known as race-track MRR, which is shown in Figure 2(a) and Figure 2(b) respectively.

Figure 1. Configuration of MRR, (a). Single waveguide-coupled, (b). Double waveguide-coupled

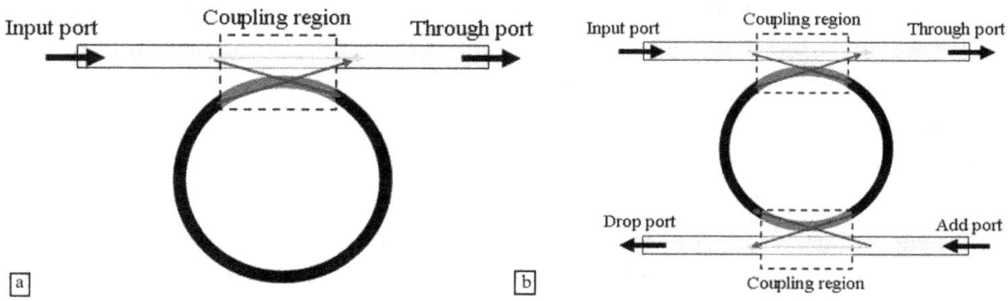

Figure 2. Race-track MRR; (a) Single waveguide-coupled; (b). Double waveguide-coupled

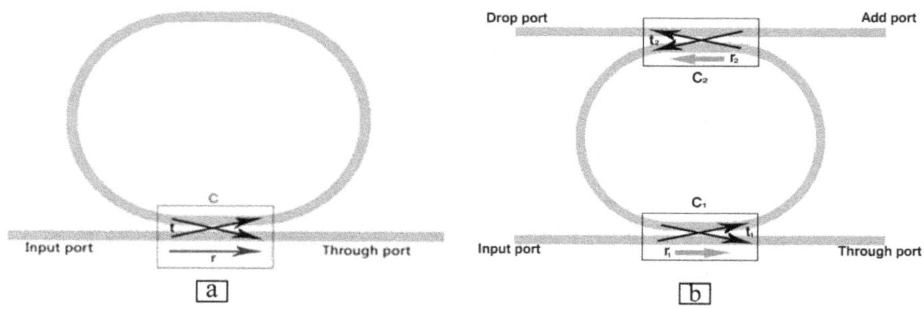

When the optical signal of the suitable wavelength is launched from the input port to the MRR, one of the resonant modes gets coupled into the ring-cavity due to constructive interference and if the mode of the wavelength match with one of the resonant wavelengths of the ring-cavity then it comes out of the drop port and the other mode of light which is not coupled to the ring-cavity comes out from the through port. The coupling of the mode of light is from the straight waveguide to the ring-cavity waveguide of the MRR is dependent on the integral multiple of the resonant wavelengths, which is given by following equation (Rakshit et al, 2014)

$$m\lambda = 2\pi r.\ n_{eff}$$

where m is an integer (m=1,2,3,.....), λ is the input wavelength, n_{eff} is the effective refractive index of the waveguide mode, r is the radius of the MRR. This is the conditions for constructive interference in the MRR.

The input-output couplers determine the amount of light couples to the ring. Then input light will comes out from the drop port and maximum transmittance will obtain at that port and minimum transmittance will obtain at the through port at resonance without pump power. The application of pump power to the ring generated free-carrier concentration through TPA (two-photon absorption) effect which causes refractive-index (RI) change and hence resonance condition. Therefore, π-phase shift occur in one complete ring so that output light signal will be directed to the through port from drop port. Thus optical switching can be realized for a signal between two output ports. The change is RI due to pump power is given below.

$n_{eff} = n_o + n_{2.} I, \; I = P/S$

$\phi = (2\pi/\lambda).\Delta n.L$

where n_o and n_2 are the linear and nonlinear refractive indexes of the material respectively. S is the effective area of the cross-section of the MRR, I and P is the intensity and power of the optical signal respectively. ϕ is the phase-shift of the MRR and Δn is the change in refractive index of the material of the MRR. L is the circumference of the ring and λ is the operating wavelength of the MRR Bianucci et al, 2007)

The output electric fields at the through port and drop port can be written as respectively,

$$E_t = \frac{\sqrt{1-\kappa_1} - \sqrt{1-\kappa_2}\exp(-\alpha L)\exp^2(j\varphi)}{1-\sqrt{1-\kappa_1}\sqrt{1-\kappa_2}\exp(-\alpha L)\exp^2(j\varphi)} E_{i1} + \frac{-\sqrt{\kappa_1}\sqrt{\kappa_2}\exp(-\alpha \frac{L}{2})\exp(j\varphi)}{1-\sqrt{1-\kappa_1}\sqrt{1-\kappa_2}\exp(-\alpha L)\exp^2(j\varphi)} E_{i2}$$

$$E_d = \frac{-\sqrt{\kappa_1}\sqrt{\kappa_2}\exp(-\alpha \frac{L}{2})\exp(j\varphi)}{1-\sqrt{1-\kappa_1}\sqrt{1-\kappa_2}\exp(-\alpha L)\exp^2(j\varphi)} E_{i1} + \frac{\sqrt{1-\kappa_2} - \sqrt{1-\kappa_1}\exp(-\alpha L)\exp^2(j\varphi)}{1-\sqrt{1-\kappa_1}\sqrt{1-\kappa_2}\exp(-\alpha L)\exp^2(j\varphi)} E_{i2}$$

where, κ_1, κ_2 are the field coupling coefficient between the ring and input and output bus respectively, κ_n is wave propagation constant, where $\kappa_n = \frac{2\pi}{\lambda} n_{eff}$, λ is the resonant wavelength of the ring, n_{eff} is effective refractive index of the material of the ring, $x = \exp(-\alpha \frac{L}{2})$, $\varphi = \frac{\kappa_n . L}{2}$, E_{i1} and E_{i2} are the input and add port field respectively. Hence the MRR can be used as an optical switch and the status of the switch is controlled by the optical pump pulse. The transfer function of the single MRR at through port (T_p) and drop port (T_d) by considering $E_{i2} = 0$ can be written as, $T_p = \frac{|E_t|^2}{|E_{i1}|^2}$ and $T_d = \frac{|E_d|^2}{|E_{i1}|^2}$ respectively. Figure 3 shows the corresponding graphs of the transfer function of through port and drop port.

Figure 3. Transmittance graph at through port (dashed line) and drop port (solid line) outputs

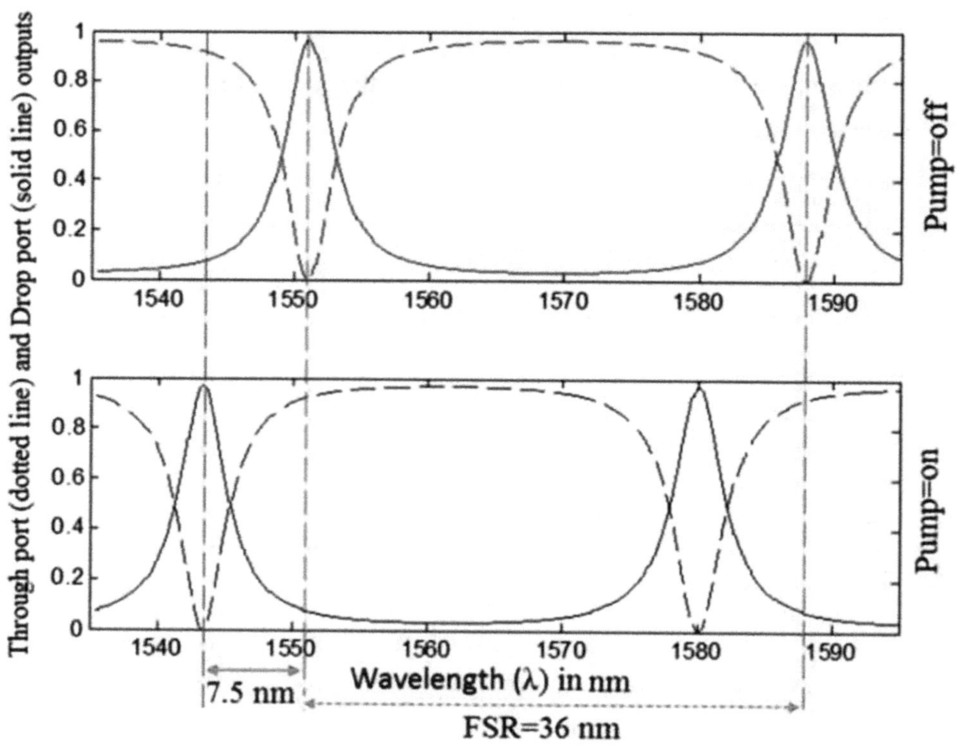

The race-track structure of MRR is preferred for mode-conversion to occur in MRR which support both TE (Transverse Electric) and TM (Transverse Magnetic) modes of light. While propagating through curved waveguides with small radius of the bend, the TE and TM modes are no longer orthogonal due to their hybridness and experience a polarization-conversion in the ring. The equation of coupler from differential coupled mode theory (CMT) is written below (Melloni et al, 2002; Melloni et al, 2004).

$$\frac{da(z)}{dz} = -i\beta_1 a(z) - iK_1 b(z)$$

$$\frac{db(z)}{dz} = -i\beta_2 a(z) - iK_2 b(z)$$

where a(z) and b(z) are the complex amplitude of the fields in the waveguides, β_1, β_2 are their propagation constants and K_1 and K_2 are the field coupling constant. By the linear coupled mode theory (LCMT), the equation is the combination of even and odd modes of the coupled waveguides. The transmission matrix ($\mathrm{T.M_c}$) of the coupler of the micro-ring resonator in the coupled mode theory (CMT) form, can be written as (Cusmai eta l, 2005).

$$\mathrm{T.M_c} = \mathrm{e}^{-i\beta \mathrm{L_c}} \begin{bmatrix} \cos(\mathrm{KL_c}) & -\mathrm{i}\sin(\mathrm{KL_c}) \\ -\mathrm{i}\sin(\mathrm{KL_c}) & \cos(\mathrm{KL_c}) \end{bmatrix}$$

$$= \mathrm{e}^{-i\beta \mathrm{L_c}} \begin{bmatrix} \mathrm{r} & -\mathrm{it} \\ -\mathrm{it} & \mathrm{r} \end{bmatrix} \tag{1}$$

where $\mathrm{e}^{-i\beta \mathrm{L_c}} = \mathrm{z}^{-1}$ and $\mathrm{L_c}$ is the coupling length of the MRR. t and r are the field coupling coefficient and can be represented as $\mathrm{t} = \sin(\mathrm{KL_c})$ and $\mathrm{r} = \cos(\mathrm{KL_c})$ respectively. Equation (1) can be rewritten in the form of the rotation matrix ($\mathrm{R.M_c}$) using Jones matrix and is given below.

$$\mathrm{R.M_c} = \begin{bmatrix} \cos\phi & -\mathrm{i}\sin\phi \\ -\mathrm{i}\sin\phi & \cos\phi \end{bmatrix} \tag{2}$$

The transmission matrix ($T.M_R$) of the whole ring-cavity waveguide in the coupled-mode complex amplitude form is written as-

$$T.M_R = e^{-i\beta L_R}\begin{bmatrix}\cos\phi - i\ R\sin\phi & -i\ S\sin\phi \\ -i\ S\sin\phi & \cos\phi + i\ R\sin\phi\end{bmatrix} \tag{3}$$

where, $e^{-i\beta L_R} = z^{-1}$, $L_R = \frac{\pi}{2\delta}$ is the length of the MRR, $\beta = (\beta_{TM} + \beta_{TE})/2$ is the average phase constant of the two modes, $R = \frac{\Delta\beta}{2\delta}$, $S^2 = 1\text{-}R^2$, $\Delta\beta = \beta_{TE} - \beta_{TM}$, $\phi = \delta L_R$, $\delta = (\frac{\Delta\beta^2}{4} + K_p^2)^{1/2}$, $K_p = S^2\sin^2\phi$ is the field coupling coefficient between the two modes. By using CMT approach, the transmission matrix of a birefringent bends waveguide ($T.M_B$) is given below.

$$T.M_B = e^{-i\beta L_B}\begin{bmatrix}\cos\phi - i\ R\sin\phi & -i\ S\sin\phi \\ -i\ S\sin\phi & \cos\phi + i\ R\sin\phi\end{bmatrix} \tag{4}$$

where $e^{-i\beta L_B} = z^{-1}$, L_B is the length of the bend waveguide, $\phi = \delta L_B$. The TE-TE (H_{EE}) or TM-TM (H_{MM}) MRR transfer function is defined below.

$$H_{EE\ or\ MM} = \frac{rz^{-2} - \left(1 - r^2\right)z^{-1}\cos\phi \pm i\ Rt^2z^{-1}\sin\phi + r}{1 + r^2z^{-2} - 2rz^{-1}\cos\phi} \tag{5}$$

Similarly, the transfer function for TE to TM or TM to TE can be written as (Cusmai et al, 2005)

$$H_{em\ or\ me} = \frac{iSt^2z^{-1}\sin\phi}{1 + r^2z^{-2} - 2rz^{-1}\cos\phi} \tag{6}$$

Above equation (5) and (6) obtained using coupled mode theory can be used to design micro-ring resonator based mode (polarization)-conversion. The Eq. (5) and Eq. (6) obtained from the CMT approach for the micro-ring resonator is also applicable in the realization of mode-conversion based all-optical switch and all-optical logic gates.

DESIGN OF MICRO-RING RESONATOR FOR POLARIZATION CONVERSION

An ultra-compact, race-track shaped MRR of Si/SiO_2 with bending radius of 5μm and coupling length of 3μm is considered to obtain mode-conversion and all-optical logic circuits. The coupling efficiency of the MRR can be enhanced by race-track shaped MRR. The high refractive-index contrast in race-track MRR reduces coupling losses in the coupling region. Figure 4(a) shows the proposed design of ring resonator for mode-conversion.

The substrate of SiO_2 (glass) ridge waveguide and the Si based ring waveguide is used to design MRR based optical switch, as shown in Figure 4 and the refractive indices of Si and SiO_2 are considered to be 3.455 and 1.445 respectively (Bharti et al, 2018) The coupling length, radius and the other parameters are optimized through finite-difference-time-domain (FDTD) solutions in such a way that it supports both quasi-TE and quasi-TM mode of light. The height-to-width ratio is also chosen in such a way that it supports both quasi-TE and quasi-TM mode for mode-conversion, unlike in case of (Dai et al, 2015). The source and pump wavelength as well as pump power is selected suitability for mode-conversion (Cusmai et al, 2005). The advantage of the proposed design is that overall losses of Si/SiO_2 based MRR have relatively low. Scattering loss is also very low, below 1dB. Absorption loss, bending loss and insertion loss are also in low levels at the wavelength nearly to 1.55 μm. All the parameters listed in Table 1 are optimized through FDTD simulation such that polarization conversion took place in a more feasible manner.

Figure 4(b) resembles the two waveguide coupled race-track MRR which is employed to obtain polarization-conversion based all-optical switching and different logic circuits in the MRR. To obtain all-switching and different logic circuits, the source and pump signal both are injected from the input port of the MRR. The

Figure 4. (a) Schematic view of MRR; (b) Proposed labelled diagram of MRR

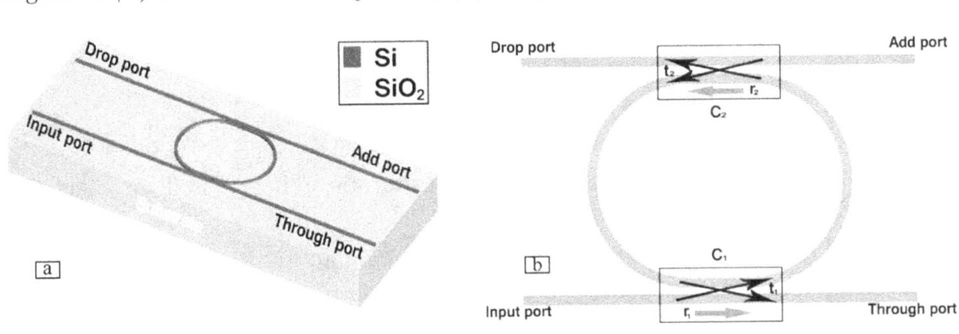

Table 1. Parameters used for mode-conversion in MRR

S. No.	Parameters	Description
1.	MRR waveguide material	Si
2.	Substrate material	SiO_2 (Glass)
3.	Gap between the ring and straight waveguide	0.1μm
4.	Coupling length (L_c)	3μm
5.	Bending radius of MRR	5μm
6.	TPA coefficient, β_2	$7.9\text{x}10^{-10}$ cm/W
7.	Base Angle	90 degree
8.	cross-section area (A_{eff})	450 x 415 nm^2
9.	Coupling Nature	Lateral coupling
10.	Overall losses (α) of the ring	<2dB
11.	Height of the waveguide (h_g)	0.22μm

outputs are obtained from drop port and through port respectively. The proposed MRR based circuit can perform NOT, OR/NOR, AND and NAND logic operations in both output ports simultaneously. The polarization-conversion occurs in the output ports of the MRR due to the application of a suitable nature of the source and pump signal and its input power respectively. Refractive index changes due to pump power and a π-phase shift occurs in one complete circle of the ring, so the all-optical polarization switching of light in through port and drop port occurs.

The design showed in Figure 4 is quite small and the low optical pump power is required for polarization (mode)-conversion. The two-photon absorption (TPA) effect plays a vital role in mode-conversion based switching and logic circuit designing and requires comparatively less pump power, small size and have ultra- fast switching speed (Cusmai et al, 2005). Furthermore, the switching power of TPA based MRR switch is three times smaller than that of Kerr based MRR switch (Chun-fei et al, 2009) is given below (Rakshit et al., 2016)

$$\Delta n = -\left[8.8 \times 10^{-22} \Delta N_e + 8.5 \times 10^{-22} (\Delta N_h)^{0.8}\right]$$

where ΔN_h and ΔN_e resembles the hole-concentration change and the electron-concentration change respectively.

The quasi polarized mode serves as an intermediate mode to exchange power between side-coupled waveguide to ring waveguide of the MRR. The height-to-width ratio of the waveguide is optimized to support both quasi-TE and quasi-TM mode. The relationship is established between the polarization state of inputs and the corresponding output signals to obtain polarization switch and logic gates in a MRR by utilizing polarization-conversion phenomena.

WORKING OF POLARIZATION CONVERSION USING MICRO-RING RESONATOR

For light transmission of the quasi-TE and quasi-TM mode for Si/SiO_2 based MRR, Figure 5(a) shows the transmission graph of TE source in the wavelength range of 1.52μm to 1.555μm. From Figure 5 (a), the non-resonant and resonant wavelengths are listed as 1.522μm (P_1), 1.552μm (P_2) and 1.53μm (I_X where x= quasi-TE or quasi-TM mode of light) are used as pump (P_1, P_2) and source (I_X) wavelengths respectively. In the proposed design, the all-optical switching and the logic gates are obtained at the resonant wavelength of 1.53μm by utilizing the above mentioned non-resonant wavelengths as an pump signals and resonant wavelength as input signal respectively.

The nature of output pulse is determined by the output electric field intensity. The high level of output field represents quasi-TE mode of polarized light while the low level of field represents quasi-TM mode of polarized light. The magnitude less than 2.5 in the electric field graph is regarded as quasi-TM while magnitude more than 2.5 resembles quasi-TE mode in the output. The quasi-TE mode regarded as "logic 1" and the quasi-TM mode is said to be "logic 0". Figure. 4(b) shows the working diagram of the proposed scheme and Eq. (1) to Eq. (4) are used to model polarization-conversion using MRR.

Figure 5. (a). Transfer function graph of proposed ring resonator; (b). Schematic for realization of polarization switch; I_{TE}= quasi-TE input, P_2=Pump signal

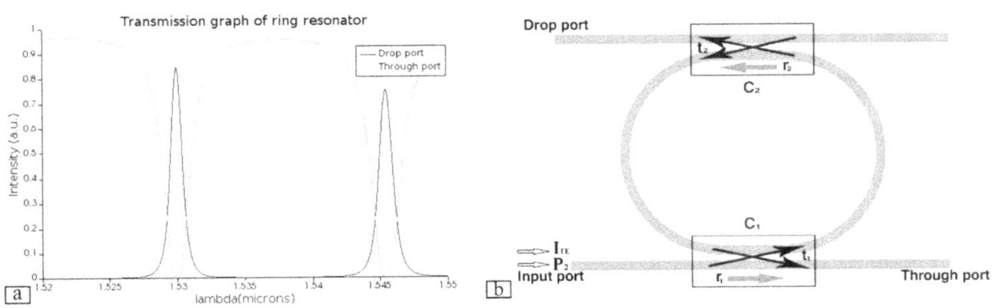

DESIGN OF A POLARIZATION SWITCH

To design polarization switch, a fixed source (I_{TE}) of quasi-TE polarization state of light is considered and the nature of different pump signal (P_2) is applied from the input port of the MRR, as highlighted in Figure 5(b). The nature of the source and pump pulses should be "mode pulse" because of its large optical spectrum. The different outputs are obtained from through port as well as from drop port, depending on the nature of pump (P_2) signal. The all-optical switching is observed if the fixed input signal (I_{TE}) of quasi-TE launched at the wavelength of 1.53μm and pump (P_2) at a wavelength of 1.552μm are applied respectively. At a wavelength of 1.53μm, the polarization-conversion based all-optical switching is realized at through port and drop port output. If the state of polarization of both source and pump are quasi-TE, the quasi-TE is obtained at output (through) port and quasi-TM is obtained at drop port as depicted in Figure 6(a). Similarly, for the same input, quasi-TM is obtained at through port and quasi-TE is obtained at drop port, if the pump signal is changed to quasi-TM at the wavelength of 1.552μm, as shown in Figure 6(b). This behavior resembles polarization-conversion based on all-optical switching in an MRR, due to pump power. The simulated results are tabulated in Table 2. The total power of 50nW/m is calculated for all-optical switching in the MRR through finite-difference-time-domain (FDTD) simulation.

Table 2. All optical polarization conversion switching in a MMR

Source nature (Fixed nature of source)	**Pump (P_2) nature (Varying nature)**	**Through port (Electric field intensity)**	**Drop port (Electric field intensity)**
quasi-TE	quasi-TE	quasi-TE	quasi-TM
quasi-TE	quasi-TM	quasi-TM	quasi-TE

Figure 6. Drop port (quasi-TM) and Through port (quasi-TE) output; (a). pump is quasi-TE, (b). pump is quasi-TM

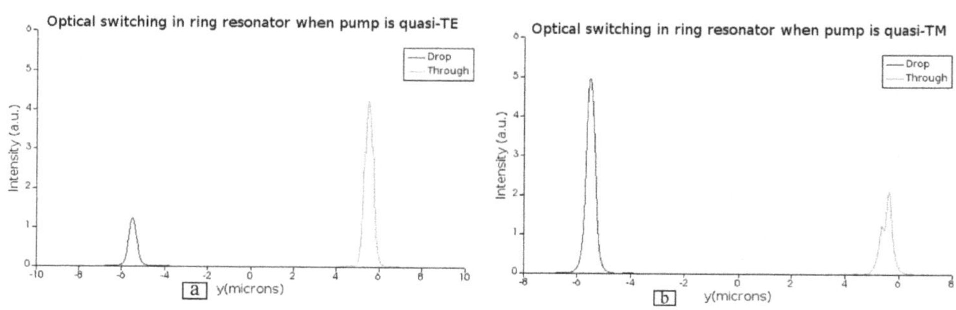

When the pump (P_2) of quasi-TE is applied, high electric field intensity (or quasi-TE nature) is obtained at the through port while drop port shows low intensity (or quasi-TM nature). When the nature of pump (P_2) is changed to quasi-TM nature, the intensity of through port output is low (or quasi-TM nature), while the drop port intensity is high (or quasi-TE nature). This behavior shows the all-optical switching of MRR. Thus, the polarization-conversion based all-optical switching occurs at the output ports of the MRR at the same instant, due to the application different pump nature.

The time-domain graph is also obtained from simulation to verify its switching behavior. Figure 7 illustrates the behavior of polarization-conversion based all-optical switch, which shows the relation of field intensity with respect to time. The high intensity shows the quasi-TE nature while low intensity shows the quasi-TM nature of the graph at a particular instant of time. When pump is quasi-TE, the output of through port is as high as nearly 3 a.u. at time 275.8 fs and drop port output is low around 0.5 a.u. at 470.8 fs, but the field-magnitude is changed if the pump (P_2) nature is altered from quasi-TE to quasi-TM. The low magnitude of about 0.026 a.u. is noted for through port at the same time of 275.8 fs and high magnitude of drop port around 2.92 a.u. is noted at 470.8 fs, when pump is quasi-TM.

DESIGN OF NOT GATE

The logic gate operations are performed by establishing a relationship between polarization nature of both input and pump source, to the polarization (mode) nature

Figure 7. Switching behavior of MRR; (a) when pump is quasi-TE; (b); when pump is quasi-TM

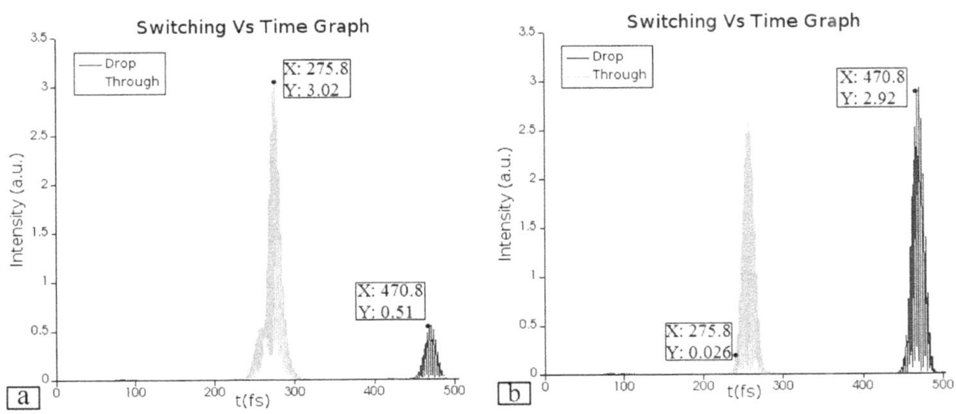

of through port output. The output pulse, O_x (where x= quasi-TE nature or quasi-TM nature of polarization) are obtained from the through port, as depicted in Figure 8.

To obtain the NOT logic gate, quasi-TE mode of fixed source (I_X) is injected at the resonant wavelength of 1.53μm. When the input pump source (P_1) of quasi-TE nature (logic "1") is applied at the non-resonant wavelength of 1.522μm, the output of quasi-TM nature (logic"0") is recognized. Figure 9(a) shows the output of through port obtained at 1.53μm. But, if the pump nature is changed (P_1) from quasi-TE to quasi-TM mode, then the quasi-TE (logic "1") mode is achieved at the output port. Figure 9(b) obtained from FDTD simulation clearly shows the nature of quasi-TE at 1.53μm. The power of 50nW/m with a single pump (P_1) is required for the realization of all-optical NOT logic gate using FDTD simulation.

The field magnitude at output port as shown in Figure 9(a) is about 1.9 which clearly shows quasi-TM nature of polarized (mode) state. While the value in Figure 9(b) is around 3.2 which shows quasi-TE mode of the polarized state. Table 3 shows the truth table of all-optical NOT gate based on the polarized states of the source and the pump input.

Figure 8. Diagram of proposed MRR for obtaining all-optical NOT, AND and NAND gate; I_X=Input source; O_x=output; P_1,P_2=Pump signal

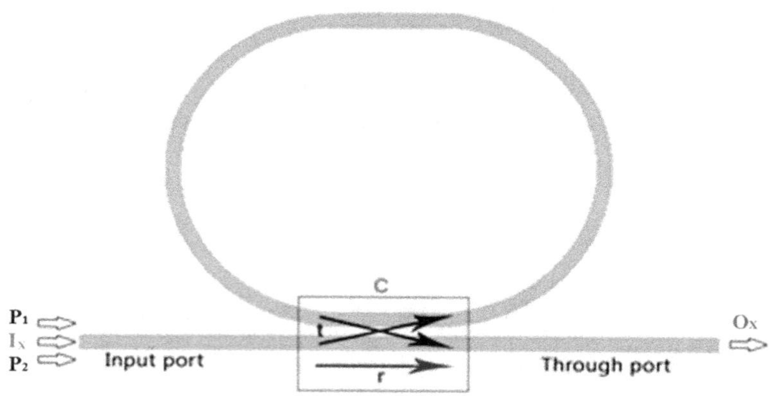

Table 3. Logic operation for the all-optical not gate

Source (Fixed state)	Pump P_1(Varying state)	Output (field intensity)
(quasi-TE)	1 (quasi-TE)	0 (quasi-TM)
(quasi-TE)	0 (quasi-TM)	1 (quasi-TE)

Figure 9. Simulation output of NOT gate; (a). "Logic 0" obtained if input is "logic 1";(b). "Logic 1" obtained if input is "logic 0"

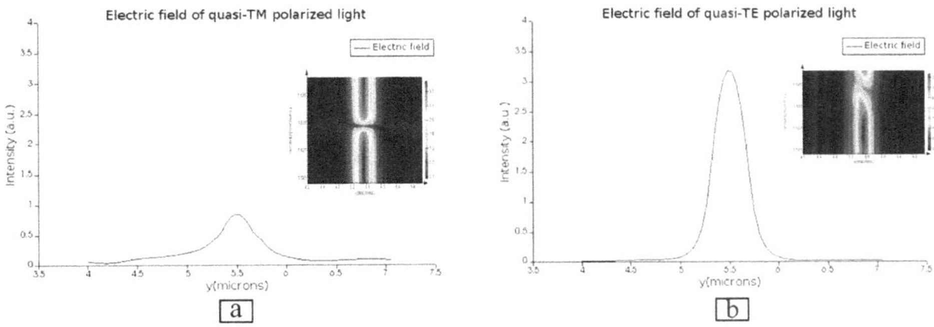

DESIGN OF OR/NOR LOGIC GATES

And for designing all-optical logic OR/NOR gates, two pump signals (P_1 and P_2) and a fixed source is injected from input port simultaneously, as illustrated in Figure 10. The nature of pump signal is varied in such a way that all-optical OR gate is obtained from through port and all-optical NOR gate is obtained from drop port of the MRR at the same operating wavelength.

The OR/NOR gate is obtained if the fixed input signal (I_{TE}) of quasi-TE launched at the wavelength of 1.53μm and pump signals (P_1, P_2) of 1.522μm and 1.552μm wavelength are applied respectively from the input port, as shown in Figure 10. At a wavelength of 1.53μm, the polarization-conversion dependent all-optical OR/NOR

Figure 10. Working diagram of MRR for obtaining OR/NOR gate

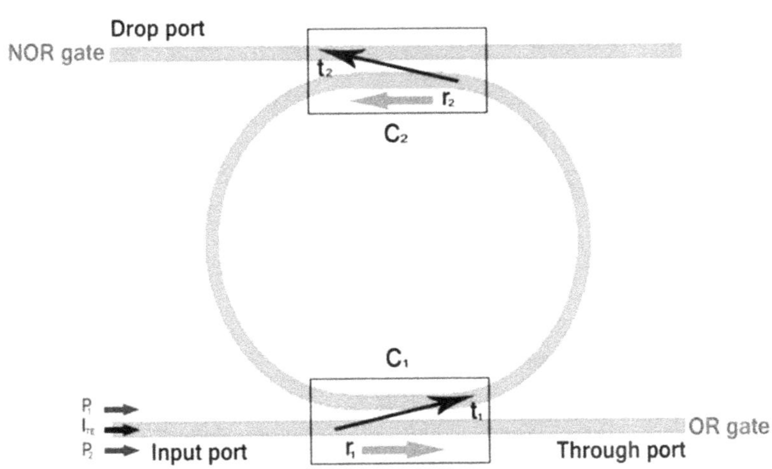

logic gates are realized at through port and drop port output respectively. If the state of polarization of both pumps is quasi-TE (logic "1"), the output of quasi-TE (logic "1") is obtained at through port and quasi-TM (logic "0") is obtained at drop port as depicted in Figure 11(a) and Figure 12(a) respectively. And if, either P_1 and P_2 of quasi-TE (logic "1") or quasi-TM (logic "0") nature is applied, the output of quasi-TE (logic "1") is obtained at through port as shown in Figure 11(b) and Figure 11(c) respectively for both the cases. Similarly, at the same input, quasi-TM (logic "0") is obtained at drop port, as shown in Figure 12(b) and Figure12(c) respectively for both the cases. But, if the polarized state of both the pump P_1 and P_2 as quasi-TM (logic "0") is selected, the output of quasi-TM (logic "0") and quasi-TE (logic "1") is obtained at through port and drop port respectively, as shown in Figure 11(d)

Figure 11. Outputs field graph obtained at through port (resembles OR logic gate); (a). "Logic 1" obtained if both inputs are "logic 1"; (b). "Logic 1" obtained if inputs are "logic 1" and "logic 0"; (c). "Logic 1" obtained if inputs are "logic 0" and "logic 1"; (d). "Logic 0" obtained at output if both inputs are "logic 0"

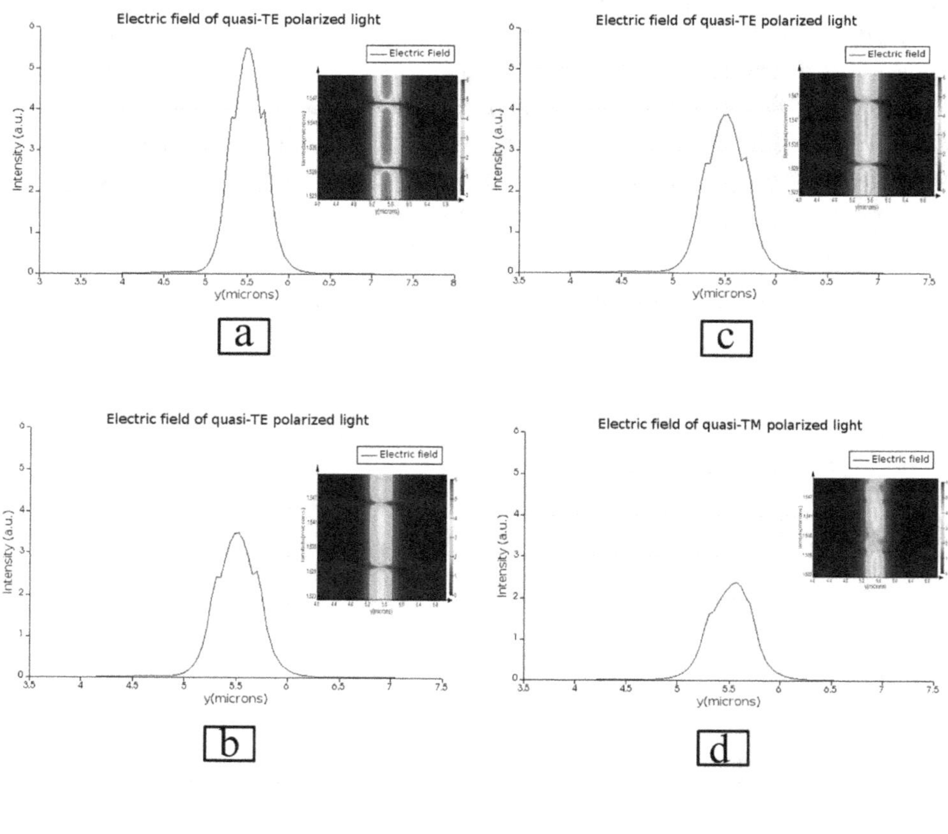

Figure 12. Outputs field graph obtained at drop port (resembles NOR logic gate); (a). "Logic 0" obtained at output" if both inputs are "logic 1; (b). "Logic 0" obtained at output if inputs are "logic 1" and "logic 0"; (c). "Logic 0" obtained at output if inputs are "logic 0" and "logic 1"; (d). "Logic 1" obtained if both inputs are "logic 0".

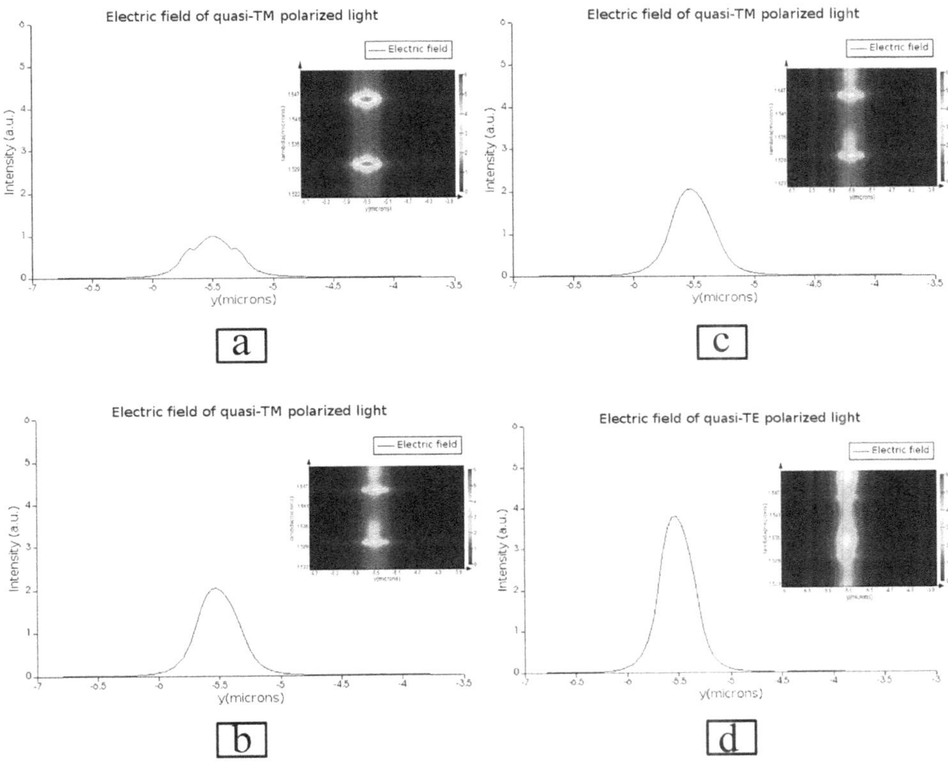

and Figure 12(d). The state of output resemble the all-optical OR/NOR logic gates and summarized in Table 4. The total power of 60 nW/m is required for all-optical OR/NOR logic gates.

DESIGN OF 'AND' AND 'NAND' GATE

'AND' Gate

The AND gate is obtained if the fixed input signal (I_X) of quasi-TE and pump signal (P_1, P_2) are applied at the same wavelength as applied for OR/NOR logic gate using

Table 4. All-optical polarization-conversion based or/nor logic gates

Source nature	Pump (P_1) nature	Pump (P_2) nature	Through port output (OR gate)	Drop port output (NOR gate)
quasi-TE	quasi-TE	quasi-TE	1 (quasi-TE)	0(quasi-TM)
quasi-TE	quasi-TE	quasi-TM	1 (quasi-TE)	0(quasi-TM)
quasi-TE	quasi-TM	quasi-TE	1 (quasi-TE)	0(quasi-TM)
quasi-TE	quasi-TM	quasi-TM	0(quasi-TM)	1(quasi-TE)

the structure of Figure 8. At a same wavelength of 1.53µm, the polarization conversion occurs. If the state of polarization of both pump is quasi-TE (logic"1") the output of quasi-TE (logic "1") is obtained, as depicted in Figure 13(a). And if, either P_1 and P_2 of quasi-TE (logic "1") or quasi-TM (logic "0") states are applied, the output (O_x) state of the signal is quasi-TM (logic "0") appears, as shown in Figure 13(b) and Figure 13(c) respectively. Also, if the polarized state of both the pump P_1 and P_2 as quasi-TM (logic "0") is selected, the output O_x obtained as quasi-TM (logic "0") at wavelength of 1.53µm, as shown in Figure 13(d). The state of output resembles the all-optical AND logic gate and summarized in Table 5. The total power 26nW/m is required for all-optical AND gate. In case of AND gate as well as NAND gate, all the outputs are obtained at the through port of MRR.

'NAND' Gate

To design NAND gate in the same configuration of Figure 8, an input source I_X, (where X= quasi-TE/quasi-TM mode of light) and pump (P_1) are applied at the wavelength of 1.53µm and 1.522µm respectively and the mode conversion occurs at the same resonant wavelength of 1.53µm. Quasi-TM mode (logic "0") obtained at output port (O_x) if both input and pump sources of quasi-TE mode (logic "1") is applied, as shown in Figure 14(a). But if I_X of quasi-TE nature (logic "1") and P_1 of

Table 5. All-optical polarization conversion based on logic gates in a MRR

Source (I_X)	Pump pulse (P_1)	Pump pulse (P_2)	AND gate output
1 (quasi-TE)	1 (quasi-TE)	1 (quasi-TE)	1 (quasi-TE)
1 (quasi-TE)	1 (quasi-TE)	0 (quasi-TM)	0 (quasi-TM)
1 (quasi-TE)	0 (quasi-TM)	1 (quasi-TE)	0 (quasi-TM)
1 (quasi-TE)	0 (quasi-TM)	0 (quasi-TM)	0 (quasi-TM)

Figure 13. Outputs field graph obtained at drop port (resembles AND logic gate); (a). "Logic 1" obtained if both inputs are "logic 1"; (b). "Logic 0" obtained if inputs are "logic 1" and "logic 0"; (c). "Logic 0" obtained if inputs are "logic 0" and "logic 1"; (d). "Logic 0" obtained if both inputs are "logic 0"

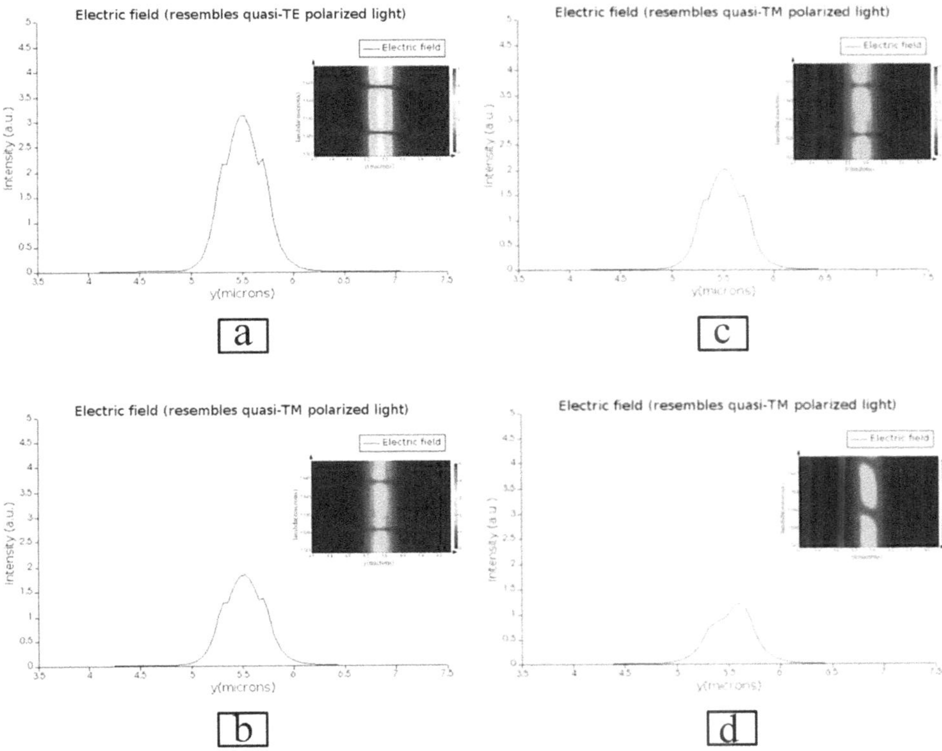

quasi-TM nature (logic "0") are applied to the input port, then the quasi-TE (logic "1") nature of the signal appears at the through port, as shown in Figure 14(b). Similarly if I_X of quasi-TM mode (logic "0") and P_1 of quasi-TE mode (logic "1") are applied, then also the quasi-TE (logic "1") nature of the signal will appear at the output port, as shown in Figure 14(c). However, if both input and pump sources are of quasi-TM nature (logic "0") are applied, the O_x of quasi-TE (logic "1") nature is obtained, as shown in Figure 14(d). The nature of input and output light resembles the all-optical NAND gate. Table 6 shows the summarized operation of the all-optical NAND gate. The total power of 90 nW/m is required for NAND gate operation using FDTD simulation.

Table 6. Logic operation of the all-optical NAND gate

Source nature (I_x)	Pump nature (P_1)	Output (Electric field) NAND gate
1 (quasi-TE)	1 (quasi-TE)	0 (quasi-TM)
1 (quasi-TE)	0 (quasi-TM)	1 (quasi-TE)
0 (quasi-TM)	1 (quasi-TE)	1 (quasi-TE)
0 (quasi-TM)	0 (quasi-TM)	1 (quasi-TE)

Figure 14. Outputs field graph obtained at drop port (resembles NAND logic gate);(a). "Logic 0"obtained if both inputs are "logic1"; (b). "Logic 1" obtained if inputs are "logic 1"and "logic 0"; (c). "Logic 1" obtained if inputs are "logic 0"and "logic 1"; (d). "Logic 1" obtained if both inputs are "logic 0".

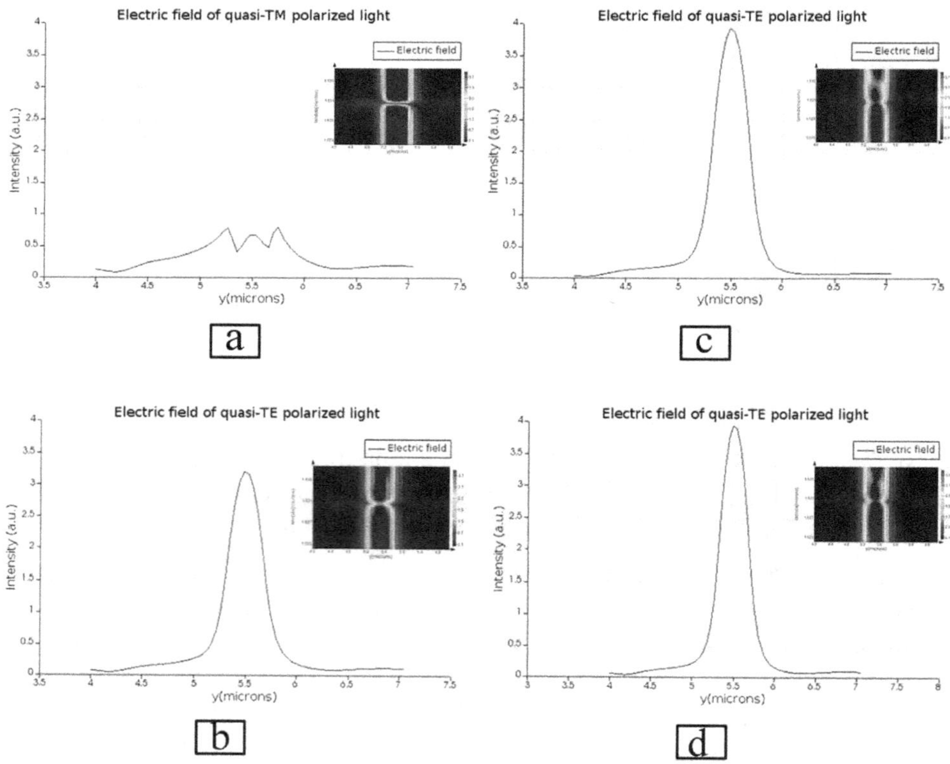

DISCUSSIONS

The race-track configuration of MRR is more suitable for mode-conversion as compared to other architecture of MRR. The optimized coupling is achieved using race-track structure of MRR and it favors more for mode-conversion phenomena (Bharti eta l, 2018). Therefore, optimized coupling length, bending radius and other parameters of the model are selected through FDTD simulation. If the bending radius of MRR is less than 1 μm then it allows only quasi-TM mode significantly. There is less losses for radius more than 4μm due to modal-mismatch in the race-track MRR. Therefore, in the proposed model, the radius of the ring is considered as 5 μm for mode-conversion (Bharti et al, 2018). And coupling length and gap between straight and bend waveguide are optimized through FDTD simulation and optimized values are 3μm and 0.1μm respectively. At optimum operating point, the maximum of 92% of input light coupled to the ring. Rectangular waveguide is selected due to the fact that it confines light in two-dimensions and suitable for silicon-on-insulator (SOI) platform. The selection for the size of the height of the waveguide for the propagation of both TE and TM mode is critical. Therefore, in the proposed device quasi-TE and quasi-TM modes are opted for propagation. The height and width of the waveguide is selected as 0.22μm and 0.4 μm respectively to support both the mode. Also, the high-index-contrast of the waveguides is selected to support small bending radius of the MRR.

In the proposed design electron beam lithography technology is preferred over complementary metal-oxide semiconductor (CMOS) fabrication technique on SOI substrate (Mookherjea et al, 2008)

Some performance parameters of the proposed model using MRR are determined and described below.

1. **Operational Speed:** The output polarization status is switched between quasi-T and quasi-TM modes by changing pump status. The operational speed can be calculated using the following formula.

$$\mathrm{T} = \frac{\mathrm{Y}^2}{1 + \mathrm{X}^2 - 2\mathrm{X}\cos\phi} \tag{7}$$

where

$$\mathrm{X} = \cos^2(\mathrm{K})\exp(-\alpha\pi\mathrm{R}),\ \mathrm{Y} = \sin^2(\mathrm{K})\exp\left(-\frac{\alpha\pi\mathrm{R}}{2}\right), \phi = \frac{4\pi^2\mathrm{R}\ \mathrm{n}_{\mathrm{eff}}}{\lambda},$$

n_{eff} is effective refractive index of the MRR. The required time of 0.2 ps to emerge output is obtained from the FDTD simulation. The switching time for TPA also depends on the carrier recombination time, and the switching time of MRR is limited by the photon lifetime on the ring cavity.

2. **Extinction Ratio (ER):** ER explains the switching performance between two output polarization states. ER can be expressed as.

$$ER\left(dB\right) = 10\log\left(\frac{P^1_{min}}{P^0_{max}}\right) \tag{8}$$

where P^0_{max} and P^1_{min} are the maximum and minimum values of the peak intensity low (0) level (quasi-TM mode) to the high (1) level (quasi-TE mode). The ER of 13.70dB is calculated for polarization-conversion based switch.

3. **Buildup Factor (B):** The improvement in optical power is determined by buildup factor. The buildup factor is evaluated as stated below (Bhrati et al, 2018)

$$B = \left|\frac{-r\exp\left(-\frac{\alpha}{2}L\right)}{1-\sqrt{1-t^2}\exp\left(-\frac{\alpha}{2}\mathrm{L}\right)}\right|^2 \tag{9}$$

It indicates the optical power magnification inside the ring-cavity with respect to the bus-waveguide which is around 26.40 in the polarization based all-optical switch using proposed MRR.

4. **Quality Factor (Q-Factor):** Q-factor is used to evaluate the spectral range of the resonance condition for a micro-ring resonator (Bhrati et al, 2018). Thus, Q-factor is defined below.

$$\mathrm{Q} = \frac{\text{Stored energy}}{\text{Power lost}}$$

The Q-value is obtained as high as 1400 from the simulation for the all-optical polarization switch and for NOT gate it is 1500 while 2400 is noted for OR/NOR logic gates. The optimum value of Q-factor is also noted as 2400 for AND gate and 1500 for NAND gate respectively.

CONCLUSION

The design and modeling of all-optical polarization-switch as well as all-optical logic-gates based on polarization-conversion using single silicon MRR, which has two different polarization (or mode) outputs based on the pump signal nature, is demonstrated and realized. The simulation of the proposed design of MRR for all-optical switch and logic gates are designed using finite-difference time-domain (FDTD) method. It simulates and optimizes the performance of photonic integrated circuits. The output status are altered by applying different "nature" of the pump signal to the MRR which changes its refractive index and hence resonance wavelength of the MRR and thus optical switching takes place. The coupled-mode theory (CMT) is employed to model polarization-conversion based MRR. The ring radius, gap between straight waveguide and ring as well as the refractive index contrast between the materials are the main parameters used for the polarization-conversion phenomena. The two pump signals are required to achieve OR/NOR logic gate, but only a single pump pulse is required for obtaining NOT logic gate and for obtaining polarization-conversion based switch. And for realizing NAND gate one pump signal varying input source is required. The design is applicable for both the mode states viz. quasi-TE and quasi-TM.

The mode state of the input source as well as the pump light is adjusted to achieve all-optical polarization switch and all-optical NOT, OR/NOR, AND-NAND logic gates. The input source is applied at resonant wavelength. The total power calculated in simulation which are required for all-optical polarization conversion based switch, NOT, OR/NOR, AND-NAND logic operations are 50 nW/m, 50 nW/m, 60 nW/m and 26nW/m-90nW/m respectively. The MRR also shows outstanding performance as its Q (quality)-factor is very high. The performance of the design is evaluated by calculating various 'figure of merits' and 'spectral parameters'. The proposed device has a good transmission as operating speed of 0.2ps is noted here for both switchings as well as for logic gate operations.

The mode-conversion based logic gates in a micro-ring resonator represent the most promising solution for the present day technology due to their less complex circuitry, ultra-compact size as single micro-ring resonator is used. The study may be used for ultra-compact and ultra-fast optical-signal processing, design of different logic-gates, adder/subtractor, multiplexer and demultiplexer circuit based on mode-conversion. The results can be utilized to design new logical devices like polarization dependent switch, logic gates etc. The design is robust, simple, stable, easy to fabricate and silicon-on-insulator (SOI) compatible. The structure is compatible for interconnects and capable for integrating in electronics as well as in plasmonics circuits. The proposed device has a potential to be used in other polarization conversion based arithmetic and logic circuits. The design is ultra-compact and capable to interconnect in other optical devices. It may also be used in optical communication networks, optical signal processing and optical time-division multiplexing (OTDM).

REFERENCES

Bharti, G. K., Rakshit, J. K., & Pal, S. S. (2018). Design and modeling of mode-conversion in ring resonator and its application in all-optical switching. *Microsystem Technologies*, 1–12.

Bianucci, P., Fietz, C. R., Robertson, J. W., Shvets, G., & Shih, C. K. (2007). Polarization conversion in a silica microsphere. *Optics Express*, *15*(11), 7000–7005. doi:10.1364/OE.15.007000 PMID:19547016

Chun-fei, L. I., & Dou, N. A. (2009). Optical switching in silicon nanowaveguide ring resonators based on Kerr effect and TPA effect. *Chinese Physics Letters*, *26*(5). doi:10.1088/0256-307X/26/5/054203

Cusmai, G., Morichetti, F., Rosotti, P., Costa, R., & Melloni, A. (2005). Circuit-oriented modeling of ring resonator. *Optical and Quantum Electronics*, *37*(1-3), 343–358. doi:10.100711082-005-1191-7

Dai, J., Zhang, M., Zhou, F., Wang, Y., Lu, L., & Liu, D. (2015). All-optical logic operation of polarized light signals in highly nonlinear silicon hybridplasmonicmicroring resonators. *Applied Optics*, *54*(14), 4471–4477. doi:10.1364/AO.54.004471 PMID:25967504

Fietz, C., & Shvets, G. (2007). Non-linear polarization conversion using microring resonators. *Optics Letters*, *32*(12), 1683–1685. doi:10.1364/OL.32.001683 PMID:17572746

Jinno, M., & Matsumoto, T. (1992). Nonlinear Sagnac interferometer switch and its applications. *IEEE Journal of Quantum Electronics*, *28*(4), 875–882. doi:10.1109/3.135205

Khorasaninejad, M., & Saini, S. S. (2009). All-optical logic gates using nonlinear effects in silicon-on-insulator waveguides. *Applied Optics*, *48*(25), F31–F36. doi:10.1364/AO.48.000F31 PMID:19724310

Kim, S. H., Takei, R., Shoji, Y., & Mizumoto, T. (2009). Single-trench waveguide TE-TM mode converter. *Optics Express*, *17*(14), 11267–11273. doi:10.1364/OE.17.011267 PMID:19582039

Lamouroux, B., Prade, B., & Orszag, A. (1982). Polarization effect in optical fiber ring resonators. *Optics Letters*, *7*(8), 391–393. doi:10.1364/OL.7.000391 PMID:19714032

Li, J., Li, L., Jin, L., & Li, C. (2006). All-optical switch and limiter based on nonlinear polarization in Mach-Zehnder interferometer coupled with a polarization-maintaining fiber ring-resonator. *Optics Communications*, *260*(1), 318–323. doi:10.1016/j.optcom.2005.10.054

Liang, T. K., Nunes, L. R., Tsuchiya, M., Abedin, K. S., Miyazaki, T., Van Thourhout, D., ... Tsang, H. K. (2006). High speed logic gate using two-photon absorption in silicon waveguides. *Optics Communications*, *265*(1), 171–174. doi:10.1016/j.optcom.2006.03.031

Little, B. E., & Chu, S. T. (2000). Theory of polarization rotation and conversion in vertically coupled microresonators. *IEEE Photonics Technology Letters*, *12*(4), 401–403. doi:10.1109/68.839032

Lui, W. W., Hirano, T., Yokoyama, K., & Huang, W. P. (1998). Polarization rotation in semiconductor in bending waveguides: A coupled mode theory formulation. *Journal of Lightwave Technology*, *16*(5), 929–936. doi:10.1109/50.669050

Melloni, A., & Martinelli, M. (2002). Synthesis of direct coupled resonators bandpass filters for WDM systems. *Journal of Lightwave Technology*, *20*(2), 296–303. doi:10.1109/50.983244

Melloni, A., Morichetti, F., & Martinelli, M. (2004). Polarization conversion in ring resonator phase shifters. *Optics Letters*, *29*(23), 2785–2787. doi:10.1364/OL.29.002785 PMID:15605505

Mookherjea, S., & Melloni, A. (2008). *Microring resonators in integrated optics*. Academic Press.

Nayar, B. K., Finlayson, N., Doran, N. J., Davey, S. T., Williams, D. L., & Arkwright, J. W. (1991). All-optical switching in a 200-m twin-core fiber nonlinear Mach–Zehnder interferometer. *Optics Letters*, *16*(6), 408–410. doi:10.1364/OL.16.000408 PMID:19773949

Rakshit, J. K., Chattopadhyay, T., & Roy, J. N. (2013). Design of ring resonator based all optical switch for logic and arithmetic operations–a theoretical study. *Optik-International Journal for Light and Electron Optics.*, *124*(23), 6048–6057. doi:10.1016/j.ijleo.2013.04.075

Rakshit, J. K., & Roy, J. N. (2014). Micro ring resonator based all optical reconfigurable logic operations. *Optics Communications*, *321*, 38–46. doi:10.1016/j.optcom.2014.01.048

Rakshit, J. K., & Roy, J. N. (2016). All-optical ultrafast switching in a silicon microring resonator and its application to design multiplex/demultiplexer, adder/subtractor and comparator circuit. *Optica Applicata*, *46*, 517–539.

Soref, R. (2006). The past, present, and future of silicon photonics. *IEEE Journal of Selected Topics in Quantum Electronics*, *12*(6), 1678–1687. doi:10.1109/JSTQE.2006.883151

Suzuki, K., Hirayama, T., & Arakawa, T. (2016). *Proposal of compact TE/TM polarization switch based on microring resonator*. OECC/PC2016.

Zoiros, K. E., Houbavlis, T., & Kalyvas, M. (2004). Ultra-high speed all-optical shift registers and their applications in OTDM networks. *Optical and Quantum Electronics*, *36*(11), 1005–1053. doi:10.100711082-004-2040-9

Compilation of References

Abbasi, F., & Bahrampour, A. R. (2008). High harmonics generation in under dense plasma from Relativistic Thomson scattering. *IEEE 35th International Conference on Plasma Science*.

Abed, A. T. (2018). Highly compact size serpentine-shaped multiple-input–multiple-output fractal antenna with CP diversity. *IET Microwaves, Antennas & Propagation*, *12*(4), 636–640. doi:10.1049/iet-map.2017.0770

Agrawall, N. P., Kumar, G., & Ray, K. P. (1998). Wide Band Planar Monopole Antennas. *IEEE Transactions on Antennas and Propagation*, *46*(2), 294–295. doi:10.1109/8.660976

Al-Fares, Loukissas, & Vahdat. (2008). A scalable, commodity data centre network architecture. *ACM SIGCOMM'08*, 63–74.

Altug, H., & Vučković, J. (2005). Photonic Crystal Nanocavity Array Laser. *Optics Express*, *13*(22), 8819–8828. doi:10.1364/OPEX.13.008819 PMID:19498914

Amiri, I. S., Alavi, S. E., Bahadoran, M., Afroozeh, A., & Ahmad, H. (2015). Nanometer bandwidth soliton generation and experimental transmission within nonlinear fiber optics using an add-drop filter system. *Journal of Computational and Theoretical Nanoscience*, *12*(2), 221–225. doi:10.1166/jctn.2015.3721

Andreani, L. C., Agio, M., Bajoni, D., Belotti, M., Galli, M., Guizzetti, G., ... Vecchi, G. (2003). Optical Properties and Photonic Mode Dispersion in Two-Dimensional and Waveguide- Embedded Photonic Crystals. *Synthetic Metals*, *139*(3), 695–700. doi:10.1016/S0379-6779(03)00330-8

Andreas, M. (2005). *Wireless Communications*. Wiley-IEEE Press.

Arafin, S., & Coldren, L. A. (2018). Advanced InP Photonic Integrated Circuits for Communication and Sensing. *IEEE Journal of Selected Topics in Quantum Electronics*, *24*(1), 1–12. doi:10.1109/JSTQE.2017.2754583

Awasthi, S. K., Mishra, A., Malaviya, U., & Ojha, S. P. (2009). Wave Propagation in a One-Dimensional Photonic Crystal with Metamaterial. *Solid State Communications*, *149*(33-34), 1379–1383. doi:10.1016/j.ssc.2009.05.017

Azuma, H. (2008). Quantum Computation with Kerr-Nonlinear Photonic Crystals. *Journal of Physics. D, Applied Physics*, *41*(025102).

Baba, T., Matsumoto, T., & Asatsuma, T. (2008). Negative Refraction in Photonic Crystals. *Advances in Science and Technology*, *55*, 91–100. doi:10.4028/www.scientific.net/AST.55.91

Bellatoni, J., Compton, R., & Levy, H. (1989). A New W-Band Coplanar Waveguide Test Fixture. *IEEE MTT-S International Microwave Symposium Digest*, 1203-1204. 10.1109/MWSYM.1989.38940

Benson, T., Akella, A., & Maltz, D. A. (2010). Network traffic characteristics of data centres in the wild. *10th ACM SIGCOMM Conference on Internet Measurement*, 267–280.

Bharti, G. K., Rakshit, J. K., & Pal, S. S. (2018). Design and modeling of mode-conversion in ring resonator and its application in all-optical switching. *Microsystem Technologies*, 1–12.

Bhattacharya, P. (2017). Comparative Study for Proposed Algorithm for All-Optical Network with Negative Acknowledgment (AO-NACK). In *Proceedings of the 7th International Conference on Computer and Communication Technology*. ACM. 10.1145/3154979.3154981

Bhattacharya, P., Tiwari, K. A., & Srivastava, R. (in press). Dual Buffers Optical Based Packet Switch incorporating Arrayed Waveguide Gratings. *Journal of Engineering Research*.

Bhavarthe, P. P., Rathod, S. S., & Reddy, K. T. V. (2018). Mutual coupling reduction in patch antenna using Electromagnetic Band Gap (EBG) structure for IoT application. *International Conference on Communication information and Computing Technology*. 10.1109/ICCICT.2018.8325867

Bianucci, P., Fietz, C. R., Robertson, J. W., Shvets, G., & Shih, C. K. (2007). Polarization conversion in a silica microsphere. *Optics Express*, *15*(11), 7000–7005. doi:10.1364/OE.15.007000 PMID:19547016

Born, M., & Wolf, E. (1999). *Principles of Optics* (7th ed.). Cambridge University Press.

Caglayan, H., Bulu, I., Loncar, M., & Ozbay, E. (2009). Experimental Observation of Subwavelength Localization using Metamaterial-Based Cavities. *Optics Letters*, *34*(1), 88–90. doi:10.1364/OL.34.000088 PMID:19109649

Çakmak, V., Ufuk, F., & Karabulut, N. (2017). Diffusion-weighted MRI of pulmonary lesions: Comparison of apparent diffusion coefficient and lesion-to-spinal cord signal intensity ratio in lesion characterization. *Journal of Magnetic Resonance Imaging*, *45*(3), 845–854. doi:10.1002/jmri.25426 PMID:27519160

Caloz, C., Sanada, A., & Itoh, T. (2003). Surface Plasmons at the Interface between Right-Handed and Left-Handed 2D Metamaterials. *IEEE Antennas and Propagation Society International Symposium*, *3*, 363 – 366.

Carver, K. R., & Coffey, E. L. (1979). *Theoretical Investigation of the Microstrip Antenna*. Technical Report PT-00929, Physical Science Laboratory, New Mexico State University, Las Cruces, NM.

Chandra, S. K., & Mukhopadhyay, S. (2012). All optical alternative approach of conducting NAND and NOR logic gates with phase encoding principle. *Optik-International Journal for Light and Electron Optics*, *123*(11), 1022–1025. doi:10.1016/j.ijleo.2011.07.022

Chang, G.-K., Yu, J., Yeo, Y.-K., Chowdhury, A., & Jia, Z. (2006, May). Enabling technologies for next generation optical packet-switching networks. *Proceedings of the IEEE*, *94*(5), 892–910. doi:10.1109/JPROC.2006.873433

Chattopadhyay, K., Das, S., & Bhadra Chaudhuri, S. R. (2012). Bandwidth Enhancement of a Microstrip line Fed Hexagonal Wide-slot Antenna Using Fork-like Tuning Stub. *International Journal of Soft Computing and Engineering*, *2*(2).

Chattopadhyay, K., RoyChoudhury, S., Das, S., & Bhadra Chaudhuri, S.R. (2013). Wideband microstrip-line-fed hexagonal slot antenna for WiMAX and Wireless Local Area Network applications. *The Journal of Engineering, IET, UK*.

Chattopadhyay, K., Mitra, T., Das, D., Das, S., & Bhadra Chaudhuri, S. R. (2013). Bandwidth Enhancement of Hexagonal Wide-slot Antenna Using a Tuning Stub. *Journal of the Institution of Electronics and Telecommunication Engineers*, *59*(6), 732–738. doi:10.4103/0377-2063.126976

Chen, S., Shi, Y., He, S., & Dai, D. (2016). Compact eight-channel thermally reconfigurable optical add/drop multiplexers on silicon. *IEEE Photonics Technology Letters*, *28*(17), 1874–1877. doi:10.1109/LPT.2016.2574459

Chen, S.-Y., & Hsu, P. (2004). CPW-fed folded-slot antenna for 5.8 GHz RFID tags. *Electronics Letters*, *40*(24), 1516–1517. doi:10.1049/el:20046780

Chen, W. F., Tsai, C. C., & Huang, C. Y. (2008). Compact Wide- slot antenna for ultra-wideband Communications. *Electronics Letters*, *44*(15), 892. doi:10.1049/el:20081563

Chou, C., Shyu, J. C., Huang, Y. C., & Yuan, C. K. (1998). Common-path optical heterodyne profilometer: A configuration. *Applied Optics*, *37*(19), 4137–4142. doi:10.1364/AO.37.004137 PMID:18285853

Chun-fei, L. I., & Dou, N. A. (2009). Optical switching in silicon nanowaveguide ring resonators based on Kerr effect and TPA effect. *Chinese Physics Letters*, *26*(5). doi:10.1088/0256-307X/26/5/054203

Cisco Global Cloud Index. (2011). *Forecast and Methodology, 2011–2016*. Cisco White Paper.

Cisco global cloud index: forecast and methodology, 2012-2017. (n.d.). Available: http://www.cisco.com/

Collin, R. E. (1966). Foundation of Microwave Engineering (2nd ed.). McGraw-Hill Book Co.

Collin, R.E. (1992). *Foundation of Microwave Engg*. McGraw-Hill Inc. International.

Collin, R. E. (1960). *Field Theory of Guided waves*. McGraw-Hill Book Co.

Cusmai, G., Morichetti, F., Rosotti, P., Costa, R., & Melloni, A. (2005). Circuit-oriented modeling of ring resonator. *Optical and Quantum Electronics*, *37*(1-3), 343–358. doi:10.100711082-005-1191-7

D'Orazio, A., De Palo, V., De Sario, M., Petruzzelli, V., & Prudenzano, F. (2003). Finite Difference Time Domain Modeling of Light Amplification in Active Photonic Bandgap Structures. *Progress in Electromagnetics Research*, *39*, 299–339. doi:10.2528/PIER02112501

Dai, J., Zhang, M., Zhou, F., Wang, Y., Lu, L., & Liu, D. (2015). All-optical logic operation of polarized light signals in highly nonlinear silicon hybridplasmonicmicroring resonators. *Applied Optics*, *54*(14), 4471–4477. doi:10.1364/AO.54.004471 PMID:25967504

Das, A., & Das, S. K. (2009). *Microwave Engineering*. McGraw Hill.

Das, A., Sarkar, M., Das, R., Das, D., & Debnath, P. (2012, March). Numerical Characterization of Rectangular Wave Guide in Free space using HFSS Software at lower microwave Frequencies. *IJETAE*, *2*(3), 60–64.

Dastranj, A., & Abiri, H. (2010). Bandwidth Enhancement of Printed E-Shaped Slot Antennas Fed by CPW and Microstrip Line. *IEEE Transactions on Antennas and Propagation*, *58*(4), 1402–1407. doi:10.1109/TAP.2010.2041164

Dastranj, A., Imani, A., & Moghaddasi, M. N. (2008). Printed Wide Slot Antenna for Wideband Applications. *IEEE Transactions on Antennas and Propagation*, *56*(10), 3097–3102. doi:10.1109/TAP.2008.929459

Davis, J. A., Evans, G. H., Crabtree, K., & Moreno, I. (2004). Programmable birefringent lenses with a liquid-crystal display. *Applied Optics*, *43*(34), 6235–6241. doi:10.1364/AO.43.006235 PMID:15619832

Debnath, P., Karmakar, A., Saha, A., & Huda, S. (2018). UWB MIMO Slot Antenna with Minkowski Fractal Shaped Isolators for Isolation Enhancement. *Progress In Electromagnetics Research M*, *75*, 69–78. doi:10.2528/PIERM18090506

Demetriadou, A. (2015). The impact of natural modes in plasmonic imaging. *Scientific Reports*, *5*, 18247.

Derneryd, A. (1976). Linearly Polarized Microstrip Antennas. *IEEE Transactions on Antennas and Propagation*, *24*(6), 846–851. doi:10.1109/TAP.1976.1141445

Derneryd, A. (1978). A Theoretical Investigation of the Rectangular Microstrip Antenna Element. *IEEE Transactions on Antennas and Propagation*, *26*(4), 532–535. doi:10.1109/TAP.1978.1141890

Despins, C., Labeau, F., Le-Ngoc, T., Labelle, R., Cheriet, M., Thibeault, C., & Gagnon, F. (2011, August). Leveraging green communications for carbon emission reductions: Techniques, testbeds, and emerging carbon footprint standards. *IEEE Communications Magazine*, *49*(8), 101–109. doi:10.1109/MCOM.2011.5978422

Deyasi, A., Paul, P., & Verma, P. (2017). Calculating impedance variation with frequency of MIM surface plasmon structure for spectroscopic applications. *IEEE International Conference on Devices for Integrated Circuit*, 81-84.

Deyasi, A., Verma, P., & Sarkar, A. (2018). *Calculating Transmittance and Field Enhancement of n-Layer MIM Surface Plasmon Structure for Detection of Biological Nano-Objects*. Springer.

Deyasi, A., & Verma, P. (2018). Computation of Skin Depth for MIM Surface Plasmon Structure at High Frequency. *Fifteenth International Conference on Wireless and Optical Communications Networks*.

Dey, S., & Mukhopadhyay, S. (2017). Approach of implementing phase encoded quantum square root of NOT gate. *Electronics Letters*, *53*(20), 1375–1377. doi:10.1049/el.2017.2500

Dey, S., & Mukhopadhyay, S. (2018). A method of generation of sharp peaked parabolic and hyper parabolic optical pulses by the use of Pockels effect for long distance communication. *Journal of Optics*, *47*(3), 272–277. doi:10.100712596-018-0461-2

Dey, S., & Mukhopadhyay, S. (2018). Implementation of all-optical Pauli-Y gate by the integrated phase and polarisation encoding. *IET Optoelectronics*, *12*(4), 176–179. doi:10.1049/iet-opt.2017.0138

Díaz, A. M. S., Galaz-Larios, M. C., & Ramírez-García, E. (2017). Analysis and design of electromagnetic band gap structures with application in planar antennas. *42nd International Conference on Infrared, Millimeter, and Terahertz Waves*. 10.1109/IRMMW-THz.2017.8067190

Díaz-Rubio, A., Carbonell, J., Torrent, D., Cervera, F., Kirleis, M. A., Piqué, A., & Sánchez-Dehesa, J. (2012). Radial Photonic Crystals for Microwave Operation. *The Sixth International Congress on Advanced Electromagnetic Materials in Microwaves and Optics*, 393-395.

Ding, S. Y., Yi, J., Li, J. F., Ren, B., Wu, D. Y., Panneerselvam, R., & Tian, Z. Q. (2016). Nanostructure-based Plasmon-enhanced Raman spectroscopy for surface analysis of materials. *Nature Reviews Materials*, *1*, 16021.

Dong, Y., & Liu, P. (2016). Applications of graphene in terahertz electromagnetic band-gap structure designs. *Asia-Pacific International Symposium on Electromagnetic Compatibility*. 10.1109/APEMC.2016.7522966

Dutta, S., & Mukhopadhyay, S. (2011). All optical frequency encoding method for converting a decimal number to its equivalent binary number using tree architecture. *Optik - International Journal for Light and Electron Optics, 122*(2), 125-127.

Dutta, S., & Mukhopadhyay, S. (2011). Alternating approach of implementing frequency encoded all-optical logic gates and flip-flop using semiconductor optical amplifier. *Optik - International Journal for Light and Electron Optics, 122*(12), 1088-1094.

Dyson, L. (1957). Common-path interferometer for testing purposes. *Journal of the Optical Society of America*, *47*(5), 386–390. doi:10.1364/JOSA.47.000386

Edwards, T. (1992). *Foundation for Microstrip circuit Design* (2nd ed.). John Wiley and Sons.

El-Khozondar, H. J., El-Khozondar, R. J., & Shabat, M. M. (2014). Metamaterial-dielectric Photonics Crystal Waveguide Structure. *Optics*, *4*(1-2), 1–4.

Elwi, T. A. (2017). Electromagnetic band gap structures based on ultra wideband microstrip antenna. *Microwave and Optical Technology Letters*, *59*(4), 827–834. doi:10.1002/mop.30397

Eshrah, I. A., Kishk, A. A., Yakovlev, A. B., & Glisson, A. W. (2005). Modal Analysis of Corrugated Rectangular Waveguides Supporting Left-Hand Propagation. *IEEE AP-Society International Symposium Digest*. 10.1109/APS.2005.1551408

Federal Communications Commission. (2002). *Federal communications commission revision of Part 15 of the Commission's rules regarding ultra-wideband (UWB) transmission system from 3.1 to 10.6 GHz*. Washington, DC: ET-Docket.

Fiala, W., Finkelmann, H., & Gfeller, W.F. (1993). *Bifocals manufactured from birefringent materials*. Prague Meeting on Macromolecules. Special Lecture SL4.

Fiala, W. (1990). Birefringent multi-focal lenses: Theory and application to the correction of refractive error. *Optometry and Vision Science*, *67*(10), 787–791. doi:10.1097/00006324-199010000-00006 PMID:2247302

Fiala, W. (1992). Multi-focal intra-ocular lenses fabricated from media exhibiting tuned birefringence. *Optometry and Vision Science*, *69*(4), 329–333. doi:10.1097/00006324-199204000-00013 PMID:1565436

Fietz, C., & Shvets, G. (2007). Non-linear polarization conversion using microring resonators. *Optics Letters*, *32*(12), 1683–1685. doi:10.1364/OL.32.001683 PMID:17572746

Finkenzeller, K. (2010). Fundamentals and applications in contactless smart cards, radio frequency identification and near-field communication. In *RFID handbook*. John Wiley & Sons.

Gallo, M., Antonino-Daviu, E., Ferrando-Bataller, M., Bozzetti, M., Molina-Garcia-Pardo, J. M., & Juan-Llacer, L. (2012). A broadband pattern diversity annular slot antenna. *IEEE Transactions on Antennas and Propagation*, *60*(3), 1596–1600. doi:10.1109/TAP.2011.2180314

Gao, P., He, S., & Wei, X. (2014). Compact printed UWB diversity slot antenna with 5.5-GHz band-notched characteristics. *IEEE Antennas and Wireless Propagation Letters*, *13*, 376–379. doi:10.1109/LAWP.2014.2305772

Gao, Y., Chen, H., Qiu, H., Lu, Q., & Huang, C. (2011). Transmission Spectra Characteristics of 1D Photonic Crystals with Complex Dielectric Constant. *Rare Metals*, *30*(2), 150–154. doi:10.100712598-011-0215-7

Garai, S. K., & Mukhopadhyay, S. (2010). A method of optical implementation of frequency encoded different logic operations using second harmonic & difference frequency generation techniques in non-linear material. *Optik-International Journal for Light and Electron Optics*, *121*(8), 715–721. doi:10.1016/j.ijleo.2008.10.011

Garai, S. K., & Mukhopadhyay, S. (2010). All-optical frequency encoded binary half subtractor using periodically poled lithium niobate waveguide and reflecting semiconductor optical amplifier. *Optics and Photonics Letters*, *3*(1), 15–22. doi:10.1142/S1793528810000074

Garcia-Vidal, F. J., Sanchez-Dehesa, J., Dechelette, A., Bustarret, E., Lopez-Rios, T., Fournier, T., & Pannetier, B. (1991). Localized surface plasmons in lamellar metallic gratings. *Journal of Lightwave Technology*, *17*(11), 2191–2195. doi:10.1109/50.803010

Garg, R., Bhartia, P., Bahl, I., & Ittipiboon, A. (2001). *Microstrip Antenna Design Handbook*. Artech House.

Gayen, D. K., Chattopadhyay, T., & Zoiros, K. E. (2015). All-optical D flip-flop using single quantum-dot semiconductor optical amplifier assisted Mach–Zehnder interferometer. *Journal of Computational Electronics*, *14*(1), 129–138. doi:10.100710825-014-0632-6

Geer, D. (2005, May). Chip makers turn to multicore processors. *IEEE Computer*, *38*(5), 11–13. doi:10.1109/MC.2005.160

Ghosh, A. K., Bhattacharya, A., Raul, M., & Basuray, A. (2012). Trinary arithmetic and logic unit (TALU) using savart plate and spatial light modulator (SLM) suitable for optical computation in multivalued logic. *Optics & Laser Technology*, *44*(5), 1583–1592. doi:10.1016/j.optlastec.2011.11.044

Ghosh, B., Hazra, S., Haldar, N., Roy, D., Patra, S. N., Swarnakar, J., ... Mukhopadhyay, S. (2018). A Novel Approach to Realize of All Optical Frequency Encoded Dibit Based XOR and XNOR Logic Gates Using Optical Switches with Simulated Verification. *Optics and Spectroscopy*, *124*(3), 337–342. doi:10.1134/S0030400X1803013X

Ghosh, B., & Mukhopadhyay, S. (2013). A novel realization of all-optical dibit represented frequency encoded Boolean and quaternary inverters without switching device. *Optik-International Journal for Light and Electron Optics*, *124*(21), 4813–4815. doi:10.1016/j.ijleo.2013.02.035

Ghosh, J., Ghosal, S., Mitra, D., & Chaudhuri, S. R. B. (2016). Mutual coupling reduction between closely placed microstrip patch antenna using meander line resonator. *Prog. Electromagn. Res. Lett*, *59*, 115–122. doi:10.2528/PIERL16012202

Gordon-II, J. G., & Swalen, J. D. (1977). The effect of thin organic films on the surface plasmaresonace on gold. *Optics Communications*, *22*(3), 374–376. doi:10.1016/S0030-4018(97)90032-8

Greenberg, Hamilton, Jain, Kandula, Kim, Lahiri, … Sengupta. (2009). VL2: a scalable and flexible data centre network. *ACM SIGCOMM'09, 39*, 51–62.

Greenberg, A., Hamilton, J., Maltz, D. A., & Patel, P. (2009, January). The cost of a cloud: Research problems in data centre networks. *Computer Communication Review*, *39*(1), 68–73. doi:10.1145/1496091.1496103

GreenDataProject. (2008). *Where does power go?* Available: http://www.greendataproject.org

Greenpeace International. (2010). *Make IT green: cloud computing and its contribution to climate change*. Available: http://www.greenpeace.org/

Gripp, Simsarian, LeGrange, Bernasconi, & Neilson. (2012). Photonic terabit routers: the IRIS project. *Optical Fiber Communication Conference*.

Gupta, K. C., Ramesh, G., & Bahl, I. J. (1979). Microstrip Lines and Slot lines. Artech House, Inc.

Habib, M. F., Tornatore, M., Leenheer, M. D., Dikbiyik, F., & Mukherjee, B. (2012, August). Design of disaster resilient optical data centre networks. *Journal of Lightwave Technology, 30*(16), 2563–2573. doi:10.1109/JLT.2012.2201696

Hajlaoui, E. A. (2018). New triple band electromagnetic band gap microstrip patch antenna with two shaped parasitic elements. *Journal of Computational Electronics, 17*(1), 452–457. doi:10.100710825-017-1100-x

Hammersted, E. O. (1975). Equations for microstrip Circuit Design. *5th European Microwave Conference.*

Hazra, L. N. (2006). Walsh filters for tailoring of resolution in microscopic imaging. *Micron (Oxford, England), 38*(2), 129–135. doi:10.1016/j.micron.2006.07.003 PMID:16908176

Hecht, E. (2002). *Optics.* Addison-Wesley.

Herminghaus, S., Riedel, M., Leiderer, P., Weiland, U., Giebel, K. F., Bechinger, C., & Bastmeyer, M. (1999). Imaging of cell/substrate contacts of living cells with surface plasmon resonance microscopy. *Journal of Biophysics, 76*(1), 509–516. doi:10.1016/S0006-3495(99)77219-X

High Frequency Structure Simulator based on the Finite Element Method. (2013). Ansoft Corporation.

Homola, J., Yee, S. S., & Gauglitz, G. (1999). Surface Plasmon resonance sensors [review]. *Sensors and Actuators. B, Chemical, 54*(1-2), 3–15. doi:10.1016/S0925-4005(98)00321-9

Huang, H., Zhang, J., Ji, X., Zhou, J., Bao, M., & Huang, Y. (2009). Photonic Crystal Biosensors Based on Surface Plasmons. *IEEE 9th International Conference on Solid-State and Integrated-Circuit Technology.*

Huang, W., & Wu, Z. (2017). Study on band gap characteristics of EBG structures with interdigital and spiral ground plane. *AIP Conference Proceedings*, 1890.

Hu, G., Xiao, D., Wang, Y., Xiang, T., & Zhou, Q. (2017). Securing image information using double random phase encoding and parallel compressive sensing with updated sampling processes. *Optics and Lasers in Engineering, 98*, 123–133. doi:10.1016/j.optlaseng.2017.06.013

Hunter, D. K., Chia, M. C., & Andonovic, I. (1998, December). Buffering in optical packet switches. *Journal of Lightwave Technology*, *16*(12), 2081–2094. doi:10.1109/50.736577

Hwang, R. B., Liu, H. W., & Chin, C. Y. (2009). A metamaterial-based E-plane Horn Antenna. *Progress in Electromagnetics Research*, *93*, 275–289. doi:10.2528/PIER09050606

Iqbal, A., Saraereh, O. A., Ahmad, A. W., & Bashir, S. (2018). Mutual Coupling Reduction Using F-Shaped Stubs in UWB-MIMO Antenna. *IEEE Access: Practical Innovations, Open Solutions*, *6*, 2755–2759. doi:10.1109/ACCESS.2017.2785232

Ishimaru, A., Jaruwatanadilok, S., & Kuga, Y. (2005). Generalized Surface Plasmon Resonance Sensors Using Metamaterials And Negative Index Materials. *Progress in Electromagnetic Research, PIER*, *51*, 139–152. doi:10.2528/PIER04020603

Iskander, M., & Lind, T. (1989). Electromagnetic Coupling of Coplanar Waveguides and Microstrip Lines to Highly Lossy Dielectric Media. *IEEE MTT*, 1910-1917.

Islam, M. T., & Alam, M. S. (2013). Compact EBG structure for alleviating mutual coupling between patch antenna array elements. *Progress in Electromagnetics Research*, *137*, 425–438. doi:10.2528/PIER12121205

Ivchenko, E. L., Voronov, M. M., Erementchouk, M. V., Deych, L. I., & Lisyansky, A. A. (2004). Multiple-quantum-well-based photonic crystals with simple and compound elementary supercells. *Physical Review B: Condensed Matter and Materials Physics*, *70*(19), 195106. doi:10.1103/PhysRevB.70.195106

Jacquinot, P., & Roizen-Dossier, B. (1964). *Apodization in Progress in Optics* (Vol. 3). Amsterdam: North-Holland.

Jang, Y. W. (2001). Experimental study of Large Bandwidth Three-Offset Microstripline-Fed Slot Antenna. *IEEE Microwave and Wireless Components Letters*, *11*(10), 425–427. doi:10.1109/7260.959318

Jan, J. Y., & Su, J. W. (2005). Bandwidth Enhancement of a Printed wide slot Antenna With a Rotated Slot. *IEEE Transactions on Antennas and Propagation*, *53*(6), 2111–2114. doi:10.1109/TAP.2005.848518

Jan, J. Y., & Wang, L. C. (2009). Printed Wideband Rhombus Slot Antenna With a Pair of Parasitic Strips for Multiband Applications. *IEEE Transactions on Antennas and Propagation*, *57*(4), 1267–1270. doi:10.1109/TAP.2009.2015859

Jetté-Charbonneau, S., Charbonneau, R., Lahoud, N., Mattiussi, G. A., & Berini, P. (2005). Bragg Gratings Based on Long-Range Surface Plasmon-Polariton Waveguides: Comparison of Theory and Experiment. *IEEE Journal of Quantum Electronics*, *41*(12), 1480–1491. doi:10.1109/JQE.2005.858466

Jiang, L., Zhao, X., Fei, Y., Yu, D., Qian, J., Tong, J., ... He, S. (2016). The effects of magnetic fields exposure on relative permittivity of saline solutions measured by a high resolution SPR system. *Scientific Reports*, *6*, 25111.

Jiao, C. Q. (2011). Selective Suppression of Electromagnetic Modes in a Rectangular Waveguide by using Distributed Wall Losses. *Progress in Electromagnetics Research Letters*, *22*, 119–128. doi:10.2528/PIERL11013102

Jinno, M., & Matsumoto, T. (1992). Nonlinear Sagnac interferometer switch and its applications. *IEEE Journal of Quantum Electronics*, *28*(4), 875–882. doi:10.1109/3.135205

Jinno, M., Takara, H., Kozicki, B., Tsukishima, Y., Sone, Y., & Matsuoka, S. (2009, November). Spectrum-efficient and scalable elastic optical path network: Architecture, benefits, and enabling technologies. *IEEE Communications Magazine*, *47*(11), 66–73. doi:10.1109/MCOM.2009.5307468

Jorgenson, R. C., & Yee, S. S. (1993). A fiber-optic chemical sensor based on surface plasmon resonance. *Sensors and Actuators. B, Chemical*, *12*(3), 213–220. doi:10.1016/0925-4005(93)80021-3

Kachris, C., Kanonakis, K., & Tomkos, I. (2013). Optical Interconnection Networks in Data Centres: Recent Trends and Future Challenges. *IEEE Communications Magazine, 51*(9), 39-45.

Kachris, C., & Tomkos, I. (2012). A survey on optical interconnects for data centres. *IEEE Communications Surveys and Tutorials*, *14*(4), 1021–1036. doi:10.1109/SURV.2011.122111.00069

Kahrizi, M., Sarkar, T. K., & Maricevic, Z. A. (1993). Analysis of a wide Radiating slot in the Ground plane of a Microstrip line. *IEEE Transactions on Microwave Theory and Techniques*, *41*(1), 29–37. doi:10.1109/22.210226

Kaiser, T., Zheng, F., & Dimitrov, E. (2009). An overview of ultra-wide-band systems with MIMO. *Proceedings of the IEEE*, *97*(2), 285–312. doi:10.1109/JPROC.2008.2008784

Kang, L., Li, H., Wang, X., & Shi, X. (2015). Compact offset microstrip-fed MIMO antenna for band-notched UWB applications. *IEEE Antennas and Wireless Propagation Letters*, *14*, 1754–1757. doi:10.1109/LAWP.2015.2422571

Kao, K. C., & Hockham, G. A. (1986). Dielectric-fibre surface waveguides for optical frequencies. *IEE Proceedings. Part J, Optoelectronics*, *133*(3), 191–198. doi:10.1049/ip-j.1986.0030

Karaomerlioglu, F., Mamedov, A. M., & Ozbay, E. (2014). Optical Properties of Metamaterial-based Devices Modulated by a Liquid Crystal. *Applied Physics. A, Materials Science & Processing*, *117*(2), 611–619. doi:10.100700339-014-8711-x

Katz, R. H. (2009, February). Tech titans building boom. *IEEE Spectrum*, *46*(2), 40–54. doi:10.1109/MSPEC.2009.4768855

Khan, B. D. B. M. S., Capobianco, A. D., Shafique, M. F., Ijaz, B., Naqvi, A., & Braaten, B. D. (2015). Isolation Enhancement of a Wideband MIMO Antenna using Floating Parasitic Elements. *Microwave and Optical Technology Letters*, *57*(7), 1677–1682. doi:10.1002/mop.29162

Khare, R., & Nema, R. (2012). Reflection Coefficient Analysis Of Chebyshev Impedance Matching Network Using Different Algorithms. *International Journal of Innovative Research in Science, Engineering and Technology*, *1*(2), 214–218.

Khorasaninejad, M., & Saini, S. S. (2009). All-optical logic gates using nonlinear effects in silicon-on-insulator waveguides. *Applied Optics*, *48*(25), F31–F36. doi:10.1364/AO.48.000F31 PMID:19724310

Kim, H., Kwon, B., Suh, M., Kang, D. S., Kim, Y., & Jeon, D. Y. (2011). Degradation characteristics of red light-emitting CuInS2/ZnS quantum dots as a wavelength converter for LEDs. *Electrochemical and Solid-State Letters*, *14*(10), K55–K57. doi:10.1149/1.3610644

Kim, H., & Yoon, Y. J. (2005). Microstrip-Fed Slot Antennas With Suppressed Harmonics. *IEEE Transactions on Antennas and Propagation*, *53*(9), 2809–2817. doi:10.1109/TAP.2005.854557

Kim, S. H., Takei, R., Shoji, Y., & Mizumoto, T. (2009). Single-trench waveguide TE-TM mode converter. *Optics Express*, *17*(14), 11267–11273. doi:10.1364/OE.17.011267 PMID:19582039

Kittel, C. (1996). *Introduction to Solid State Physics* (7th ed.). New York: Wiley.

Knoll, W. (2000). Interfaces and thin films as seen by bound electromagnetic waves. *Annual Review of Physical Chemistry*, *49*(1), 569–638. doi:10.1146/annurev.physchem.49.1.569 PMID:15012436

Kolb, D. M. (1982). *The study of solid-liquid interfaces by surface plasmon polariton excitation.* Surface Polaritons. doi:10.1016/B978-0-444-86165-8.50014-1

Koley, B., Vusirikala, V., Lam, C., & Gill, V. (2010). 100GbE and beyond for warehouse scale computing. *15th IEEE Opto electronics and Communications Conference (OECC),* 106–107.

Kossel, M. A., Kung, R., Benedickter, H., & B¨achtold, W. (1999). An active tagging system using circular-polarization modulation. *IEEE Transactions on Microwave Theory and Techniques*, *47*(12), 2242–2248. doi:10.1109/22.808966

Kubota, H., & Inoue, S. (1959). Diffraction images in the polarizing microscope. *Journal of the Optical Society of America*, *49*(2), 191–198. doi:10.1364/JOSA.49.000191 PMID:13621312

Kubota, H., & Saito, H. (1960). Further study on diffraction Images in the polarizing microscope. *Journal of the Optical Society of America*, *50*(10), 1020–1024. doi:10.1364/JOSA.50.001020

Kumar & Nath. (2017). *Introducing Multiband and Wideband Microstrip Patch Antennas Using Fractal Geometries: Development in Last Decade*. Springer.

Kumar Naik, K., & Vijaya Sri, A. (2017). Design of Concentric Circular Ring Patch with DGS for Dual-Band at Satellite Communication and Radar Applications. In *Wireless Personal Communications.* Springer.

Kumar Naik, K., & Amala Vijaya Sri, P. (2018). Design of hexadecagon circular patch antenna with dgs at ku band for satellite communications. *Progress In Electromagnetics Research M*, *63*, 163–173. doi:10.2528/PIERM17092205

Kuo, C. H., & Moghaddam, M. (2006). Backscattering enhancement of surface plasmons from multilayer rough surfaces. *IEEE Antennas and Propagation Society International Symposium*, 749 – 752.

Lam. (2010). Optical network technologies for data centre networks. *OFC/NFOEC 2010*, NWA3.

Lam, C. F., Liu, H., Koley, B., Zhao, X., Kamalov, V., & Gill, V. (2010, July). Fiber optic communication technologies: What's needed for datacentre network operations? *IEEE Communications Magazine*, *48*(7), 32–39. doi:10.1109/MCOM.2010.5496876

Lamouroux, B., Prade, B., & Orszag, A. (1982). Polarization effect in optical fiber ring resonators. *Optics Letters*, *7*(8), 391–393. doi:10.1364/OL.7.000391 PMID:19714032

Lee, C. S., Lee, S. W., & Chuang, S. L. (1985). Plot of Modal Field Distribution in Rectangular and Circular Waveguides. *IEEE Transactions on Microwave Theory and Techniques*, *33*(3), 271–274. doi:10.1109/TMTT.1985.1132998

Lee, R. Q., Lee, K. F., & Bobinchak, J. (1987, September). Characteristics of a two-layer electro-magnetically coupled rectangular patch antenna. *Electronics Letters*, *23*(20), 1070–1072. doi:10.1049/el:19870748

Lee, S. W., & Sung, Y. J. (2015). Reconfigurable Rhombus-Shaped Patch Antenna With Y-Shaped Feed for Polarization Diversity. *IEEE Antennas and Wireless Propagation Letters*, *14*, 163–166. doi:10.1109/LAWP.2014.2358651

Lee, Y., Yeo, J., & Mittra, R. (2003). Investigation of electromagnetic bandgap (EBG) structures for antenna pattern control. *Proceedings of the IEEE International Antennas and Propagation Symposium*, 2, 1115-1118.

Li, H., Peng, W., Wang, Y., & Hu, L. (2011). Theoretical analyses of localized surface Plasmon resonance spectrum with nanoparticles imprinted polymers. *IEEE 2011 Asia Communications and Photonics Conference and Exhibition*, 1-8.

Liang, T. K., Nunes, L. R., Tsuchiya, M., Abedin, K. S., Miyazaki, T., Van Thourhout, D., ... Tsang, H. K. (2006). High speed logic gate using two-photon absorption in silicon waveguides. *Optics Communications*, *265*(1), 171–174. doi:10.1016/j.optcom.2006.03.031

Liboff, R. L., & Dalmon, G. C. (1985). *Transmission Lines, Waveguides and Smith Chart*. New York: McMillan Publishing Co.

Lier, E. (1982). Improved Formulas for Input Impedance of Coax-Fed Microstrip Patch Antennas. *IEE Proc.*, *129*(H).

Li, H., Zhou, Y., Mou, X., Ji, Z., Yu, H., & Wang, L. (2014). Miniature four-band CPW-fed antenna for RFID/WiMAX/WLAN applications. *IEEE Antennas and Wireless Propagation Letters*, *13*, 1684–1688. doi:10.1109/LAWP.2014.2345564

Li, J., Li, L., Jin, L., & Li, C. (2006). All-optical switch and limiter based on nonlinear polarization in Mach-Zehnder interferometer coupled with a polarization-maintaining fiber ring-resonator. *Optics Communications*, *260*(1), 318–323. doi:10.1016/j.optcom.2005.10.054

Li, L., Zhou, Z. L., Hong, J. S., & Wang, B. Z. (2011). Compact dual-band-notched UWB planar monopole antenna with modified SRR. *Electronics Letters*, *47*(17), 950–951. doi:10.1049/el.2011.1874

Limpert, J., Schreiber, T., Nolte, S., Zellmer, H., Tunnermann, T., Iliew, R., ... Jakobsen, C. (2003). High Power Air-Clad Large-Mode-Area Photonic Crystal Fiber Laser. *Optics Express*, *11*(7), 818–823. doi:10.1364/OE.11.000818 PMID:19461794

Lin, S. Y., Hietala, V. M., Lyo, S. K., & Zaslavsky, A. (1996). Photonic Bandgap Quantum Well and Quantum Box Structures: A high-Q Resonant Cavity. *Applied Physics Letters*, *68*(23), 3233–3235. doi:10.1063/1.116558

Li, P., Liang, J., & Chen, X. (2006). Study of printed elliptical/circular slot antennas for ultrawideband applications. *IEEE Transactions on Antennas and Propagation*, *54*(6), 1670–1675. doi:10.1109/TAP.2006.875499

Li, Q., Feresidis, A. P., Mavridou, M., & Hall, P. S. (2015). Miniaturized double-layer EBG structures for broadband mutual coupling reduction between UWB monopoles. *IEEE Transactions on Antennas and Propagation*, *63*(3), 1168–1171. doi:10.1109/TAP.2014.2387871

Lit, J. W. Y. (1971). Effects of a half wave filter in a portion of an aperture of a perfect lens. *Journal of the Optical Society of America*, *61*(3), 297–302. doi:10.1364/JOSA.61.000297

Little, B. E., & Chu, S. T. (2000). Theory of polarization rotation and conversion in vertically coupled microresonators. *IEEE Photonics Technology Letters*, *12*(4), 401–403. doi:10.1109/68.839032

Liu, Cheung, & Yuk. (2013). Compact MIMO antenna for portable devices in UWB applications. *IEEE Transactions on Antennas and Propagation, 8*, 4257–4264.

Liu, Y.F., Lau, K.L., Xue, Q., & Chan, C.H. (2004). Experimental studies of Printed Wide-Slot Antenna for Wide-Band Applications. *IEEE Antennas and wireless propagation Letters, 3*, 273–275.

Liu, F. (2018, July). High Purcell factor generation of indistinguishable on-chip single photons. *Nature Nanotechnology*. PMID:30013218

Liu, H.-W., Ku, C.-H., & Yang, C.-F. (2010). Novel CPW-fed planar monopole antenna for WiMAX/WLAN applications. *IEEE Antennas and Wireless Propagation Letters*, *9*, 240–243. doi:10.1109/LAWP.2010.2044860

Liu, W. C., & Hu, Z.-K. (2005). Broadband CPW-fed folded-slot monopole antenna for 5.8 GHz RFID application. *Electronics Letters*, *41*(17), 937–939. doi:10.1049/el:20052222

Liu, Y., Liu, Q., Chen, S., Cheng, F., Wang, H., & Peng, W. (2016). Surface Plasmon Resonance Biosensor Based on Smart Phone Platforms. *Scientific Reports*, *5*, 12864.

Liu, Y., & Salemink, H. W. M. (2012). Photonic Crystal-based All-Optical On-Chip Sensor. *Optics Express*, *20*(18), 19912–19920. doi:10.1364/OE.20.019912 PMID:23037043

Li, Z., Du, Z., Takahashi, M., Saito, K., & Ito, K. (2012). Reducing Mutual Coupling of MIMO Antennas With Parasitic Elements for Mobile Terminals. *IEEE Transactions on Antennas and Propagation*, *60*(2), 473–481. doi:10.1109/TAP.2011.2173432

Lo, Y. T. (1979). Theory and Experiment on Microstrip Antennas. *IEEE Transactions on Antennas and Propagation*, *AP-27*.

Lo, Y. T., & Richards, W. F. (1981). Perturbation Approach to Design of Circularly Polarized Microstrip Antennas. *Electronics Letters*, *17*(11), 383. doi:10.1049/el:19810269

Luberto, M., & Fano, W. G. (2015). Microstrip antenna design using EBG (Electromagnetic Band Gap) structures at 2.4GHz. *XVI Workshop on Information Processing and Control*. 10.1109/RPIC.2015.7497097

Lui, W. W., Hirano, T., Yokoyama, K., & Huang, W. P. (1998). Polarization rotation in semiconductor in bending waveguides: A coupled mode theory formulation. *Journal of Lightwave Technology*, *16*(5), 929–936. doi:10.1109/50.669050

Luneberg, R. K. (1944). *Mathematical Theory of Optics*. Brown University Press.

Maity, Chottopadhyay, Banerjee, & Deyasi. (n.d.). Novel band-pass filter design using photonic multiple quantum well structure with p-polarized incident wave at 1550 nm. *Journal of Electron Devices, 17*, 1400-1405.

Maity, A., Chottopadhyay, B., Banerjee, U., & Deyasi, A. (2013). Novel Band-Pass Filter Design using Photonic Multiple Quantum Well Structure with p-Polarized Incident Wave at 1550 nm. *Journal of Electron Devices*, *17*, 1400–1405.

Mandal, K., & Sarkar, P. P. (2013, April). High Gain Wide-Band U-Shaped Patch Antennas with Modified Ground Planes. *IEEE Transactions and Antenna Propagations*, *61*(4), 2279–2282. doi:10.1109/TAP.2012.2233455

Mandal, S. (2014). Evaluation of polychromatic image quality of a birefringent lens by means of optical transfer function. *Optik-International Journal for Light and Electron Optics*, *125*(1), 121–125. doi:10.1016/j.ijleo.2013.06.028

Mandal, S., & Ghosh, A. (2012). Polychromatic intensity point spread function of a birefringent lens. *Optik-International Journal for Light and Electron Optics*, *123*(18), 1623–1626. doi:10.1016/j.ijleo.2011.08.045

Mandal, S., Sanyal, S., & Ghosh, A. (2007). Imaging characteristics of a birefringent lens under broadband illumination. *Optik-International Journal for Light and Electron Optics*, *118*(7), 335–339. doi:10.1016/j.ijleo.2006.04.005

Mansoor, R., Koziel, S., Sasse, H., & Duffy, A. (2015). Crosstalk suppression bandwidth optimisation of a vertically coupled ring resonator add/drop filter. *IET Optoelectronics*, *9*(2), 30–36. doi:10.1049/iet-opt.2014.0059

Marathay, A. S. (1969). Realization of complex filters with polarized light. *Journal of the Optical Society of America*, *59*(6), 748–752. doi:10.1364/JOSA.59.000748

Marcueitz, N. (1951). *Waveguide Handbook, Rad. Lab. Series* (Vol. 10). New York: McGraw Hill Book Co.

Marcuvitz, N. (1986). *Waveguide Hand Book. Peter Peregrinus Ltd.*

Maxwell, J. C. (1865). A dynamical theory of the electromagnetic field. *Philosophical Transactions of the Royal Society of London*, *155*, 459–512. doi:10.1098/rstl.1865.0008

Meisner, D., Gold, B. T., & Wenisch, T. F. (2009). PowerNap: eliminating server idle power. *14th International Conference on Architectural Support for Programming Languages and Operating Systems*, 205–216.

Melloni, A., & Martinelli, M. (2002). Synthesis of direct coupled resonators bandpass filters for WDM systems. *Journal of Lightwave Technology*, *20*(2), 296–303. doi:10.1109/50.983244

Melloni, A., Morichetti, F., & Martinelli, M. (2004). Polarization conversion in ring resonator phase shifters. *Optics Letters*, *29*(23), 2785–2787. doi:10.1364/OL.29.002785 PMID:15605505

Mesquita, J. M., Pires, N., & Moreira, A. A. (2013). Influence of deformations on the matching of a flexible dual-band antenna. *Antennas and Propagation (EuCAP), 2013 7th European Conference on*, 3283-3285.

Mezouari, S., & Harvey, A. R. (2003). Phase pupil functions for reduction of defocus and spherical aberrations. *Optics Letters*, *28*(10), 771–773. doi:10.1364/OL.28.000771 PMID:12779141

Miller, D. A. B. (2009, July). Device requirements for optical interconnects to silicon chips. *Proceedings of the IEEE*, *97*(7), 1166–1185. doi:10.1109/JPROC.2009.2014298

Mookherjea, S., & Melloni, A. (2008). *Microring resonators in integrated optics*. Academic Press.

Mukhopadhyay, S. (1992). Binary optical data subtraction by using a ternary dibit representation technique in optical arithmetic problems. *Applied Optics*, *31*(23), 4622–4623. doi:10.1364/AO.31.004622 PMID:20725467

Munson, R. (1974). Conformal Microstrip Antennas and Microstrip Phased Arrays. *IEEE Transactions on Antennas and Propagation*, *22*(1), 74–78. doi:10.1109/TAP.1974.1140723

Nagaraj, K., Khandelwal, H., Killian, C., & Kompella, R. R. (2012). Hierarchy-aware distributed overlays in data centres using DC2. *Fourth International Conference on Communication Systems and Networks (COMSNETS)*, 1–10.

Namdar, A., Entezar, S. R., Rahimi, H., & Tajalli, H. (2010). Backward TAMM States In 1D Single-Negative Metamaterial Photonic Crystals. *Progress In Electromagnetics Research Letters*, *13*, 149–159. doi:10.2528/PIERL09121305

Nayar, B. K., Finlayson, N., Doran, N. J., Davey, S. T., Williams, D. L., & Arkwright, J. W. (1991). All-optical switching in a 200-m twin-core fiber nonlinear Mach–Zehnder interferometer. *Optics Letters*, *16*(6), 408–410. doi:10.1364/OL.16.000408 PMID:19773949

Nicholes, S. C. (2010, February). An 8×8 InP monolithic tunable optical router (MOTOR) packet forwarding chip. *Journal of Lightwave Technology*, *28*(4), 641–650. doi:10.1109/JLT.2009.2030145

Nikitin, P. V., Rao, K. V. S., Lam, S. F., Pillai, V., Martinez, R., & Heinrich, H. (2005). Power Reflection Coefficient Analysis for Complex Impedances in RFID Tag Design. *IEEE Transactions on Microwave Theory and Techniques*, *53*(9), 2721–2725. doi:10.1109/TMTT.2005.854191

Niranjan, V., Saxena, A. K., & Srivastava, K. V. (2012). CPW-fed slot patch antenna for 5.2/5.8 ghz wlan application. *PIERS Proceedings,* 1350-1352.

Nozaki, K., Matsuo, S., Fujii, T., Takeda, K., Ono, M., Shakoor, A., ... Notomi, M. (2016). Photonic-Crystal Nano-Photodetector with Ultrasmall Capacitance for On-Chip Light-To-Voltage Conversion without an Amplifier. *Optica*, *3*(5), 483–492. doi:10.1364/OPTICA.3.000483

Nylander, C., Liedberg, B., & Lundstrom, I. (1983). Surface plasmon resonance for gas and biosensing. *Sensors and Actuators*, *4*, 299–304. doi:10.1016/0250-6874(83)85036-7

Ojeda-Castaneda, J., Andres, P., & Diaz, A. (1986). Annular apodizers for low sensitivity to defocus and to spherical aberration. *Optics Letters*, *11*(8), 487–489. doi:10.1364/OL.11.000487 PMID:19738664

Ojeda-Castaneda, J., & Berriel-Valdos, L. R. (1988). Arbitrarily high focal depth with finite apertures. *Optics Letters*, *13*(3), 183–185. doi:10.1364/OL.13.000183 PMID:19742021

Ojeda-Castaneda, J., & Berriel-Valdos, L. R. (1990). Zone plate for arbitrarily high focal depth. *Applied Optics*, *29*(7), 994–997. doi:10.1364/AO.29.000994 PMID:20562947

Ojeda-Castaneda, J., Berriel-Valdos, L. R., & Montes, E. (1985). Spatial filter for increasing the depth of focus. *Optics Letters*, *10*(11), 520–522. doi:10.1364/OL.10.000520 PMID:19730471

Ojeda-Castaneda, J., & Gomez-Sarabia, C. M. (1989). Focal depth: Optimum annular apodizer. *Applied Optics*, *28*(20), 4263–4264. doi:10.1364/AO.28.004263 PMID:20555858

Osterberg, H., & Wilkins, J. E. Jr. (1949). The resolving power of a coated objective. *Journal of the Optical Society of America*, *39*(7), 553–557. doi:10.1364/JOSA.39.000553

Otto, A. (2000). Excitation of surface plasma waves in silver by the method of frustrated totalreflection. *Zeitschrift fur Physik, 216*(4), 398–410. doi:10.1007/BF01391532

Pechprasarn, S., & Somekh, M. (2011). *Analyzing Surface Plasmon Microscopy with Rigorous Diffraction Theory. IEEE Functional Optical Imaging.*

Pendry, J. B. (2000). Negative refraction makes a perfect lens. *Physical Review Letters, 85*(18), 3966–3969. doi:10.1103/PhysRevLett.85.3966 PMID:11041972

Pendry, J. B., Holden, A. J., Robbins, D. J., & Stewart, W. J. (1999). Magnetism from conductors and enhanced nonlinear phenomena. *IEEE Transactions on Microwave Theory and Techniques, 47*(11), 2075–2084. doi:10.1109/22.798002

Popov, A. K., Myslivets, S. A., George, T. F., & Shalaev, V. M. (2007). *2007 Compensating losses in positive- and negative-index metamaterials through nonlinear-optical quantum switching, IEEE paper on Lasers and Electro-Optics - Pacific Rim, 2007*. CLEO/Pacific Rim.

Povinelli, M. L., Johnson, S. G., Joannopoulos, J. D., & Pendry, J. B. (2003). Toward Photonic-Crystal Metamaterials: Creating Magnetic Emitters in Photonic Crystals. *Applied Physics Letters, 82*(7), 1069–1071. doi:10.1063/1.1544428

Pozar, D. M. (1998). *Microwave Engineering* (2nd ed.). New York: John Wiley &sons Inc.

Proietti, Yin, Yu, Ye, Nitta, Akella, & Yoo. (2012). All-Optical Physical Layer NACK in AWGR-Based Optical Interconnects. *IEEE Photonics Technology Letters, 24*(5), 410-412.

Pues, H., & Capelle, A. V. D. (1984). Accurate Transmission Line Model for the Rectangular MicrostripAntenna. *IEE Proc., 131*(H).

Pues, H. F., & Van De Capelle, A. R. (1989, November). An impedance-matching technique for increasing the bandwidth of microstrip antennas. *IEEE Transactions and Antenna Propagations, AP-37*(11), 1345–1354. doi:10.1109/8.43553

Qu, S. W., Li, J.-L., Chen, J. X., & Xue, Q. (2007). Ultrawideband Strip-Loaded Circular Slot Antenna With Improved Radiation Patterns. *IEEE Transactions on Antennas and Propagation, 55*(11), 3348–3353. doi:10.1109/TAP.2007.908847

Qu, S.-W., & Xue, Q. (2007). A Y-Shaped Stub Proximity Coupled V-Slot Microstrip Patch Antenna. *IEEE Antennas and Wireless Propagation Letters*, *6*, 40–42. doi:10.1109/LAWP.2007.891515

Raether, H., & Kretschmann, E. (1968). Radaitive decay of non-radiative surface plasmons excited by light. *Zeitschrift fur Naturforschung. Teil B. Anorganische Chemie, Organische Chemie, Biochemie, Biophysik, Biologie*, *23A*, 2135–2136.

Rakshit, J. K., Chattopadhyay, T., & Roy, J. N. (2013). Design of ring resonator based all optical switch for logic and arithmetic operations–a theoretical study. *Optik-International Journal for Light and Electron Optics.*, *124*(23), 6048–6057. doi:10.1016/j.ijleo.2013.04.075

Rakshit, J. K., & Roy, J. N. (2014). Micro ring resonator based all optical reconfigurable logic operations. *Optics Communications*, *321*, 38–46. doi:10.1016/j.optcom.2014.01.048

Rakshit, J. K., & Roy, J. N. (2016). All-optical ultrafast switching in a silicon microring resonator and its application to design multiplex/demultiplexer, adder/subtractor and comparator circuit. *Optica Applicata*, *46*, 517–539.

Rashed, A. N. Z. (2012). Optical Add Drop Multiplexer (OADM) Based on Dense Wavelength Division Multiplexing Technology in Next Generation Optical Networks. *International Journal of Multimedia & Ubiquitous Engineering*, *7*(1), 1–14.

Rastegarfar, Garcia, Laroachelle, & Rusch. (n.d.). Cross Layer Performance Analysis of Recirculation Buffers for Optical Data Centres. *Journal of Lightwave Technology*, *31*(3), 432–445. doi:10.1109/JLT.2012.2226865

Ray, M., Sarkar, S. K., Chakraborty, R. N., & Basuray, A. (2008). Birefringence measurement using a confocal setup with a pair of birefringent lenses. *Optical Engineering (Redondo Beach, Calif.)*, *47*(3), 033604–033608. doi:10.1117/1.2899095

Reininger, P., Kalchmair, S., Gansch, R., Andrews, A. M., Detz, H., Zederbauer, T., ... Strasser, G. (2012). Optimized Photonic Crystal Design for Quantum Well Infrared Photodetectors. *Proceedings of the Society for Photo-Instrumentation Engineers*, *8425*, 84250A. doi:10.1117/12.923045

Ren, J., Hu, W., Yin, Y., & Fan, R. (2014). Compact printed MIMO antenna for UWB applications. *IEEE Antennas and Wireless Propagation Letters*, *13*, 1517–1520. doi:10.1109/LAWP.2014.2343454

Richards, W. F. (1981). An Improved Theory for Microstrip Antennas and Applications. *IEEE Transactions on Antennas and Propagation*, *AP-29*.

Riordan, M., & Hoddeson, L. (1998). *Crystal Fire: The Birth of the Information Age* (1st ed.). New York: Sloan Technology Series.

Ritchie, R. H. (1957). Plasma losses by fast electrons in thin films. *Physical Review*, *106*(5), 874–881. doi:10.1103/PhysRev.106.874

Robinson, S., & Nakkeeran, R. (2011). Photonic Crystal Ring Resonator based Add-Drop Filter using Hexagonal Rods for CWDM Systems. *Optoelectronics Letters*, *7*(3), 164–166. doi:10.100711801-011-0172-2

Rudziński, A. (2007). Analytic Expressions for Electromagnetic Field Envelopes in a 1D Photonic Crystal. *Acta Physica Polonica A*, *111*(3), 323–333. doi:10.12693/APhysPolA.111.323

Ruppin, R. (2000). Surface polaritons of a left-handed medium. *Physics Letters. [Part A]*, *277*(1), 61–64. doi:10.1016/S0375-9601(00)00694-0

Russell, P. S. J. (2006). Photonic-Crystal Fibers. *Journal of Lightwave Technology*, *24*(12), 472–474. doi:10.1109/JLT.2006.885258

Saha, S., Biswas, S., & Mukhopadhyay, S. (2017). Optical scheme of conversion of a positionally encoded decimal digit to frequency encoded Boolean form using Mach–Zehnder interferometer-based semiconductor optical amplifier. *IET Optoelectronics*, *11*(5), 201–207. doi:10.1049/iet-opt.2016.0078

Santiago, A. G., Costa, J. R., & Fernandes, C. A. (2013). Broadband UHF RFID passive tag antenna for near-body applications. *IEEE Antennas and Wireless Propagation Letters*, *12*, 136–139. doi:10.1109/LAWP.2013.2243400

Sanyal, S., Bandyopadhyay, P., & Ghosh, A. (1998). Vector wave imagery using a birefringent lens. *Optical Engineering (Redondo Beach, Calif.)*, *37*(2), 592–599. doi:10.1117/1.601859

Sanyal, S., & Ghosh, A. (1999). Imaging characteristics of birefringent lenses under focused and defocused condition. *Optik-International Journal for Light and Electron Optics.*, *110*(11), 513–520.

Sanyal, S., & Ghosh, A. (2000). High focal depth with a quasi-bifocus birefringent lens. *Applied Optics*, *39*(14), 2321–2325. doi:10.1364/AO.39.002321 PMID:18345140

Sanyal, S., & Ghosh, A. (2002). Frequency response characteristics of a birefringent lens. *Optical Engineering (Redondo Beach, Calif.)*, *41*(3), 592–597. doi:10.1117/1.1430420

Sanyal, S., & Ghosh, A. (2003). Image assessment of a birefringent lens based on the factor of encircled energy. *Optical Engineering (Redondo Beach, Calif.)*, *42*(4), 1058–1064. doi:10.1117/1.1555736

Sanyal, S., Kawata, Y., Ghosh, A., & Mandal, S. (2004). Frequency response characteristics of a birefringent lens with off-axis aberrations. *Applied Optics*, *43*(19), 3838–3847. doi:10.1364/AO.43.003838 PMID:15250550

Sarkar, B., & Mukhopadhyay, S. (2017). An all optical scheme for implementing an integrated Pauli's X, Y and Z quantum gates with optical switches. *Journal of Optics*, *46*(2), 143–148. doi:10.100712596-017-0398-x

Sarkar, P. P., Ghosh, B., & Patra, S. N. (2016). Simulative study of all optical frequency encoded dibit based universal NAND and NOR logic gates using a reflective semiconductor optical amplifier and an add/drop multiplexer. *Journal of Optical Technology*, *83*(4), 257–262. doi:10.1364/JOT.83.000257

Sarkar, P. P., Ghosh, B., Patra, S. N., & Mukhopadhyay, S. (2017). A new scheme of an all optical frequency encoded dibit based latch with its simulated result. *Journal of Optical Technology*, *84*(9), 631–634. doi:10.1364/JOT.84.000631

Sarkar, P. P., & Mukhopadhyay, S. (2014). All optical frequency encoded NAND logic operation along with the simulated result. *Journal of Optics*, *43*(3), 177–182. doi:10.100712596-014-0197-6

Sarkar, P. P., Satpati, B., & Mukhopadhyay, S. (2013). New simulative studies on performance of semiconductor optical amplifier based optical switches like frequency converter and add-drop multiplexer for optical data processors. *Journal of Optics*, *42*(4), 360–366. doi:10.100712596-013-0133-1

Sarkar, T. K. (2006). *"Who Was James Clerk Maxwell and What Is/Was His Electromagnetic Theory", Seminar paper on IEEE AP/MTT HK Joint Chapter.* City University of Hong Kong.

Satija, J., Shukla, G. M., & Mukherji, S. (2011). Potential of dendrimeric architecture in surface Plasmon resonance biosensor. *IEEE 2010 International Conference on Systems in Medicine and Biology.*

Sato, K.-I., & Hasegawa, H. (2009, July). Optical networking technologies that will create future bandwidth abundant networks. *Journal of Optical Communications and Networking*, *1*(2), A81–A93. doi:10.1364/JOCN.1.000A81

Schaubert, D. H., Pozar, D. M., & Adrian, A. (1989, June). Effect of microstrip antenna substrate thickness and permittivity: Comparison of theories and experiment. *IEEE Transactions and Antennas Propagations*, *AP-37*(6), 677–682. doi:10.1109/8.29353

Scheumann, V., Zizlsperger, Z., Mack, J., Jung, G., Badia, A., Arnold, S., & Knoll, W. (1999). Probing the elecrochemical deposition and/or desorption of self-assembled and electropolymerizableorganic thin film by surface plasmon spectroscopy and atomic force microscopy. *Sensors and Actuators. B, Chemical*, *54*(1-2), 145–165. doi:10.1016/S0925-4005(98)00333-5

Selvan, K. T. (2007). Engineering Education: Presentation of Maxwell's Equations in Historical Perspective and the Likely Desirable Outcomes. *IEEE Antennas & Propagation Magazine*, *49*(5), 155–160. doi:10.1109/MAP.2007.4395329

Sengupta, D. L., & Sarkar, T. K. (2003). Maxwell, Hertz, the Maxwellians, and the Early History of Electromagnetic Waves. *IEEE Antennas & Propagation Magazine*, *45*(2), 13–19. doi:10.1109/MAP.2003.1203114

Shacham, A., & Bergman, K. (2007, July/August). Building ultralow-latency interconnection networks using photonic integration. *IEEE Micro*, *27*(4), 6–20. doi:10.1109/MM.2007.64

Shacham, A., Bergman, K., & Carloni, L. P. (2008, September). Photonic networks-on-chip for future generations of chip multiprocessors. *IEEE Transactions on Computers*, *57*(9), 1246–1260. doi:10.1109/TC.2008.78

Sharawi, M. (2017, April). Current Misuses and Future Prospects for Printed Multiple-Input, Multiple-Output Antenna Systems. *IEEE Antennas & Propagation Magazine*, 162–170.

Sharawi, M. S. (2017). Current Misuses and Future Prospects for Printed Multiple-Input, Multiple-Output Antenna Systems. *IEEE Antennas & Propagation Magazine*, *59*(2), 162–170. doi:10.1109/MAP.2017.2658346

Sharma, A. K., Jha, R., & Gupta, B. D. (2007). Fiber-Optic Sensors Based on Surface Plasmon Resonance: A Comprehensive Review. *IEEE Sensors Journal*, *7*(8), 1118–1129. doi:10.1109/JSEN.2007.897946

Shi, F., Chen, Y., Han, P., & Soukoulis, C. M. (2013). Investigation of One-Dimensional Photonic Bandgap Structures containing Lossy Double-Negative Metamaterials through the Bloch Impedance. *Journal of the Optical Society of America. B, Optical Physics*, *30*(6), 1473. doi:10.1364/JOSAB.30.001473

Shukla, Jain, & Srivastava. (2014). Physical Layer Analysis of Arrayed Waveguide Based Optical Switch. *International Journal of Applied Engineering Research*, *9*(21), 10035-10050.

Shukla, Jain, & Srivastava. (2015). Performance Evaluation of an AWG based Optical Router. *Optical and Quantum Electronics*, *48*(1).

Shukla, Jain, & Srivastava. (2016). Design of an Arrayed Waveguide Gratings based Optical Packet Switch. *Journal of Engineering Science and Technology*, *11*(12), 1705-1721.

Shukla, V. (2016). Design of AWG based Optical Switch for High Speed Optical Networks. *IJE Transactions A: Basics*, *29*(7), 909–915.

Shukla, V., & Jain, A. (2018). Design and analysis of high-speed optical routers for next generation data centre network. *Journal of Engineering Research*, *6*(2), 122–137.

Shukla, V., & Singh, O. P., Khare, M., Sharma, S., & Srivastava, R. (2018). Estimation of burst length in optical burst switching networks based on early release of control packet. *Journal of Engineering Research*, *6*(1).

Shukla, V., & Srivastava, R. (2015). WDM fiber delay lines and AWG based optical packet switch architecture. *Proceedings of National Conference on Innovative Trends in Computer Science Engineering (ITCSE-2015)*.

Simon, W., Werthen, M., & Wolff, I. (1998). A Novel Coplanar Transmission Line to Rectangular Waveguide Transition. *IEEE MTT-Society International Symposium Digest*, 257-260. 10.1109/MWSYM.1998.689369

Singh, Tiwari, & Srivastava. (2018). Design and analysis of hybrid optical and electronic buffer based optical packet switch. *Sādhanā*, *43*(2), 19.

Singh, S. P., Kar, S., & Jain, V. K. (2007). Performance of all-optical WDM network in presence of four-wave mixing, optical amplifier noise, and wavelength converter noise. *Fiber and Integrated Optics*, *26*(2), 79–97. doi:10.1080/01468030601131477

Singh, S., Kaur, S., Kaur, R., & Kaler, R. S. (2016). Photonic processing of all-optical Johnson counter using semiconductor optical amplifiers. *IET Optoelectronics*, *11*(1), 8–14. doi:10.1049/iet-opt.2015.0106

Soref, R. (2006). The past, present, and future of silicon photonics. *IEEE Journal of Selected Topics in Quantum Electronics*, *12*(6), 1678–1687. doi:10.1109/JSTQE.2006.883151

Southworth, G. C. (1980). *Principles and Application of Waveguide Transmission*. D Van Nostrand Co.

Sreekanth, K. V., Alapan, Y., ElKabbash, M., Ilker, E., Hinczewski, M., Gurkan, U. A., ... Strangi, G. (2016). Extreme sensitivity biosensing platform based on hyperbolic metamaterials. *Nature Materials*, *15*(6), 621–627. doi:10.1038/nmat4609 PMID:27019384

Srivastava, R. (n.d.). *WDM fiber delay lines based optical packet switch architectures* (Ph.D. dissertation). IIT Kanpur.

Sung, Y. (2011). A Printed Wide-Slot Antenna With a Modified L-Shaped Microstrip Line for Wideband Applications. *IEEE Transactions on Antennas and Propagation*, *59*(10), 3918–3922. doi:10.1109/TAP.2011.2163748

Sung, Y. (2012). Bandwidth Enhancement of a Microstrip Line-Fed Printed Wide-Slot Antenna with a Parasitic Centre Patch. *IEEE Transactions on Antennas and Propagation*, *60*(4), 1712–1716. doi:10.1109/TAP.2012.2186224

Suzuki, K., Hirayama, T., & Arakawa, T. (2016). *Proposal of compact TE/TM polarization switch based on microring resonator*. OECC/PC2016.

Sze, J. Y., & Wong, K. L. (2001). Bandwidth Enhancement of a Microstrip line fed Printed wide slot Antenna. *IEEE Transactions on Antennas and Propagation*, *49*(7), 1020–1024. doi:10.1109/8.933480

Taubenblatt, M. A. (2012, February). Optical interconnects for high-performance computing. *Journal of Lightwave Technology*, *30*(4), 448–457. doi:10.1109/JLT.2011.2172989

Tripathi, S., Mohan, A., & Yadav, S. (2015). A compact Koch fractal UWB MIMO antenna with WLAN band-rejection. *IEEE Antennas and Wireless Propagation Letters*, *14*, 1565–1568. doi:10.1109/LAWP.2015.2412659

Tseng, C.-F., Huang, C.-L., & Hsu, C.-H. (2006). A wideband planar inverted-F dielectric resonator antenna for RFID system applications. *Microwave and Optical Technology Letters*, *48*(7), 1302–1305. doi:10.1002/mop.21616

Tucker, R. S. (2006, December). The role of optics and electronics in high-capacity routers. *Journal of Lightwave Technology*, *24*(12), 4655–4673. doi:10.1109/JLT.2006.885774

Usha, S. P., Mishra, S. K., & Gupta, B. D. (2016). Fabrication and Characterization of a SPR Based Fiber Optic Sensor for the Detection of Chlorine Gas Using Silver and Zinc Oxide. *Materials (Basel)*, *8*(5), 2204–2216. doi:10.3390/ma8052204

Vallini, F., Gu, Q., Kats, M., Fainman, Y., & Frateschi, N. C. (2013). Carrier saturation in multiple quantum well metallo-dielectric semiconductor nanolaser: Is bulk material a better choice for gain media? *Optics Express*, *21*(22), 25985–25998. doi:10.1364/OE.21.025985 PMID:24216824

Verma, P., Deyasi, A., & Paul, P. (2018). Computing Reflectance of Three-Layer Surface Plasmon-Based Sensor at Visible Spectra. In Lecture Notes in Networks and Systems: Industry Innovative Innovations in Science, Engineering and Technology. Springer.

Verma, P., Paul, P., & Deyasi, A. (2017). Wavelength dependence of impedance and quality factor for MIM surface plasmon structure. *1st IEEE International Conference on Electronics, Materials Engineering and Nano-Technology*. 10.1109/IEMENTECH.2017.8076946

Verma, R., & Daya, K. S. (2011). Effect of Forbidden Bands of Electromagnetic Bandgap Engineered Ground Plane on the Response of Half Wave Length Linear Microwave Resonator. *Journal of Applied Physics*, 109.

Veselago, V. G. (1968). The electrodynamics of substances withsimultaneously negative values of ε and μ. *Soviet Physics - Uspekhi*, *10*(4), 509–514. doi:10.1070/PU1968v010n04ABEH003699

Veselago, V., Braginsky, L., Shklover, V., & Hafner, C. (2006). Negative Refractive Index Materials. *Journal of Computational and Theoretical Nanoscience*, *3*(2), 1–30. doi:10.1166/jctn.2006.3000

Wang, Y., Knoll, W., & Dostalek, J. (2011). Long range surface Plasmon resonance bacterial pathogen biosensor with magnetic nanoparticle assay. *IEEE 2011 International Workshop on Biophotonics*.

Wang, J., & Chen, L. R. (2015). Low crosstalk Bragg grating/Mach-Zehnder interferometer optical add-drop multiplexer in silicon photonics. *Optics Express*, *23*(20), 26450–26459. doi:10.1364/OE.23.026450 PMID:26480158

Wang, K. Y., & Foster, A. C. (2012). Ultralow power continuous-wave frequency conversion in hydrogenated amorphous silicon waveguides. *Optics Letters*, *37*(8), 1331–1333. doi:10.1364/OL.37.001331 PMID:22513676

Wang, L., Johannessen, E. A., Hammond, P. A., Cui, L., Reid, S. W. J., Cooper, J. M., & Cumming, D. R. S. (2005). A programmable microsystem using system-on-chip for real-time biotelemetry. *IEEE Transactions on Biomedical Engineering*, *52*(7), 1251–1260. doi:10.1109/TBME.2005.847562 PMID:16041988

Whiting, A. I., Abouraddy, A. F., Saleh, B. E. A., Teich, M. C., & Fourkas, J. T. (2003). Polarization-assisted transverse and axial optical superresolution. *Optics Express*, *11*(15), 1714–1723. doi:10.1364/OE.11.001714 PMID:19466051

Wilkins, J. E. Jr. (1950). The resolving power of a coated objective II. *Journal of the Optical Society of America*, *40*(4), 222–224. doi:10.1364/JOSA.40.000222

Wi, S.-H., Lee, Y.-S., & Yook, J.-G. (2007, April). Wideband Microstrip Patch Antenna With U-Shaped Parasitic Elements. *IEEE Transactions and Antenna Propagations*, *55*(4), 1196–1199. doi:10.1109/TAP.2007.893427

Wong, E. W. M., & Zukerman, M. (2006, December). Bandwidth and buffer trade offs in optical packet switching. *Journal of Lightwave Technology*, *24*(12), 4790–4798. doi:10.1109/JLT.2006.885249

Wood, R. W. (1902). On a remarkable case of uneven distribution of light in a diffraction grating spectrum. *Phil. Mag. Ser. 6*, *4*(21), 396–402. doi:10.1080/14786440209462857

Xi, K., Kao, Y.-H., & Chao, H. J. (2013). *A Petabit buffer less optical switch for data center networks. In Optical Interconnects for Future Data Center Networks* (pp. 135–154). Springer New York.

Xiong, H., Hong, J. S., & Peng, Y. H. (2012). Impedance Bandwidth and Gain Improvement for Microstrip Antenna using Metamaterials. *Radioenginnering*, *21*, 993–998.

Xu, L., Perros, H. G., & Rouskas, G. (2001, January). Techniques for optical packet switching and optical burst switching. *IEEE Communications Magazine*, *39*(1), 136–142. doi:10.1109/35.894388

Xu, X., Chen, H., Xiong, Z., Jin, A., Gu, C., Cheng, B., & Zhang, D. (2007). Fabrication of Photonic Crystals on Several Kinds of Semiconductor Materials by using Focused-Ion Beam Method. *Thin Solid Films*, *515*(22), 8297–8300. doi:10.1016/j.tsf.2007.02.061

Yablonovitch, E. (1987). Inhibited Spontaneous Emission in Solid-State Physics and Electronics. *Physical Review Letters*, *58*(20), 2059–2061. doi:10.1103/PhysRevLett.58.2059 PMID:10034639

Yamamoto, M. (2008). Surface Plasmon (SPR) Theory, Tutorial. *Rev Polarogr.*, *48*. doi:10.5189/revpolarography.48.209

Yang & Samii. (2009). *Electromagnetic Band Gap Structures in Antenna Engineering*. Cambridge University Press.

Yang, Fan, & Feng. (2005). A spiral electromagnetic bandgap (EBG) structure and its application in microstrip antenna arrays. *Asia-Pacific Microwave Conference Proceedings*. doi:10.1063/1.5005254

Yang, L., Fan, M., Chen, F., She, J., & Feng, Z. (2005). A Novel Compact Electromagnetic–Bandgap (EBG) Structure and Its Applications for Microwave Circuits. *IEEE Transactions on Microwave Theory and Techniques*, *53*(1), 183–190. doi:10.1109/TMTT.2004.839322

Yang, X., Xiang, C., & Zhang, G. (1993). Birefringent common-path interferometer for testing large convex spherical surfaces. *Optical Engineering (Redondo Beach, Calif.)*, *32*(5), 1080–1083. doi:10.1117/12.133348

Yin, Y. (2013, March). LIONS: An AWGR-based low-latency optical switch for high-performance computing and data centres. *Selected Topics in Quantum Electronics. IEEE Journal of*, *19*(2), 3600409–3600409.

Yokota, H. (2015). *Control of localized surface plasmon resonance energy in monolayer structures of gold and silver nanoparticles* (Vol. 40). Physical Chemistry Chemical Physics.

Yoo, S. J. B. (2006, December). Optical packet and burst switching technologies for the future photonic Internet. *Journal of Lightwave Technology*, *24*(12), 4468–4492. doi:10.1109/JLT.2006.886060

Yoo, S. J. B. (2011, March/April). Energy efficiency in the future Internet: The role of optical packet switching and optical-label switching. *IEEE Journal of Selected Topics in Quantum Electronics*, *17*(2), 406–418. doi:10.1109/JSTQE.2010.2076793

Yoshimura, Y. (1972). A Microstripline slot Antenna. *IEEE Transactions on Microwave Theory and Techniques*, 760–762.

Zavats, A. V., Darmanyad, S. A., Gcrard, D., Salomnn, L., & De Fornel, F. (2005). Surface Plasmon Polaritons on Nanostructured Surfaces and Thin Films. *IEEE 10th International Conference on Mathematical Methods in Electromagnetic Theory*, 73 – 78.

Zhang, J.-Y., Zhang, F., Tian, W.-P., & Luo, Y.-L. (2015). Acs-fed UWB-MIMO antenna with shared radiator. *Electronics Letters*, *51*(7), 1301–1302. doi:10.1049/el.2015.1327

Zhang, S., Lau, B. K., Sunesson, A., & He, S. (2012). Closely-packed UWB MIMO/diversity antenna with different patterns and polarizations for USB dongle applications. *IEEE Transactions on Antennas and Propagation*, *60*(9), 4372–4380. doi:10.1109/TAP.2012.2207049

Zhang, S., Ying, Z., Xiong, J., & He, S. (2009). Ultrawideband MIMO/diversity antennas with a tree-like structure to enhance wideband isolation. *IEEE Antennas and Wireless Propagation Letters*, *8*, 1279–1282. doi:10.1109/LAWP.2009.2037027

Zhang, W., Wang, H., & Bergman, K. (2012, December). Next-generation optically-interconnected high performance data centres. *Journal of Lightwave Technology*, *30*(24), 3836–3844. doi:10.1109/JLT.2012.2212696

Zhao, D., Ding, Y. G., Wang, Y., & Ruan, C. J. (2010). Linear Analysis of a Rectangular Waveguide Cyclotron Maser with a Sheet Electron Beam. *Physics of Plasmas*, *17*(11), 113110. doi:10.1063/1.3514596

Zhu, J., Zhou, P., Su, X., & You, Z. (2016). Accurate and fast 3D surface measurement with temporal-spatial binary encoding structured illumination. *Optics Express*, *24*(25), 28549–28560. doi:10.1364/OE.24.028549 PMID:27958499

Ziolkowski, R. W., & Heyman, E. (2001). Wave propagation in media having negative permittivity and permeability. *Physical Review. E*, *64*(5), 056625. doi:10.1103/PhysRevE.64.056625 PMID:11736134

Zoiros, K. E., Houbavlis, T., & Kalyvas, M. (2004). Ultra-high speed all-optical shift registers and their applications in OTDM networks. *Optical and Quantum Electronics*, *36*(11), 1005–1053. doi:10.100711082-004-2040-9

Related References

To continue our tradition of advancing academic research, we have compiled a list of recommended IGI Global readings. These references will provide additional information and guidance to further enrich your knowledge and assist you with your own research and future publications.

Abed, S., Khir, T., & Ben Brahim, A. (2016). Thermodynamic and Energy Study of a Regenerator in Gas Turbine Cycle and Optimization of Performances. *International Journal of Energy Optimization and Engineering*, *5*(2), 25–44. doi:10.4018/IJEOE.2016040102

Abu Bakar, W. A., Abdullah, W. N., Ali, R., & Mokhtar, W. N. (2016). Polymolybdate Supported Nano Catalyst for Desulfurization of Diesel. In T. Saleh (Ed.), *Applying Nanotechnology to the Desulfurization Process in Petroleum Engineering* (pp. 263–280). Hershey, PA: IGI Global. doi:10.4018/978-1-4666-9545-0.ch009

Addo-Tenkorang, R., Helo, P., & Kantola, J. (2016). Engineer-To-Order Product Development: A Communication Network Analysis for Supply-Chain's Sustainable Competitive Advantage. In R. Addo-Tenkorang, J. Kantola, P. Helo, & A. Shamsuzzoha (Eds.), *Supply Chain Strategies and the Engineer-to-Order Approach* (pp. 43–59). Hershey, PA: IGI Global. doi:10.4018/978-1-5225-0021-6.ch003

Adebiyi, I. D., Popoola, P. A., & Pityana, S. (2016). Mitigation of Wear Damage by Laser Surface Alloying Technique. In E. Akinlabi, R. Mahamood, & S. Akinlabi (Eds.), *Advanced Manufacturing Techniques Using Laser Material Processing* (pp. 172–196). Hershey, PA: IGI Global. doi:10.4018/978-1-5225-0329-3.ch007

Ahmad, W. (2016). Sulfur in Petroleum: Petroleum Desulfurization Techniques. In T. Saleh (Ed.), *Applying Nanotechnology to the Desulfurization Process in Petroleum Engineering* (pp. 1–52). Hershey, PA: IGI Global. doi:10.4018/978-1-4666-9545-0.ch001

Ahmed, I., Ahmad, N., Mehmood, I., Haq, I. U., Hassan, M., & Khan, M. U. (2016). Applications of Nanotechnology in Transportation Engineering. In A. Khitab & W. Anwar (Eds.), *Advanced Research on Nanotechnology for Civil Engineering Applications* (pp. 180–207). Hershey, PA: IGI Global. doi:10.4018/978-1-5225-0344-6.ch006

Aikhuele, D. (2018). A Study of Product Development Engineering and Design Reliability Concerns. *International Journal of Applied Industrial Engineering*, *5*(1), 79–89. doi:10.4018/IJAIE.2018010105

Al-Najar, B. T., & Bououdina, M. (2016). Bioinspired Nanoparticles for Efficient Drug Delivery System. In M. Bououdina (Ed.), *Emerging Research on Bioinspired Materials Engineering* (pp. 69–103). Hershey, PA: IGI Global. doi:10.4018/978-1-4666-9811-6.ch003

Al-Shebeeb, O. A., Rangaswamy, S., Gopalakrishan, B., & Devaru, D. G. (2017). Evaluation and Indexing of Process Plans Based on Electrical Demand and Energy Consumption. *International Journal of Manufacturing, Materials, and Mechanical Engineering*, *7*(3), 1–19. doi:10.4018/IJMMME.2017070101

Alexakis, H., & Makris, N. (2016). Validation of the Discrete Element Method for the Limit Stability Analysis of Masonry Arches. In V. Sarhosis, K. Bagi, J. Lemos, & G. Milani (Eds.), *Computational Modeling of Masonry Structures Using the Discrete Element Method* (pp. 292–325). Hershey, PA: IGI Global. doi:10.4018/978-1-5225-0231-9.ch012

AlMegren, H. A., Gonzalez-Cortes, S., Huang, Y., Chen, H., Qian, Y., Alkinany, M., ... Xiao, T. (2016). Preparation of Deep Hydrodesulfurzation Catalysts for Diesel Fuel using Organic Matrix Decomposition Method. In H. Al-Megren & T. Xiao (Eds.), *Petrochemical Catalyst Materials, Processes, and Emerging Technologies* (pp. 216–253). Hershey, PA: IGI Global. doi:10.4018/978-1-4666-9975-5.ch009

Alshammari, A., Kalevaru, V. N., Bagabas, A., & Martin, A. (2016). Production of Ethylene and its Commercial Importance in the Global Market. In H. Al-Megren & T. Xiao (Eds.), *Petrochemical Catalyst Materials, Processes, and Emerging Technologies* (pp. 82–115). Hershey, PA: IGI Global. doi:10.4018/978-1-4666-9975-5.ch004

Amel, M. (2016). Synthesis, Characterizations, and Biological Effects Study of Some Quinoline Family. In M. Bououdina (Ed.), *Emerging Research on Bioinspired Materials Engineering* (pp. 160–196). Hershey, PA: IGI Global. doi:10.4018/978-1-4666-9811-6.ch006

Amna, T., Haasan, M. S., Khil, M., & Hwang, I. (2016). Impact of Electrospun Biomimetic Extracellular Environment on Proliferation and Intercellular Communication of Muscle Precursor Cells: An Overview – Intercellular Communication of Muscle Precursor Cells with Extracellular Environment. In M. Bououdina (Ed.), *Emerging Research on Bioinspired Materials Engineering* (pp. 247–265). Hershey, PA: IGI Global. doi:10.4018/978-1-4666-9811-6.ch009

Amuda, M. O., Lawal, T. F., & Akinlabi, E. T. (2017). Research Progress on Rheological Behavior of AA7075 Aluminum Alloy During Hot Deformation. *International Journal of Materials Forming and Machining Processes*, *4*(1), 53–96. doi:10.4018/IJMFMP.2017010104

An, M., & Qin, Y. (2016). Challenges of Railway Safety Risk Assessment and Maintenance Decision Making. In B. Rai (Ed.), *Handbook of Research on Emerging Innovations in Rail Transportation Engineering* (pp. 173–211). Hershey, PA: IGI Global. doi:10.4018/978-1-5225-0084-1.ch009

Anil, M., Ayyildiz-Tamis, D., Tasdemir, S., Sendemir-Urkmez, A., & Gulce-Iz, S. (2016). Bioinspired Materials and Biocompatibility. In M. Bououdina (Ed.), *Emerging Research on Bioinspired Materials Engineering* (pp. 294–322). Hershey, PA: IGI Global. doi:10.4018/978-1-4666-9811-6.ch011

Armutlu, H. (2018). Intelligent Biomedical Engineering Operations by Cloud Computing Technologies. In U. Kose, G. Guraksin, & O. Deperlioglu (Eds.), *Nature-Inspired Intelligent Techniques for Solving Biomedical Engineering Problems* (pp. 297–317). Hershey, PA: IGI Global. doi:10.4018/978-1-5225-4769-3.ch015

Arokiyaraj, S., Saravanan, M., Bharanidharan, R., Islam, V. I., Bououdina, M., & Vincent, S. (2016). Green Synthesis of Metallic Nanoparticles Using Plant Compounds and Their Applications: Metallic Nanoparticles Synthesis Using Plants. In M. Bououdina (Ed.), *Emerging Research on Bioinspired Materials Engineering* (pp. 1–34). Hershey, PA: IGI Global. doi:10.4018/978-1-4666-9811-6.ch001

Atik, M., Sadek, M., & Shahrour, I. (2017). Single-Run Adaptive Pushover Procedure for Shear Wall Structures. In V. Plevris, G. Kremmyda, & Y. Fahjan (Eds.), *Performance-Based Seismic Design of Concrete Structures and Infrastructures* (pp. 59–83). Hershey, PA: IGI Global. doi:10.4018/978-1-5225-2089-4.ch003

Aydin, A., Akyol, E., Gungor, M., Kaya, A., & Tasdelen, S. (2018). Geophysical Surveys in Engineering Geology Investigations With Field Examples. In N. Ceryan (Ed.), *Handbook of Research on Trends and Digital Advances in Engineering Geology* (pp. 257–280). Hershey, PA: IGI Global. doi:10.4018/978-1-5225-2709-1.ch007

Azevedo, N. M., Lemos, J. V., & Rocha de Almeida, J. (2016). Discrete Element Particle Modelling of Stone Masonry. In V. Sarhosis, K. Bagi, J. Lemos, & G. Milani (Eds.), *Computational Modeling of Masonry Structures Using the Discrete Element Method* (pp. 146–170). Hershey, PA: IGI Global. doi:10.4018/978-1-5225-0231-9.ch007

Bamufleh, H. S., Noureldin, M. M., & El-Halwagi, M. M. (2016). Sustainable Process Integration in the Petrochemical Industries. In H. Al-Megren & T. Xiao (Eds.), *Petrochemical Catalyst Materials, Processes, and Emerging Technologies* (pp. 150–163). Hershey, PA: IGI Global. doi:10.4018/978-1-4666-9975-5.ch006

Banerjee, S., Gautam, R. K., Gautam, P. K., Jaiswal, A., & Chattopadhyaya, M. C. (2016). Recent Trends and Advancement in Nanotechnology for Water and Wastewater Treatment: Nanotechnological Approach for Water Purification. In A. Khitab & W. Anwar (Eds.), *Advanced Research on Nanotechnology for Civil Engineering Applications* (pp. 208–252). Hershey, PA: IGI Global. doi:10.4018/978-1-5225-0344-6.ch007

Bas, T. G. (2017). Nutraceutical Industry with the Collaboration of Biotechnology and Nutrigenomics Engineering: The Significance of Intellectual Property in the Entrepreneurship and Scientific Research Ecosystems. In T. Bas & J. Zhao (Eds.), *Comparative Approaches to Biotechnology Development and Use in Developed and Emerging Nations* (pp. 1–17). Hershey, PA: IGI Global. doi:10.4018/978-1-5225-1040-6.ch001

Beale, R., & André, J. (2017). *Design Solutions and Innovations in Temporary Structures*. Hershey, PA: IGI Global. doi:10.4018/978-1-5225-2199-0

Behnam, B. (2017). Simulating Post-Earthquake Fire Loading in Conventional RC Structures. In P. Samui, S. Chakraborty, & D. Kim (Eds.), *Modeling and Simulation Techniques in Structural Engineering* (pp. 425–444). Hershey, PA: IGI Global. doi:10.4018/978-1-5225-0588-4.ch015

Ben Hamida, I., Salah, S. B., Msahli, F., & Mimouni, M. F. (2018). Distribution Network Reconfiguration Using SPEA2 for Power Loss Minimization and Reliability Improvement. *International Journal of Energy Optimization and Engineering*, *7*(1), 50–65. doi:10.4018/IJEOE.2018010103

Benjamin, S. R., de Lima, F., & Rathoure, A. K. (2016). Genetically Engineered Microorganisms for Bioremediation Processes: GEMs for Bioremediaton. In A. Rathoure & V. Dhatwalia (Eds.), *Toxicity and Waste Management Using Bioremediation* (pp. 113–140). Hershey, PA: IGI Global. doi:10.4018/978-1-4666-9734-8.ch006

Bhaskar, S. V., & Kudal, H. N. (2017). Effect of TiCN and AlCrN Coating on Tribological Behaviour of Plasma-nitrided AISI 4140 Steel. *International Journal of Surface Engineering and Interdisciplinary Materials Science*, *5*(2), 1–17. doi:10.4018/IJSEIMS.2017070101

Bhowmik, S., Sahoo, P., Acharyya, S. K., Dhar, S., & Chattopadhyay, J. (2016). Effect of Microstructure Degradation on Fracture Toughness of 20MnMoNi55 Steel in DBT Region. *International Journal of Manufacturing, Materials, and Mechanical Engineering*, *6*(3), 11–27. doi:10.4018/IJMMME.2016070102

Bhutto, A. W., Abro, R., Abbas, T., Yu, G., & Chen, X. (2016). Desulphurization of Fuel Oils Using Ionic Liquids. In H. Al-Megren & T. Xiao (Eds.), *Petrochemical Catalyst Materials, Processes, and Emerging Technologies* (pp. 254–284). Hershey, PA: IGI Global. doi:10.4018/978-1-4666-9975-5.ch010

Bhuyan, D. (2018). Designing of a Twin Tube Shock Absorber: A Study in Reverse Engineering. In K. Kumar & J. Davim (Eds.), *Design and Optimization of Mechanical Engineering Products* (pp. 83–104). Hershey, PA: IGI Global. doi:10.4018/978-1-5225-3401-3.ch005

Bouloudenine, M., & Bououdina, M. (2016). Toxic Effects of Engineered Nanoparticles on Living Cells. In M. Bououdina (Ed.), *Emerging Research on Bioinspired Materials Engineering* (pp. 35–68). Hershey, PA: IGI Global. doi:10.4018/978-1-4666-9811-6.ch002

Brunetti, A., Sellaro, M., Drioli, E., & Barbieri, G. (2016). Membrane Engineering and its Role in Oil Refining and Petrochemical Industry. In H. Al-Megren & T. Xiao (Eds.), *Petrochemical Catalyst Materials, Processes, and Emerging Technologies* (pp. 116–149). Hershey, PA: IGI Global. doi:10.4018/978-1-4666-9975-5.ch005

Bügler, M., & Borrmann, A. (2016). Simulation Based Construction Project Schedule Optimization: An Overview on the State-of-the-Art. In F. Miranda & C. Abreu (Eds.), *Handbook of Research on Computational Simulation and Modeling in Engineering* (pp. 482–507). Hershey, PA: IGI Global. doi:10.4018/978-1-4666-8823-0.ch016

Calderon, F. A., Giolo, E. G., Frau, C. D., Rengel, M. G., Rodriguez, H., Tornello, M., ... Gallucci, R. (2018). Seismic Microzonation and Site Effects Detection Through Microtremors Measures: A Review. In N. Ceryan (Ed.), *Handbook of Research on Trends and Digital Advances in Engineering Geology* (pp. 326–349). Hershey, PA: IGI Global. doi:10.4018/978-1-5225-2709-1.ch009

Carmona-Murillo, J., & Valenzuela-Valdés, J. F. (2016). Motivation on Problem Based Learning. In D. Fonseca & E. Redondo (Eds.), *Handbook of Research on Applied E-Learning in Engineering and Architecture Education* (pp. 179–203). Hershey, PA: IGI Global. doi:10.4018/978-1-4666-8803-2.ch009

Ceryan, N. (2016). A Review of Soft Computing Methods Application in Rock Mechanic Engineering. In P. Samui (Ed.), *Handbook of Research on Advanced Computational Techniques for Simulation-Based Engineering* (pp. 1–70). Hershey, PA: IGI Global. doi:10.4018/978-1-4666-9479-8.ch001

Ceryan, N., & Can, N. K. (2018). Prediction of The Uniaxial Compressive Strength of Rocks Materials. In N. Ceryan (Ed.), *Handbook of Research on Trends and Digital Advances in Engineering Geology* (pp. 31–96). Hershey, PA: IGI Global. doi:10.4018/978-1-5225-2709-1.ch002

Ceryan, S. (2018). Weathering Indices Used in Evaluation of the Weathering State of Rock Material. In N. Ceryan (Ed.), *Handbook of Research on Trends and Digital Advances in Engineering Geology* (pp. 132–186). Hershey, PA: IGI Global. doi:10.4018/978-1-5225-2709-1.ch004

Chandrasekaran, S., Silva, B., Patil, A., Oo, A. M., & Campbell, M. (2016). Evaluating Engineering Students' Perceptions: The Impact of Team-Based Learning Practices in Engineering Education. *International Journal of Quality Assurance in Engineering and Technology Education*, *5*(4), 42–59. doi:10.4018/IJQAETE.2016100103

Chen, H., Padilla, R. V., & Besarati, S. (2017). Supercritical Fluids and Their Applications in Power Generation. In L. Chen & Y. Iwamoto (Eds.), *Advanced Applications of Supercritical Fluids in Energy Systems* (pp. 369–402). Hershey, PA: IGI Global. doi:10.4018/978-1-5225-2047-4.ch012

Chen, L. (2017). Principles, Experiments, and Numerical Studies of Supercritical Fluid Natural Circulation System. In L. Chen & Y. Iwamoto (Eds.), *Advanced Applications of Supercritical Fluids in Energy Systems* (pp. 136–187). Hershey, PA: IGI Global. doi:10.4018/978-1-5225-2047-4.ch005

Clementi, F., Di Sciascio, G., Di Sciascio, S., & Lenci, S. (2017). Influence of the Shear-Bending Interaction on the Global Capacity of Reinforced Concrete Frames: A Brief Overview of the New Perspectives. In V. Plevris, G. Kremmyda, & Y. Fahjan (Eds.), *Performance-Based Seismic Design of Concrete Structures and Infrastructures* (pp. 84–111). Hershey, PA: IGI Global. doi:10.4018/978-1-5225-2089-4.ch004

Cortés-Polo, D., Calle-Cancho, J., Carmona-Murillo, J., & González-Sánchez, J. (2017). Future Trends in Mobile-Fixed Integration for Next Generation Networks: Classification and Analysis. *International Journal of Vehicular Telematics and Infotainment Systems*, *1*(1), 33–53. doi:10.4018/IJVTIS.2017010103

Cui, X., Zeng, S., Li, Z., Zheng, Q., Yu, X., & Han, B. (2018). Advanced Composites for Civil Engineering Infrastructures. In K. Kumar & J. Davim (Eds.), *Composites and Advanced Materials for Industrial Applications* (pp. 212–248). Hershey, PA: IGI Global. doi:10.4018/978-1-5225-5216-1.ch010

Dalgıç, S., & Kuşku, İ. (2018). Geological and Geotechnical Investigations in Tunneling. In N. Ceryan (Ed.), *Handbook of Research on Trends and Digital Advances in Engineering Geology* (pp. 482–529). Hershey, PA: IGI Global. doi:10.4018/978-1-5225-2709-1.ch014

de la Varga, D., Soto, M., Arias, C. A., van Oirschot, D., Kilian, R., Pascual, A., & Álvarez, J. A. (2017). Constructed Wetlands for Industrial Wastewater Treatment and Removal of Nutrients. In Á. Val del Río, J. Campos Gómez, & A. Mosquera Corral (Eds.), *Technologies for the Treatment and Recovery of Nutrients from Industrial Wastewater* (pp. 202–230). Hershey, PA: IGI Global. doi:10.4018/978-1-5225-1037-6.ch008

del Valle-Zermeño, R., Chimenos, J. M., & Formosa, J. (2016). Flue Gas Desulfurization: Processes and Technologies. In T. Saleh (Ed.), *Applying Nanotechnology to the Desulfurization Process in Petroleum Engineering* (pp. 337–377). Hershey, PA: IGI Global. doi:10.4018/978-1-4666-9545-0.ch011

Delgado, J. M., Henriques, A. A., & Delgado, R. M. (2016). Structural Non-Linear Models and Simulation Techniques: An Efficient Combination for Safety Evaluation of RC Structures. In F. Miranda & C. Abreu (Eds.), *Handbook of Research on Computational Simulation and Modeling in Engineering* (pp. 540–584). Hershey, PA: IGI Global. doi:10.4018/978-1-4666-8823-0.ch018

Delgado, P. S., Arêde, A., Pouca, N. V., & Costa, A. (2016). Numerical Modeling of RC Bridges for Seismic Risk Analysis. In F. Miranda & C. Abreu (Eds.), *Handbook of Research on Computational Simulation and Modeling in Engineering* (pp. 457–481). Hershey, PA: IGI Global. doi:10.4018/978-1-4666-8823-0.ch015

Deng, Y., & Liu, S. (2016). Catalysis with Room Temperature Ionic Liquids Mediated Metal Nanoparticles. In H. Al-Megren & T. Xiao (Eds.), *Petrochemical Catalyst Materials, Processes, and Emerging Technologies* (pp. 285–329). Hershey, PA: IGI Global. doi:10.4018/978-1-4666-9975-5.ch011

Deperlioglu, O. (2018). Intelligent Techniques Inspired by Nature and Used in Biomedical Engineering. In U. Kose, G. Guraksin, & O. Deperlioglu (Eds.), *Nature-Inspired Intelligent Techniques for Solving Biomedical Engineering Problems* (pp. 51–77). Hershey, PA: IGI Global. doi:10.4018/978-1-5225-4769-3.ch003

Dias, G. L., Magalhães, R. R., Ferreira, D. D., & Vitoriano, F. A. (2016). The Use of a Robotic Arm for Displacement Measurements in a Cantilever beam. *International Journal of Manufacturing, Materials, and Mechanical Engineering*, *6*(3), 45–57. doi:10.4018/IJMMME.2016070104

Dimitratos, N., Villa, A., Chan-Thaw, C. E., Hammond, C., & Prati, L. (2016). Valorisation of Glycerol to Fine Chemicals and Fuels. In H. Al-Megren & T. Xiao (Eds.), *Petrochemical Catalyst Materials, Processes, and Emerging Technologies* (pp. 352–384). Hershey, PA: IGI Global. doi:10.4018/978-1-4666-9975-5.ch013

Dixit, A. (2018). Application of Silica-Gel-Reinforced Aluminium Composite on the Piston of Internal Combustion Engine: Comparative Study of Silica-Gel-Reinforced Aluminium Composite Piston With Aluminium Alloy Piston. In K. Kumar & J. Davim (Eds.), *Composites and Advanced Materials for Industrial Applications* (pp. 63–98). Hershey, PA: IGI Global. doi:10.4018/978-1-5225-5216-1.ch004

Drei, A., Milani, G., & Sincraian, G. (2016). Application of DEM to Historic Masonries, Two Case-Studies in Portugal and Italy: Aguas Livres Aqueduct and Arch-Tympana of a Church. In V. Sarhosis, K. Bagi, J. Lemos, & G. Milani (Eds.), *Computational Modeling of Masonry Structures Using the Discrete Element Method* (pp. 326–366). Hershey, PA: IGI Global. doi:10.4018/978-1-5225-0231-9.ch013

Dutta, S., Roy, P. K., & Nandi, D. (2016). Optimal Allocation of Static Synchronous Series Compensator Controllers using Chemical Reaction Optimization for Reactive Power Dispatch. *International Journal of Energy Optimization and Engineering*, *5*(3), 43–62. doi:10.4018/IJEOE.2016070103

Dutta, S., Roy, P. K., & Nandi, D. (2016). Quasi Oppositional Teaching-Learning based Optimization for Optimal Power Flow Incorporating FACTS. *International Journal of Energy Optimization and Engineering*, *5*(2), 64–84. doi:10.4018/IJEOE.2016040104

Eloy, S., Dias, M. S., Lopes, P. F., & Vilar, E. (2016). Digital Technologies in Architecture and Engineering: Exploring an Engaged Interaction within Curricula. In D. Fonseca & E. Redondo (Eds.), *Handbook of Research on Applied E-Learning in Engineering and Architecture Education* (pp. 368–402). Hershey, PA: IGI Global. doi:10.4018/978-1-4666-8803-2.ch017

Elsayed, A. M., Dakkama, H. J., Mahmoud, S., Al-Dadah, R., & Kaialy, W. (2017). Sustainable Cooling Research Using Activated Carbon Adsorbents and Their Environmental Impact. In T. Kobayashi (Ed.), *Applied Environmental Materials Science for Sustainability* (pp. 186–221). Hershey, PA: IGI Global. doi:10.4018/978-1-5225-1971-3.ch009

Ercanoglu, M., & Sonmez, H. (2018). General Trends and New Perspectives on Landslide Mapping and Assessment Methods. In N. Ceryan (Ed.), *Handbook of Research on Trends and Digital Advances in Engineering Geology* (pp. 350–379). Hershey, PA: IGI Global. doi:10.4018/978-1-5225-2709-1.ch010

Erinosho, M. F., Akinlabi, E. T., & Pityana, S. (2016). Enhancement of Surface Integrity of Titanium Alloy with Copper by Means of Laser Metal Deposition Process. In E. Akinlabi, R. Mahamood, & S. Akinlabi (Eds.), *Advanced Manufacturing Techniques Using Laser Material Processing* (pp. 60–91). Hershey, PA: IGI Global. doi:10.4018/978-1-5225-0329-3.ch004

Farag, H., & Kishida, M. (2016). Kinetic Models for Complex Parallel–Consecutive Reactions Assessment of Reaction Network and Product Selectivity. In H. Al-Megren & T. Xiao (Eds.), *Petrochemical Catalyst Materials, Processes, and Emerging Technologies* (pp. 330–351). Hershey, PA: IGI Global. doi:10.4018/978-1-4666-9975-5.ch012

Faroz, S. A., Pujari, N. N., Rastogi, R., & Ghosh, S. (2017). Risk Analysis of Structural Engineering Systems Using Bayesian Inference. In P. Samui, S. Chakraborty, & D. Kim (Eds.), *Modeling and Simulation Techniques in Structural Engineering* (pp. 390–424). Hershey, PA: IGI Global. doi:10.4018/978-1-5225-0588-4.ch014

Fernando, P. R., Hamigah, T., Disne, S., Wickramasingha, G. G., & Sutharshan, A. (2018). The Evaluation of Engineering Properties of Low Cost Concrete Blocks by Partial Doping of Sand with Sawdust: Low Cost Sawdust Concrete Block. *International Journal of Strategic Engineering*, *1*(2), 26–42. doi:10.4018/IJoSE.2018070103

Fragiadakis, M., Stefanou, I., & Psycharis, I. N. (2016). Vulnerability Assessment of Damaged Classical Multidrum Columns. In V. Sarhosis, K. Bagi, J. Lemos, & G. Milani (Eds.), *Computational Modeling of Masonry Structures Using the Discrete Element Method* (pp. 235–253). Hershey, PA: IGI Global. doi:10.4018/978-1-5225-0231-9.ch010

Gaines, T. W., Williams, K. R., & Wagener, K. B. (2016). ADMET: Functionalized Polyolefins. In H. Al-Megren & T. Xiao (Eds.), *Petrochemical Catalyst Materials, Processes, and Emerging Technologies* (pp. 1–21). Hershey, PA: IGI Global. doi:10.4018/978-1-4666-9975-5.ch001

Garg, H. (2016). Bi-Criteria Optimization for Finding the Optimal Replacement Interval for Maintaining the Performance of the Process Industries. In P. Vasant, G. Weber, & V. Dieu (Eds.), *Handbook of Research on Modern Optimization Algorithms and Applications in Engineering and Economics* (pp. 643–675). Hershey, PA: IGI Global. doi:10.4018/978-1-4666-9644-0.ch025

Gaspar, P. D., Dinho da Silva, P., Gonçalves, J. P., & Carneiro, R. (2016). Computational Modelling and Simulation to Assist the Improvement of Thermal Performance and Energy Efficiency in Industrial Engineering Systems: Application to Cold Stores. In F. Miranda & C. Abreu (Eds.), *Handbook of Research on Computational Simulation and Modeling in Engineering* (pp. 1–68). Hershey, PA: IGI Global. doi:10.4018/978-1-4666-8823-0.ch001

Ge, H., Tang, M., & Wen, X. (2016). Ni/ZnO Nano Sorbent for Reactive Adsorption Desulfurization of Refinery Oil Streams. In T. Saleh (Ed.), *Applying Nanotechnology to the Desulfurization Process in Petroleum Engineering* (pp. 216–239). Hershey, PA: IGI Global. doi:10.4018/978-1-4666-9545-0.ch007

Ghosh, S., Mitra, S., Ghosh, S., & Chakraborty, S. (2017). Seismic Reliability Analysis in the Framework of Metamodelling Based Monte Carlo Simulation. In P. Samui, S. Chakraborty, & D. Kim (Eds.), *Modeling and Simulation Techniques in Structural Engineering* (pp. 192–208). Hershey, PA: IGI Global. doi:10.4018/978-1-5225-0588-4.ch006

Gil, M., & Otero, B. (2017). Learning Engineering Skills through Creativity and Collaboration: A Game-Based Proposal. In R. Alexandre Peixoto de Queirós & M. Pinto (Eds.), *Gamification-Based E-Learning Strategies for Computer Programming Education* (pp. 14–29). Hershey, PA: IGI Global. doi:10.4018/978-1-5225-1034-5.ch002

Gill, J., Ayre, M., & Mills, J. (2017). Revisioning the Engineering Profession: How to Make It Happen! In M. Gray & K. Thomas (Eds.), *Strategies for Increasing Diversity in Engineering Majors and Careers* (pp. 156–175). Hershey, PA: IGI Global. doi:10.4018/978-1-5225-2212-6.ch008

Gopal, S., & Al-Hazmi, M. H. (2016). Advances in Catalytic Technologies for Selective Oxidation of Lower Alkanes. In H. Al-Megren & T. Xiao (Eds.), *Petrochemical Catalyst Materials, Processes, and Emerging Technologies* (pp. 22–52). Hershey, PA: IGI Global. doi:10.4018/978-1-4666-9975-5.ch002

Goyal, N., Ram, M., Bhardwaj, A., & Kumar, A. (2016). Thermal Power Plant Modelling with Fault Coverage Stochastically. *International Journal of Manufacturing, Materials, and Mechanical Engineering*, *6*(3), 28–44. doi:10.4018/IJMMME.2016070103

Goyal, N., Ram, M., & Kumar, P. (2017). Welding Process under Fault Coverage Approach for Reliability and MTTF. In M. Ram & J. Davim (Eds.), *Mathematical Concepts and Applications in Mechanical Engineering and Mechatronics* (pp. 222–245). Hershey, PA: IGI Global. doi:10.4018/978-1-5225-1639-2.ch011

Gray, M., & Lundy, C. (2017). Engineering Study Abroad: High Impact Strategy for Increasing Access. In M. Gray & K. Thomas (Eds.), *Strategies for Increasing Diversity in Engineering Majors and Careers* (pp. 42–59). Hershey, PA: IGI Global. doi:10.4018/978-1-5225-2212-6.ch003

Guha, D., Roy, P. K., & Banerjee, S. (2016). Application of Modified Biogeography Based Optimization in AGC of an Interconnected Multi-Unit Multi-Source AC-DC Linked Power System. *International Journal of Energy Optimization and Engineering*, *5*(3), 1–18. doi:10.4018/IJEOE.2016070101

Guha, D., Roy, P. K., & Banerjee, S. (2016). Grey Wolf Optimization to Solve Load Frequency Control of an Interconnected Power System: GWO Used to Solve LFC Problem. *International Journal of Energy Optimization and Engineering*, *5*(4), 62–83. doi:10.4018/IJEOE.2016100104

Gupta, A. K., Dey, A., & Mukhopadhyay, A. K. (2016). Micromechanical and Finite Element Modeling for Composites. In S. Datta & J. Davim (Eds.), *Computational Approaches to Materials Design: Theoretical and Practical Aspects* (pp. 101–162). Hershey, PA: IGI Global. doi:10.4018/978-1-5225-0290-6.ch005

Guraksin, G. E. (2018). Internet of Things and Nature-Inspired Intelligent Techniques for the Future of Biomedical Engineering. In U. Kose, G. Guraksin, & O. Deperlioglu (Eds.), *Nature-Inspired Intelligent Techniques for Solving Biomedical Engineering Problems* (pp. 263–282). Hershey, PA: IGI Global. doi:10.4018/978-1-5225-4769-3.ch013

Hansman, C. A. (2016). Developing Mentoring Programs in Engineering and Technology Education. *International Journal of Quality Assurance in Engineering and Technology Education, 5*(2), 1–15. doi:10.4018/IJQAETE.2016040101

Hasan, U., Chegenizadeh, A., & Nikraz, H. (2016). Nanotechnology Future and Present in Construction Industry: Applications in Geotechnical Engineering. In A. Khitab & W. Anwar (Eds.), *Advanced Research on Nanotechnology for Civil Engineering Applications* (pp. 141–179). Hershey, PA: IGI Global. doi:10.4018/978-1-5225-0344-6.ch005

Hejazi, T., & Akbari, L. (2017). A Multiresponse Optimization Model for Statistical Design of Processes with Discrete Variables. In M. Ram & J. Davim (Eds.), *Mathematical Concepts and Applications in Mechanical Engineering and Mechatronics* (pp. 17–37). Hershey, PA: IGI Global. doi:10.4018/978-1-5225-1639-2.ch002

Hejazi, T., & Hejazi, A. (2017). Monte Carlo Simulation for Reliability-Based Design of Automotive Complex Subsystems. In M. Ram & J. Davim (Eds.), *Mathematical Concepts and Applications in Mechanical Engineering and Mechatronics* (pp. 177–200). Hershey, PA: IGI Global. doi:10.4018/978-1-5225-1639-2.ch009

Hejazi, T., & Poursabbagh, H. (2017). Reliability Analysis of Engineering Systems: An Accelerated Life Testing for Boiler Tubes. In M. Ram & J. Davim (Eds.), *Mathematical Concepts and Applications in Mechanical Engineering and Mechatronics* (pp. 154–176). Hershey, PA: IGI Global. doi:10.4018/978-1-5225-1639-2.ch008

Henao, J., & Sotelo, O. (2018). Surface Engineering at High Temperature: Thermal Cycling and Corrosion Resistance. In A. Pakseresht (Ed.), *Production, Properties, and Applications of High Temperature Coatings* (pp. 131–159). Hershey, PA: IGI Global. doi:10.4018/978-1-5225-4194-3.ch006

Huirache-Acuña, R., Alonso-Nuñez, G., Rivera-Muñoz, E. M., Gutierrez, O., & Pawelec, B. (2016). Trimetallic Sulfide Catalysts for Hydrodesulfurization. In T. Saleh (Ed.), *Applying Nanotechnology to the Desulfurization Process in Petroleum Engineering* (pp. 240–262). Hershey, PA: IGI Global. doi:10.4018/978-1-4666-9545-0.ch008

Ilori, O. O., Adetan, D. A., & Umoru, L. E. (2017). Effect of Cutting Parameters on the Surface Residual Stress of Face-Milled Pearlitic Ductile Iron. *International Journal of Materials Forming and Machining Processes*, *4*(1), 38–52. doi:10.4018/IJMFMP.2017010103

Imam, M. H., Tasadduq, I. A., Ahmad, A., Aldosari, F., & Khan, H. (2017). Automated Generation of Course Improvement Plans Using Expert System. *International Journal of Quality Assurance in Engineering and Technology Education*, *6*(1), 1–12. doi:10.4018/IJQAETE.2017010101

Injeti, S. K., & Kumar, T. V. (2018). A WDO Framework for Optimal Deployment of DGs and DSCs in a Radial Distribution System Under Daily Load Pattern to Improve Techno-Economic Benefits. *International Journal of Energy Optimization and Engineering*, *7*(2), 1–38. doi:10.4018/IJEOE.2018040101

Ishii, N., Anami, K., & Knisely, C. W. (2018). *Dynamic Stability of Hydraulic Gates and Engineering for Flood Prevention*. Hershey, PA: IGI Global. doi:10.4018/978-1-5225-3079-4

J., J., Chowdhury, S., Goyal, P., Samui, P., & Dalkiliç, Y. (2016). Determination of Bearing Capacity of Shallow Foundation Using Soft Computing. In P. Saxena, D. Singh, & M. Pant (Eds.), *Problem Solving and Uncertainty Modeling through Optimization and Soft Computing Applications* (pp. 292-328). Hershey, PA: IGI Global. doi:10.4018/978-1-4666-9885-7.ch014

Jagan, J., Gundlapalli, P., & Samui, P. (2016). Utilization of Classification Techniques for the Determination of Liquefaction Susceptibility of Soils. In S. Bhattacharyya, P. Banerjee, D. Majumdar, & P. Dutta (Eds.), *Handbook of Research on Advanced Hybrid Intelligent Techniques and Applications* (pp. 124–160). Hershey, PA: IGI Global. doi:10.4018/978-1-4666-9474-3.ch005

Jayapalan, S. (2018). A Review of Chemical Treatments on Natural Fibers-Based Hybrid Composites for Engineering Applications. In K. Kumar & J. Davim (Eds.), *Composites and Advanced Materials for Industrial Applications* (pp. 16–37). Hershey, PA: IGI Global. doi:10.4018/978-1-5225-5216-1.ch002

Jeet, K., & Dhir, R. (2016). Software Module Clustering Using Bio-Inspired Algorithms. In P. Vasant, G. Weber, & V. Dieu (Eds.), *Handbook of Research on Modern Optimization Algorithms and Applications in Engineering and Economics* (pp. 445–470). Hershey, PA: IGI Global. doi:10.4018/978-1-4666-9644-0.ch017

Joshi, S. D., & Talange, D. B. (2016). Fault Tolerant Control for a Fractional Order AUV System. *International Journal of Energy Optimization and Engineering*, *5*(2), 1–24. doi:10.4018/IJEOE.2016040101

Julião, D., Ribeiro, S., de Castro, B., Cunha-Silva, L., & Balula, S. S. (2016). Polyoxometalates-Based Nanocatalysts for Production of Sulfur-Free Diesel. In T. Saleh (Ed.), *Applying Nanotechnology to the Desulfurization Process in Petroleum Engineering* (pp. 426–458). Hershey, PA: IGI Global. doi:10.4018/978-1-4666-9545-0.ch014

Kamthan, P. (2016). On the Nature of Collaborations in Agile Software Engineering Course Projects. *International Journal of Quality Assurance in Engineering and Technology Education*, *5*(2), 42–59. doi:10.4018/IJQAETE.2016040104

Karaman, O., Celik, C., & Urkmez, A. S. (2016). Self-Assembled Biomimetic Scaffolds for Bone Tissue Engineering. In M. Bououdina (Ed.), *Emerging Research on Bioinspired Materials Engineering* (pp. 104–132). Hershey, PA: IGI Global. doi:10.4018/978-1-4666-9811-6.ch004

Karkalos, N. E., Markopoulos, A. P., & Dossis, M. F. (2017). Optimal Model Parameters of Inverse Kinematics Solution of a 3R Robotic Manipulator Using ANN Models. *International Journal of Manufacturing, Materials, and Mechanical Engineering*, *7*(3), 20–40. doi:10.4018/IJMMME.2017070102

Kesimal, A., Karaman, K., Cihangir, F., & Ercikdi, B. (2018). Excavatability Assessment of Rock Masses for Geotechnical Studies. In N. Ceryan (Ed.), *Handbook of Research on Trends and Digital Advances in Engineering Geology* (pp. 231–256). Hershey, PA: IGI Global. doi:10.4018/978-1-5225-2709-1.ch006

Khanh, D. V., Vasant, P. M., Elamvazuthi, I., & Dieu, V. N. (2016). Multi-Objective Optimization of Two-Stage Thermo-Electric Cooler Using Differential Evolution: MO Optimization of TEC Using DE. In F. Miranda & C. Abreu (Eds.), *Handbook of Research on Computational Simulation and Modeling in Engineering* (pp. 139–170). Hershey, PA: IGI Global. doi:10.4018/978-1-4666-8823-0.ch004

Kim, D., Hassan, M. K., Chang, S., & Bigdeli, Y. (2016). Nonlinear Vibration Control of 3D Irregular Structures Subjected to Seismic Loads. In P. Samui (Ed.), *Handbook of Research on Advanced Computational Techniques for Simulation-Based Engineering* (pp. 103–119). Hershey, PA: IGI Global. doi:10.4018/978-1-4666-9479-8.ch003

Knoflacher, H. (2017). The Role of Engineers and Their Tools in the Transport Sector after Paradigm Change: From Assumptions and Extrapolations to Science. In H. Knoflacher & E. Ocalir-Akunal (Eds.), *Engineering Tools and Solutions for Sustainable Transportation Planning* (pp. 1–29). Hershey, PA: IGI Global. doi:10.4018/978-1-5225-2116-7.ch001

Kose, U. (2018). Towards an Intelligent Biomedical Engineering With Nature-Inspired Artificial Intelligence Techniques. In U. Kose, G. Guraksin, & O. Deperlioglu (Eds.), *Nature-Inspired Intelligent Techniques for Solving Biomedical Engineering Problems* (pp. 1–26). Hershey, PA: IGI Global. doi:10.4018/978-1-5225-4769-3.ch001

Kostić, S. (2018). A Review on Enhanced Stability Analyses of Soil Slopes Using Statistical Design. In N. Ceryan (Ed.), *Handbook of Research on Trends and Digital Advances in Engineering Geology* (pp. 446–481). Hershey, PA: IGI Global. doi:10.4018/978-1-5225-2709-1.ch013

Kumar, A., Patil, P. P., & Prajapati, Y. K. (2018). *Advanced Numerical Simulations in Mechanical Engineering*. Hershey, PA: IGI Global. doi:10.4018/978-1-5225-3722-9

Kumar, G. R., Rajyalakshmi, G., & Manupati, V. K. (2017). Surface Micro Patterning of Aluminium Reinforced Composite through Laser Peening. *International Journal of Manufacturing, Materials, and Mechanical Engineering*, *7*(4), 15–27. doi:10.4018/IJMMME.2017100102

Kumari, N., & Kumar, K. (2018). Fabrication of Orthotic Calipers With Epoxy-Based Green Composite. In K. Kumar & J. Davim (Eds.), *Composites and Advanced Materials for Industrial Applications* (pp. 157–176). Hershey, PA: IGI Global. doi:10.4018/978-1-5225-5216-1.ch008

Kuppusamy, R. R. (2018). Development of Aerospace Composite Structures Through Vacuum-Enhanced Resin Transfer Moulding Technology (VERTMTy): Vacuum-Enhanced Resin Transfer Moulding. In K. Kumar & J. Davim (Eds.), *Composites and Advanced Materials for Industrial Applications* (pp. 99–111). Hershey, PA: IGI Global. doi:10.4018/978-1-5225-5216-1.ch005

Lemos, J. V. (2016). The Basis for Masonry Analysis with UDEC and 3DEC. In V. Sarhosis, K. Bagi, J. Lemos, & G. Milani (Eds.), *Computational Modeling of Masonry Structures Using the Discrete Element Method* (pp. 61–89). Hershey, PA: IGI Global. doi:10.4018/978-1-5225-0231-9.ch003

Loy, J., Howell, S., & Cooper, R. (2017). Engineering Teams: Supporting Diversity in Engineering Education. In M. Gray & K. Thomas (Eds.), *Strategies for Increasing Diversity in Engineering Majors and Careers* (pp. 106–129). Hershey, PA: IGI Global. doi:10.4018/978-1-5225-2212-6.ch006

Macher, G., Armengaud, E., Kreiner, C., Brenner, E., Schmittner, C., Ma, Z., ... Krammer, M. (2018). Integration of Security in the Development Lifecycle of Dependable Automotive CPS. In N. Druml, A. Genser, A. Krieg, M. Menghin, & A. Hoeller (Eds.), *Solutions for Cyber-Physical Systems Ubiquity* (pp. 383–423). Hershey, PA: IGI Global. doi:10.4018/978-1-5225-2845-6.ch015

Maghsoodlou, S., & Poreskandar, S. (2016). Controlling Electrospinning Jet Using Microscopic Model for Ideal Tissue Engineering Scaffolds. *International Journal of Chemoinformatics and Chemical Engineering*, *5*(2), 1–16. doi:10.4018/IJCCE.2016070101

Mahendramani, G., & Lakshmana Swamy, N. (2018). Effect of Weld Groove Area on Distortion of Butt Welded Joints in Submerged Arc Welding. *International Journal of Manufacturing, Materials, and Mechanical Engineering*, *8*(2), 33–44. doi:10.4018/IJMMME.2018040103

Maiti, S. (2016). Engineered Gellan Polysaccharides in the Design of Controlled Drug Delivery Systems. In M. Bououdina (Ed.), *Emerging Research on Bioinspired Materials Engineering* (pp. 266–293). Hershey, PA: IGI Global. doi:10.4018/978-1-4666-9811-6.ch010

Majumdar, J. D., Weisheit, A., & Manna, I. (2016). Laser Surface Processing for Tailoring of Properties by Optimization of Microstructure. In E. Akinlabi, R. Mahamood, & S. Akinlabi (Eds.), *Advanced Manufacturing Techniques Using Laser Material Processing* (pp. 121–171). Hershey, PA: IGI Global. doi:10.4018/978-1-5225-0329-3.ch006

Maldonado-Macías, A. A., García-Alcaraz, J. L., Hernández-Arellano, J. L., & Cortes-Robles, G. (2016). An Ergonomic Compatibility Perspective on the Selection of Advanced Manufacturing Technology: A Case Study for CNC Vertical Machining Centers. In G. Alor-Hernández, C. Sánchez-Ramírez, & J. García-Alcaraz (Eds.), *Handbook of Research on Managerial Strategies for Achieving Optimal Performance in Industrial Processes* (pp. 137–165). Hershey, PA: IGI Global. doi:10.4018/978-1-5225-0130-5.ch008

Mamaghani, I. H. (2016). Application of Discrete Finite Element Method for Analysis of Unreinforced Masonry Structures. In V. Sarhosis, K. Bagi, J. Lemos, & G. Milani (Eds.), *Computational Modeling of Masonry Structures Using the Discrete Element Method* (pp. 440–458). Hershey, PA: IGI Global. doi:10.4018/978-1-5225-0231-9.ch017

Mansor, M. R., Sapuan, S. M., Salim, M. A., Akop, M. Z., Musthafah, M. T., & Shaharuzaman, M. A. (2016). Concurrent Design of Green Composites. In D. Verma, S. Jain, X. Zhang, & P. Gope (Eds.), *Green Approaches to Biocomposite Materials Science and Engineering* (pp. 48–75). Hershey, PA: IGI Global. doi:10.4018/978-1-5225-0424-5.ch003

Mansouri, I., & Esmaeili, E. (2016). Nanotechnology Applications in the Construction Industry. In A. Khitab & W. Anwar (Eds.), *Advanced Research on Nanotechnology for Civil Engineering Applications* (pp. 111–140). Hershey, PA: IGI Global. doi:10.4018/978-1-5225-0344-6.ch004

Manzoor, A. (2016). MOOCs for Enhancing Engineering Education. In D. Fonseca & E. Redondo (Eds.), *Handbook of Research on Applied E-Learning in Engineering and Architecture Education* (pp. 204–223). Hershey, PA: IGI Global. doi:10.4018/978-1-4666-8803-2.ch010

Martin, A., Kalevaru, V. N., & Radnik, J. (2016). Palladium in Heterogeneous Oxidation Catalysis. In H. Al-Megren & T. Xiao (Eds.), *Petrochemical Catalyst Materials, Processes, and Emerging Technologies* (pp. 53–81). Hershey, PA: IGI Global. doi:10.4018/978-1-4666-9975-5.ch003

Melnyczuk, J. M., & Palchoudhury, S. (2016). Introduction to Bio-Inspired Hydrogel and Their Application: Hydrogels. In M. Bououdina (Ed.), *Emerging Research on Bioinspired Materials Engineering* (pp. 133–159). Hershey, PA: IGI Global. doi:10.4018/978-1-4666-9811-6.ch005

Mitra-Kirtley, S., Mullins, O. C., & Pomerantz, A. E. (2016). Sulfur and Nitrogen Chemical Speciation in Crude Oils and Related Carbonaceous Materials. In T. Saleh (Ed.), *Applying Nanotechnology to the Desulfurization Process in Petroleum Engineering* (pp. 53–83). Hershey, PA: IGI Global. doi:10.4018/978-1-4666-9545-0.ch002

Moalosi, R., Uziak, J., & Oladiran, M. T. (2016). Using Blended Learning Approach to Deliver Courses in An Engineering Programme. *International Journal of Quality Assurance in Engineering and Technology Education, 5*(1), 23–39. doi:10.4018/IJQAETE.2016010103

Mohammadzadeh, S., & Kim, Y. (2017). Nonlinear System Identification of Smart Buildings. In P. Samui, S. Chakraborty, & D. Kim (Eds.), *Modeling and Simulation Techniques in Structural Engineering* (pp. 328–347). Hershey, PA: IGI Global. doi:10.4018/978-1-5225-0588-4.ch011

Mohanty, I., & Bhattacherjee, D. (2016). Artificial Neural Network and Its Application in Steel Industry. In S. Datta & J. Davim (Eds.), *Computational Approaches to Materials Design: Theoretical and Practical Aspects* (pp. 267–300). Hershey, PA: IGI Global. doi:10.4018/978-1-5225-0290-6.ch010

Mohebkhah, A., & Sarhosis, V. (2016). Discrete Element Modeling of Masonry-Infilled Frames. In V. Sarhosis, K. Bagi, J. Lemos, & G. Milani (Eds.), *Computational Modeling of Masonry Structures Using the Discrete Element Method* (pp. 200–234). Hershey, PA: IGI Global. doi:10.4018/978-1-5225-0231-9.ch009

Molina, G. J., Aktaruzzaman, F., Soloiu, V., & Rahman, M. (2017). Design and Testing of a Jet-Impingement Instrument to Study Surface-Modification Effects by Nanofluids. *International Journal of Surface Engineering and Interdisciplinary Materials Science*, *5*(2), 43–61. doi:10.4018/IJSEIMS.2017070104

Montalvan-Sorrosa, D., de los Cobos-Vasconcelos, D., & Gonzalez-Sanchez, A. (2016). Nanotechnology Applied to the Biodesulfurization of Fossil Fuels and Spent Caustic Streams. In T. Saleh (Ed.), *Applying Nanotechnology to the Desulfurization Process in Petroleum Engineering* (pp. 378–389). Hershey, PA: IGI Global. doi:10.4018/978-1-4666-9545-0.ch012

Montillet, J., Yu, K., Bonenberg, L. K., & Roberts, G. W. (2016). Optimization Algorithms in Local and Global Positioning. In P. Vasant, G. Weber, & V. Dieu (Eds.), *Handbook of Research on Modern Optimization Algorithms and Applications in Engineering and Economics* (pp. 1–53). Hershey, PA: IGI Global. doi:10.4018/978-1-4666-9644-0.ch001

Moreira, F., & Ferreira, M. J. (2016). Teaching and Learning Requirements Engineering Based on Mobile Devices and Cloud: A Case Study. In D. Fonseca & E. Redondo (Eds.), *Handbook of Research on Applied E-Learning in Engineering and Architecture Education* (pp. 237–262). Hershey, PA: IGI Global. doi:10.4018/978-1-4666-8803-2.ch012

Mukherjee, A., Saeed, R. A., Dutta, S., & Naskar, M. K. (2017). Fault Tracking Framework for Software-Defined Networking (SDN). In C. Singhal & S. De (Eds.), *Resource Allocation in Next-Generation Broadband Wireless Access Networks* (pp. 247–272). Hershey, PA: IGI Global. doi:10.4018/978-1-5225-2023-8.ch011

Mukhopadhyay, A., Barman, T. K., & Sahoo, P. (2018). Electroless Nickel Coatings for High Temperature Applications. In K. Kumar & J. Davim (Eds.), *Composites and Advanced Materials for Industrial Applications* (pp. 297–331). Hershey, PA: IGI Global. doi:10.4018/978-1-5225-5216-1.ch013

Náprstek, J., & Fischer, C. (2017). Dynamic Stability and Post-Critical Processes of Slender Auto-Parametric Systems. In V. Plevris, G. Kremmyda, & Y. Fahjan (Eds.), *Performance-Based Seismic Design of Concrete Structures and Infrastructures* (pp. 128–171). Hershey, PA: IGI Global. doi:10.4018/978-1-5225-2089-4.ch006

Nautiyal, L., Shivach, P., & Ram, M. (2018). Optimal Designs by Means of Genetic Algorithms. In M. Ram & J. Davim (Eds.), *Soft Computing Techniques and Applications in Mechanical Engineering* (pp. 151–161). Hershey, PA: IGI Global. doi:10.4018/978-1-5225-3035-0.ch007

Nazir, R. (2017). Advanced Nanomaterials for Water Engineering and Treatment: Nano-Metal Oxides and Their Nanocomposites. In T. Saleh (Ed.), *Advanced Nanomaterials for Water Engineering, Treatment, and Hydraulics* (pp. 84–126). Hershey, PA: IGI Global. doi:10.4018/978-1-5225-2136-5.ch005

Nogueira, A. F., Ribeiro, J. C., Fernández de Vega, F., & Zenha-Rela, M. A. (2018). Evolutionary Approaches to Test Data Generation for Object-Oriented Software: Overview of Techniques and Tools. In M. Khosrow-Pour, D.B.A. (Ed.), Incorporating Nature-Inspired Paradigms in Computational Applications (pp. 162-194). Hershey, PA: IGI Global. doi:10.4018/978-1-5225-5020-4.ch006

Nunes, J. F., Moreira, P. M., & Tavares, J. M. (2016). Human Motion Analysis and Simulation Tools: A Survey. In F. Miranda & C. Abreu (Eds.), *Handbook of Research on Computational Simulation and Modeling in Engineering* (pp. 359–388). Hershey, PA: IGI Global. doi:10.4018/978-1-4666-8823-0.ch012

Ogunlaja, A. S., & Tshentu, Z. R. (2016). Molecularly Imprinted Polymer Nanofibers for Adsorptive Desulfurization. In T. Saleh (Ed.), *Applying Nanotechnology to the Desulfurization Process in Petroleum Engineering* (pp. 281–336). Hershey, PA: IGI Global. doi:10.4018/978-1-4666-9545-0.ch010

Ong, P., & Kohshelan, S. (2016). Performances of Adaptive Cuckoo Search Algorithm in Engineering Optimization. In P. Vasant, G. Weber, & V. Dieu (Eds.), *Handbook of Research on Modern Optimization Algorithms and Applications in Engineering and Economics* (pp. 676–699). Hershey, PA: IGI Global. doi:10.4018/978-1-4666-9644-0.ch026

Osho, M. B. (2018). Industrial Enzyme Technology: Potential Applications. In S. Bharati & P. Chaurasia (Eds.), *Research Advancements in Pharmaceutical, Nutritional, and Industrial Enzymology* (pp. 375–394). Hershey, PA: IGI Global. doi:10.4018/978-1-5225-5237-6.ch017

Padmaja, P., & Marutheswar, G. (2017). Certain Investigation on Secured Data Transmission in Wireless Sensor Networks. *International Journal of Mobile Computing and Multimedia Communications*, *8*(1), 48–61. doi:10.4018/IJMCMC.2017010104

Paixão, S. M., Silva, T. P., Arez, B. F., & Alves, L. (2016). Advances in the Reduction of the Costs Inherent to Fossil Fuels' Biodesulfurization towards Its Potential Industrial Application. In T. Saleh (Ed.), *Applying Nanotechnology to the Desulfurization Process in Petroleum Engineering* (pp. 390–425). Hershey, PA: IGI Global. doi:10.4018/978-1-4666-9545-0.ch013

Palmer, S., & Hall, W. (2017). An Evaluation of Group Work in First-Year Engineering Design Education. In R. Tucker (Ed.), *Collaboration and Student Engagement in Design Education* (pp. 145–168). Hershey, PA: IGI Global. doi:10.4018/978-1-5225-0726-0.ch007

Panneer, R. (2017). Effect of Composition of Fibers on Properties of Hybrid Composites. *International Journal of Manufacturing, Materials, and Mechanical Engineering*, *7*(4), 28–43. doi:10.4018/IJMMME.2017100103

Parker, J. (2016). Hubble's Expanding Universe: A Model for Quality in Technology Infused engineering and Technology Education. *International Journal of Quality Assurance in Engineering and Technology Education*, *5*(2), 16–29. doi:10.4018/IJQAETE.2016040102

Paul, S., & Roy, P. (2018). Optimal Design of Power System Stabilizer Using a Novel Evolutionary Algorithm. *International Journal of Energy Optimization and Engineering*, *7*(3), 24–46. doi:10.4018/IJEOE.2018070102

Pavaloiu, A. (2018). Artificial Intelligence Ethics in Biomedical-Engineering-Oriented Problems. In U. Kose, G. Guraksin, & O. Deperlioglu (Eds.), *Nature-Inspired Intelligent Techniques for Solving Biomedical Engineering Problems* (pp. 219–231). Hershey, PA: IGI Global. doi:10.4018/978-1-5225-4769-3.ch010

Peña, F. (2016). A Semi-Discrete Approach for the Numerical Simulation of Freestanding Blocks. In V. Sarhosis, K. Bagi, J. Lemos, & G. Milani (Eds.), *Computational Modeling of Masonry Structures Using the Discrete Element Method* (pp. 416–439). Hershey, PA: IGI Global. doi:10.4018/978-1-5225-0231-9.ch016

Penchovsky, R., & Traykovska, M. (2016). Synthetic Approaches to Biology: Engineering Gene Control Circuits, Synthesizing, and Editing Genomes. In M. Bououdina (Ed.), *Emerging Research on Bioinspired Materials Engineering* (pp. 323–351). Hershey, PA: IGI Global. doi:10.4018/978-1-4666-9811-6.ch012

Pieroni, A., & Iazeolla, G. (2016). Engineering QoS and Energy Saving in the Delivery of ICT Services. In P. Vasant & N. Voropai (Eds.), *Sustaining Power Resources through Energy Optimization and Engineering* (pp. 208–226). Hershey, PA: IGI Global. doi:10.4018/978-1-4666-9755-3.ch009

Pioro, I., Mahdi, M., & Popov, R. (2017). Application of Supercritical Pressures in Power Engineering. In L. Chen & Y. Iwamoto (Eds.), *Advanced Applications of Supercritical Fluids in Energy Systems* (pp. 404–457). Hershey, PA: IGI Global. doi:10.4018/978-1-5225-2047-4.ch013

Plaksina, T., & Gildin, E. (2017). Rigorous Integrated Evolutionary Workflow for Optimal Exploitation of Unconventional Gas Assets. *International Journal of Energy Optimization and Engineering*, *6*(1), 101–122. doi:10.4018/IJEOE.2017010106

Puppala, A. J., Bheemasetti, T. V., Zou, H., Yu, X., Pedarla, A., & Cai, G. (2016). Spatial Variability Analysis of Soil Properties using Geostatistics. In P. Samui (Ed.), *Handbook of Research on Advanced Computational Techniques for Simulation-Based Engineering* (pp. 195–226). Hershey, PA: IGI Global. doi:10.4018/978-1-4666-9479-8.ch008

Ramdani, N., & Azibi, M. (2018). Polymer Composite Materials for Microelectronics Packaging Applications: Composites for Microelectronics Packaging. In K. Kumar & J. Davim (Eds.), *Composites and Advanced Materials for Industrial Applications* (pp. 177–211). Hershey, PA: IGI Global. doi:10.4018/978-1-5225-5216-1.ch009

Ramesh, M., Garg, R., & Subrahmanyam, G. V. (2017). Investigation of Influence of Quenching and Annealing on the Plane Fracture Toughness and Brittle to Ductile Transition Temperature of the Zinc Coated Structural Steel Materials. *International Journal of Surface Engineering and Interdisciplinary Materials Science*, *5*(2), 33–42. doi:10.4018/IJSEIMS.2017070103

Razavi, A. M., & Ahmad, R. (2016). Agile Software Development Challenges in Implementation and Adoption: Focusing on Large and Distributed Settings – Past Experiences, Emergent Topics. In I. Ghani, D. Jawawi, S. Dorairaj, & A. Sidky (Eds.), *Emerging Innovations in Agile Software Development* (pp. 175–207). Hershey, PA: IGI Global. doi:10.4018/978-1-4666-9858-1.ch010

Reccia, E., Cecchi, A., & Milani, G. (2016). FEM/DEM Approach for the Analysis of Masonry Arch Bridges. In V. Sarhosis, K. Bagi, J. Lemos, & G. Milani (Eds.), *Computational Modeling of Masonry Structures Using the Discrete Element Method* (pp. 367–392). Hershey, PA: IGI Global. doi:10.4018/978-1-5225-0231-9.ch014

Ro, H. K., & McIntosh, K. (2016). Constructing Conducive Environment for Women of Color in Engineering Undergraduate Education. In U. Thomas & J. Drake (Eds.), *Critical Research on Sexism and Racism in STEM Fields* (pp. 23–48). Hershey, PA: IGI Global. doi:10.4018/978-1-5225-0174-9.ch002

Rodulfo-Baechler, S. M. (2016). Dual Role of Perovskite Hollow Fiber Membrane in the Methane Oxidation Reactions. In H. Al-Megren & T. Xiao (Eds.), *Petrochemical Catalyst Materials, Processes, and Emerging Technologies* (pp. 385–430). Hershey, PA: IGI Global. doi:10.4018/978-1-4666-9975-5.ch014

Rudolf, S., Biryuk, V. V., & Volov, V. (2018). Vortex Effect, Vortex Power: Technology of Vortex Power Engineering. In V. Kharchenko & P. Vasant (Eds.), *Handbook of Research on Renewable Energy and Electric Resources for Sustainable Rural Development* (pp. 500–533). Hershey, PA: IGI Global. doi:10.4018/978-1-5225-3867-7.ch021

Sah, A., Bhadula, S. J., Dumka, A., & Rawat, S. (2018). A Software Engineering Perspective for Development of Enterprise Applications. In A. Elçi (Ed.), *Handbook of Research on Contemporary Perspectives on Web-Based Systems* (pp. 1–23). Hershey, PA: IGI Global. doi:10.4018/978-1-5225-5384-7.ch001

Sahoo, P., & Roy, S. (2017). Tribological Behavior of Electroless Ni-P, Ni-P-W and Ni-P-Cu Coatings: A Comparison. *International Journal of Surface Engineering and Interdisciplinary Materials Science, 5*(1), 1–15. doi:10.4018/IJSEIMS.2017010101

Sahoo, S. (2018). Laminated Composite Hypar Shells as Roofing Units: Static and Dynamic Behavior. In K. Kumar & J. Davim (Eds.), *Composites and Advanced Materials for Industrial Applications* (pp. 249–269). Hershey, PA: IGI Global. doi:10.4018/978-1-5225-5216-1.ch011

Sahu, H., & Hungyo, M. (2018). Introduction to SDN and NFV. In A. Dumka (Ed.), *Innovations in Software-Defined Networking and Network Functions Virtualization* (pp. 1–25). Hershey, PA: IGI Global. doi:10.4018/978-1-5225-3640-6.ch001

Saikia, P., Bharadwaj, S. K., & Miah, A. T. (2016). Peroxovanadates and Its Bio-Mimicking Relation with Vanadium Haloperoxidases. In M. Bououdina (Ed.), *Emerging Research on Bioinspired Materials Engineering* (pp. 197–219). Hershey, PA: IGI Global. doi:10.4018/978-1-4666-9811-6.ch007

Saladino, R., Botta, G., & Crucianelli, M. (2016). Advances in Nanotechnology Transition Metal Catalysts in Oxidative Desulfurization (ODS) Processes: Nanotechnology Applied to ODS Processing. In T. Saleh (Ed.), *Applying Nanotechnology to the Desulfurization Process in Petroleum Engineering* (pp. 180–215). Hershey, PA: IGI Global. doi:10.4018/978-1-4666-9545-0.ch006

Saleh, T. A., Danmaliki, G. I., & Shuaib, T. D. (2016). Nanocomposites and Hybrid Materials for Adsorptive Desulfurization. In T. Saleh (Ed.), *Applying Nanotechnology to the Desulfurization Process in Petroleum Engineering* (pp. 129–153). Hershey, PA: IGI Global. doi:10.4018/978-1-4666-9545-0.ch004

Saleh, T. A., Shuaib, T. D., Danmaliki, G. I., & Al-Daous, M. A. (2016). Carbon-Based Nanomaterials for Desulfurization: Classification, Preparation, and Evaluation. In T. Saleh (Ed.), *Applying Nanotechnology to the Desulfurization Process in Petroleum Engineering* (pp. 154–179). Hershey, PA: IGI Global. doi:10.4018/978-1-4666-9545-0.ch005

Salem, A. M., & Shmelova, T. (2018). Intelligent Expert Decision Support Systems: Methodologies, Applications, and Challenges. In T. Shmelova, Y. Sikirda, N. Rizun, A. Salem, & Y. Kovalyov (Eds.), *Socio-Technical Decision Support in Air Navigation Systems: Emerging Research and Opportunities* (pp. 215–242). Hershey, PA: IGI Global. doi:10.4018/978-1-5225-3108-1.ch007

Samal, M. (2017). FE Analysis and Experimental Investigation of Cracked and Un-Cracked Thin-Walled Tubular Components to Evaluate Mechanical and Fracture Properties. In P. Samui, S. Chakraborty, & D. Kim (Eds.), *Modeling and Simulation Techniques in Structural Engineering* (pp. 266–293). Hershey, PA: IGI Global. doi:10.4018/978-1-5225-0588-4.ch009

Samal, M., & Balakrishnan, K. (2017). Experiments on a Ring Tension Setup and FE Analysis to Evaluate Transverse Mechanical Properties of Tubular Components. In P. Samui, S. Chakraborty, & D. Kim (Eds.), *Modeling and Simulation Techniques in Structural Engineering* (pp. 91–115). Hershey, PA: IGI Global. doi:10.4018/978-1-5225-0588-4.ch004

Santhanakumar, M., Adalarasan, R., & Rajmohan, M. (2016). An Investigation in Abrasive Waterjet Cutting of Al6061/SiC/Al2O3 Composite Using Principal Component Based Response Surface Methodology. *International Journal of Manufacturing, Materials, and Mechanical Engineering*, *6*(4), 30–47. doi:10.4018/IJMMME.2016100103

Sareen, N., & Bhattacharya, S. (2016). Cleaner Energy Fuels: Hydrodesulfurization and Beyond. In T. Saleh (Ed.), *Applying Nanotechnology to the Desulfurization Process in Petroleum Engineering* (pp. 84–128). Hershey, PA: IGI Global. doi:10.4018/978-1-4666-9545-0.ch003

Sarhosis, V. (2016). Micro-Modeling Options for Masonry. In V. Sarhosis, K. Bagi, J. Lemos, & G. Milani (Eds.), *Computational Modeling of Masonry Structures Using the Discrete Element Method* (pp. 28–60). Hershey, PA: IGI Global. doi:10.4018/978-1-5225-0231-9.ch002

Sarhosis, V., Oliveira, D. V., & Lourenco, P. B. (2016). On the Mechanical Behavior of Masonry. In V. Sarhosis, K. Bagi, J. Lemos, & G. Milani (Eds.), *Computational Modeling of Masonry Structures Using the Discrete Element Method* (pp. 1–27). Hershey, PA: IGI Global. doi:10.4018/978-1-5225-0231-9.ch001

Satyam, N. (2016). Liquefaction Modelling of Granular Soils using Discrete Element Method. In P. Samui (Ed.), *Handbook of Research on Advanced Computational Techniques for Simulation-Based Engineering* (pp. 381–441). Hershey, PA: IGI Global. doi:10.4018/978-1-4666-9479-8.ch015

Sawant, S. (2018). Deep Learning and Biomedical Engineering. In U. Kose, G. Guraksin, & O. Deperlioglu (Eds.), *Nature-Inspired Intelligent Techniques for Solving Biomedical Engineering Problems* (pp. 283–296). Hershey, PA: IGI Global. doi:10.4018/978-1-5225-4769-3.ch014

Sezgin, H., & Berkalp, O. B. (2018). Textile-Reinforced Composites for the Automotive Industry. In K. Kumar & J. Davim (Eds.), *Composites and Advanced Materials for Industrial Applications* (pp. 129–156). Hershey, PA: IGI Global. doi:10.4018/978-1-5225-5216-1.ch007

Shah, M. Z., Gazder, U., Bhatti, M. S., & Hussain, M. (2018). Comparative Performance Evaluation of Effects of Modifier in Asphaltic Concrete Mix. *International Journal of Strategic Engineering*, *1*(2), 13–25. doi:10.4018/IJoSE.2018070102

Shah, V. S., Shah, H. R., & Samui, P. (2016). Application of Meta-Models (MPMR and ELM) for Determining OMC, MDD and Soaked CBR Value of Soil. In S. Bhattacharyya, P. Banerjee, D. Majumdar, & P. Dutta (Eds.), *Handbook of Research on Advanced Hybrid Intelligent Techniques and Applications* (pp. 454–482). Hershey, PA: IGI Global. doi:10.4018/978-1-4666-9474-3.ch015

Sharma, N., & Kumar, K. (2018). Fabrication of Porous NiTi Alloy Using Organic Binders. In K. Kumar & J. Davim (Eds.), *Composites and Advanced Materials for Industrial Applications* (pp. 38–62). Hershey, PA: IGI Global. doi:10.4018/978-1-5225-5216-1.ch003

Sharma, T. K. (2016). Application of Shuffled Frog Leaping Algorithm in Software Project Scheduling. In P. Saxena, D. Singh, & M. Pant (Eds.), *Problem Solving and Uncertainty Modeling through Optimization and Soft Computing Applications* (pp. 225–238). Hershey, PA: IGI Global. doi:10.4018/978-1-4666-9885-7.ch011

Shivach, P., Nautiyal, L., & Ram, M. (2018). Applying Multi-Objective Optimization Algorithms to Mechanical Engineering. In M. Ram & J. Davim (Eds.), *Soft Computing Techniques and Applications in Mechanical Engineering* (pp. 287–301). Hershey, PA: IGI Global. doi:10.4018/978-1-5225-3035-0.ch014

Shmelova, T. (2018). Stochastic Methods for Estimation and Problem Solving in Engineering: Stochastic Methods of Decision Making in Aviation. In S. Kadry (Ed.), *Stochastic Methods for Estimation and Problem Solving in Engineering* (pp. 139–160). Hershey, PA: IGI Global. doi:10.4018/978-1-5225-5045-7.ch006

Shukla, R., Anapagaddi, R., Singh, A. K., Allen, J. K., Panchal, J. H., & Mistree, F. (2016). Integrated Computational Materials Engineering for Determining the Set Points of Unit Operations for Production of a Steel Product Mix. In S. Datta & J. Davim (Eds.), *Computational Approaches to Materials Design: Theoretical and Practical Aspects* (pp. 163–191). Hershey, PA: IGI Global. doi:10.4018/978-1-5225-0290-6.ch006

Siero González, L. R., & Romo Vázquez, A. (2017). Didactic Sequences Teaching Mathematics for Engineers With Focus on Differential Equations. In M. Ramírez-Montoya (Ed.), *Handbook of Research on Driving STEM Learning With Educational Technologies* (pp. 129–151). Hershey, PA: IGI Global. doi:10.4018/978-1-5225-2026-9.ch007

Singh, R., & Dutta, S. (2018). Visible Light Active Nanocomposites for Photocatalytic Applications. In K. Kumar & J. Davim (Eds.), *Composites and Advanced Materials for Industrial Applications* (pp. 270–296). Hershey, PA: IGI Global. doi:10.4018/978-1-5225-5216-1.ch012

Singh, R., & Lou, H. H. (2016). Safety and Efficiency Enhancement in LNG Terminals. In H. Al-Megren & T. Xiao (Eds.), *Petrochemical Catalyst Materials, Processes, and Emerging Technologies* (pp. 164–176). Hershey, PA: IGI Global. doi:10.4018/978-1-4666-9975-5.ch007

Sözbilir, H., Özkaymak, Ç., Uzel, B., & Sümer, Ö. (2018). Criteria for Surface Rupture Microzonation of Active Faults for Earthquake Hazards in Urban Areas. In N. Ceryan (Ed.), *Handbook of Research on Trends and Digital Advances in Engineering Geology* (pp. 187–230). Hershey, PA: IGI Global. doi:10.4018/978-1-5225-2709-1.ch005

Stanciu, I. (2018). Stochastic Methods in Microsystems Engineering. In S. Kadry (Ed.), *Stochastic Methods for Estimation and Problem Solving in Engineering* (pp. 161–176). Hershey, PA: IGI Global. doi:10.4018/978-1-5225-5045-7.ch007

Strebkov, D., Nekrasov, A., Trubnikov, V., & Nekrasov, A. (2018). Single-Wire Resonant Electric Power Systems for Renewable-Based Electric Grid. In V. Kharchenko & P. Vasant (Eds.), *Handbook of Research on Renewable Energy and Electric Resources for Sustainable Rural Development* (pp. 449–474). Hershey, PA: IGI Global. doi:10.4018/978-1-5225-3867-7.ch019

Subburaman, D., Jagan, J., Dalkiliç, Y., & Samui, P. (2016). Reliability Analysis of Slope Using MPMR, GRNN and GPR. In F. Miranda & C. Abreu (Eds.), *Handbook of Research on Computational Simulation and Modeling in Engineering* (pp. 208–224). Hershey, PA: IGI Global. doi:10.4018/978-1-4666-8823-0.ch007

Sun, J., Wan, S., Lin, J., & Wang, Y. (2016). Advances in Catalytic Conversion of Syngas to Ethanol and Higher Alcohols. In H. Al-Megren & T. Xiao (Eds.), *Petrochemical Catalyst Materials, Processes, and Emerging Technologies* (pp. 177–215). Hershey, PA: IGI Global. doi:10.4018/978-1-4666-9975-5.ch008

Tüdeş, Ş., Kumlu, K. B., & Ceryan, S. (2018). Integration Between Urban Planning and Natural Hazards For Resilient City. In N. Ceryan (Ed.), *Handbook of Research on Trends and Digital Advances in Engineering Geology* (pp. 591–630). Hershey, PA: IGI Global. doi:10.4018/978-1-5225-2709-1.ch017

Tyukhov, I., Rezk, H., & Vasant, P. (2016). Modern Optimization Algorithms and Applications in Solar Photovoltaic Engineering. In P. Vasant & N. Voropai (Eds.), *Sustaining Power Resources through Energy Optimization and Engineering* (pp. 390–445). Hershey, PA: IGI Global. doi:10.4018/978-1-4666-9755-3.ch016

Ulamis, K. (2018). Soil Liquefaction Assessment by Anisotropic Cyclic Triaxial Test. In N. Ceryan (Ed.), *Handbook of Research on Trends and Digital Advances in Engineering Geology* (pp. 631–664). Hershey, PA: IGI Global. doi:10.4018/978-1-5225-2709-1.ch018

Umar, M. A., Tenuche, S. S., Yusuf, S. A., Abdulsalami, A. O., & Kufena, A. M. (2016). Usability Engineering in Agile Software Development Processes. In I. Ghani, D. Jawawi, S. Dorairaj, & A. Sidky (Eds.), *Emerging Innovations in Agile Software Development* (pp. 208–221). Hershey, PA: IGI Global. doi:10.4018/978-1-4666-9858-1.ch011

Üzüm, O., & Çakır, Ö. A. (2016). A Bio-Inspired Phenomena in Cementitious Materials: Self-Healing. In M. Bououdina (Ed.), *Emerging Research on Bioinspired Materials Engineering* (pp. 220–246). Hershey, PA: IGI Global. doi:10.4018/978-1-4666-9811-6.ch008

Valente, M., & Milani, G. (2017). Seismic Assessment and Retrofitting of an Under-Designed RC Frame Through a Displacement-Based Approach. In V. Plevris, G. Kremmyda, & Y. Fahjan (Eds.), *Performance-Based Seismic Design of Concrete Structures and Infrastructures* (pp. 36–58). Hershey, PA: IGI Global. doi:10.4018/978-1-5225-2089-4.ch002

Vasant, P. (2018). A General Medical Diagnosis System Formed by Artificial Neural Networks and Swarm Intelligence Techniques. In U. Kose, G. Guraksin, & O. Deperlioglu (Eds.), *Nature-Inspired Intelligent Techniques for Solving Biomedical Engineering Problems* (pp. 130–145). Hershey, PA: IGI Global. doi:10.4018/978-1-5225-4769-3.ch006

Vergara, D., Lorenzo, M., & Rubio, M. (2016). On the Use of Virtual Environments in Engineering Education. *International Journal of Quality Assurance in Engineering and Technology Education*, *5*(2), 30–41. doi:10.4018/IJQAETE.2016040103

Verrollot, J., Tolonen, A., Harkonen, J., & Haapasalo, H. J. (2018). Challenges and Enablers for Rapid Product Development. *International Journal of Applied Industrial Engineering*, *5*(1), 25–49. doi:10.4018/IJAIE.2018010102

Wagner, C., & Ryan, C. (2016). Physical and Digital Integration Strategies of Electronic Device Supply Chains and Their Applicability to ETO Supply Chains. In R. Addo-Tenkorang, J. Kantola, P. Helo, & A. Shamsuzzoha (Eds.), *Supply Chain Strategies and the Engineer-to-Order Approach* (pp. 224–245). Hershey, PA: IGI Global. doi:10.4018/978-1-5225-0021-6.ch011

Wang, Z., Wu, P., Lan, L., & Ji, S. (2016). Preparation, Characterization and Desulfurization of the Supported Nickel Phosphide Catalysts. In H. Al-Megren & T. Xiao (Eds.), *Petrochemical Catalyst Materials, Processes, and Emerging Technologies* (pp. 431–458). Hershey, PA: IGI Global. doi:10.4018/978-1-4666-9975-5.ch015

Yardimci, A. G., & Karpuz, C. (2018). Fuzzy Rock Mass Rating: Soft-Computing-Aided Preliminary Stability Analysis of Weak Rock Slopes. In N. Ceryan (Ed.), *Handbook of Research on Trends and Digital Advances in Engineering Geology* (pp. 97–131). Hershey, PA: IGI Global. doi:10.4018/978-1-5225-2709-1.ch003

Zhang, L., Ding, S., Sun, S., Han, B., Yu, X., & Ou, J. (2016). Nano-Scale Behavior and Nano-Modification of Cement and Concrete Materials. In A. Khitab & W. Anwar (Eds.), *Advanced Research on Nanotechnology for Civil Engineering Applications* (pp. 28–79). Hershey, PA: IGI Global. doi:10.4018/978-1-5225-0344-6.ch002

Zindani, D., & Kumar, K. (2018). Industrial Applications of Polymer Composite Materials. In K. Kumar & J. Davim (Eds.), *Composites and Advanced Materials for Industrial Applications* (pp. 1–15). Hershey, PA: IGI Global. doi:10.4018/978-1-5225-5216-1.ch001

Zindani, D., Maity, S. R., & Bhowmik, S. (2018). A Decision-Making Approach for Material Selection of Polymeric Composite Bumper Beam. In K. Kumar & J. Davim (Eds.), *Composites and Advanced Materials for Industrial Applications* (pp. 112–128). Hershey, PA: IGI Global. doi:10.4018/978-1-5225-5216-1.ch006

About the Contributors

Siddhartha Bhattacharyya did his Bachelors in Physics, Bachelors in Optics and Optoelectronics and Masters in Optics and Optoelectronics from University of Calcutta, India in 1995, 1998 and 2000 respectively. He completed PhD in Computer Science and Engineering from Jadavpur University, India in 2008. He is currently serving as the Principal of RCC Institute of Information Technology, Kolkata, India. In addition, he is also serving as the Professor of Computer Application and Dean (Research and Development and Academic Affairs) of the institute. Prior to this, he was the Professor of Information Technology of RCC Institute of Information Technology, Kolkata, India. He served as a Senior Research Scientist in the Faculty of Electrical Engineering and Computer Science of VSB Technical University of Ostrava, Czech Republic from October, 2018 to April, 2019. He served as the Head of the Department from March, 2014 to December, 2016. Prior to this, he was an Associate Professor of Information Technology of RCC Institute of Information Technology, Kolkata, India from 2011-2014. Before that, he served as an Assistant Professor in Computer Science and Information Technology of University Institute of Technology, The University of Burdwan, India from 2005-2011. He was a Lecturer in Information Technology of Kalyani Government Engineering College, India during 2001-2005. He is a co-author of 4 books and the co-editor of 20 books and has more than 200 research publications in international journals and conference proceedings to his credit. He has got two PCTs to his credit. He has been the member of the organizing and technical program committees of several national and international conferences. He served as the Editor-In-Chief of International Journal of Ambient Computing and Intelligence (IJACI) published by IGI Global, Hershey, PA. He was the General Chair of several international conferences like ICCICN 2014, ICRCICN 2015, ICRCICN 2016 and ICRCICN 2018 organized by the Department of Information Technology, RCC Institute of Information Technology, Kolkata and held at Kolkata, India. He is the founding Editor in Chief of International Journal of Hybrid Intelligence; Publisher: Inderscience. He is the member of the editorial board of several reputed international journals and acts as the Associate Editor for several others. He is serving as the Series Editors of IGI Global Book Series Advances in

Information Quality and Management (AIQM), De Gruyter Book Series Frontiers in Computational Intelligence (FCI) , CRC Press Book Series Computational Intelligence and Applications, Wiley Book Series Intelligent Signal and Data Processing and Elsevier Book Series Hybrid Computational Intelligence for Pattern Analysis and Understanding. His research interests include soft computing, pattern recognition, multimedia data processing, hybrid intelligence and quantum computing.

Pampa Debnath is presently working as Assistant Professor in the Department of Electronics and Communication Engineering in RCC Institute of Information Technology, Kolkata, INDIA. She has more than 11 years of professional teaching experience in academics. She received B.Tech, M.Tech Degree from the University of Burdwan. Her research interest covers the area of Microwave devices, Microstrip Patch antennas, SIW based circuit. She has published several research papers in IEEE Xpore, Microsystem Technologies- Springer, CRC Press and in some National and International conferences. Her major teaching subjects are Electromagnetics, RF & Microwave and Antenna. She has already served as Editor as well as Technical chair of One International Conference (ICCSE 2016) which is published by CRC Press, coordinated a few Faculty Development Programmes, Workshops, Laboratory and Industrial visits, seminars and technical events under the banner of The Institution of Engineers (INDIA) Kolkata section. She is also associated with a few National and International conference. She is a reviewer of few journals of repute and some national and International conferences. She has conducted hands on session on Photonics, Electromagnetics and Microwaves in various FDP's, workshops, seminars. She is a member of The Institution of Electronics and Telecommunication Engineers (IETE), Indian Society for Technical Education (ISTE), International Association for Engineers (IAENG).

Arpan Deyasi is presently working as Assistant Professor in the Department of Electronics and Communication Engineering in RCC Institute of Information Technology, Kolkata, INDIA. He has 13 years of professional experience in academics and industry. He received B.Sc (Hons), B.Tech, M.Tech Degree from University of Calcutta. He is working in the area of semiconductor nanostructure and semiconductor photonics. He has published more than 150 peer-reviewed research papers. His major teaching subjects are Solid State Device, Electromagnetics, Photonics. He has already organized International and National Conferences, Faculty Development Programmes, Workshops, Laboratory and Industrial Visits, Seminars and Technical Events for students under the banner of IE(I) Kolkata section. He is also associated with a few reputed conferences. He is reviewer of a few journals of repute and some prestigious conferences in INDIA and abroad. He has delivered a few talks and conducted hands-on sessions on Nanolelectronics, Photonics and

Electromagnetics in various FDP's, workshops, seminars. He is the editor of various conference proceedings and edited volumes. He is a member of IEEE Electron Device Society, IE(I), Optical Society of India, IETE, ISTE etc. He is working as SPOC of RCCIIT Local Chapter (NPTEL course), Nodal Coordinator of e-outreach programme and Faculty Advisor of the student chapter of Institution of Engineers (INDIA) in ECE Department.

Nilanjan Dey, PhD., is an Asst. Professor in the Department of Computer Science and Engineering in Techno India College of Technology, Rajarhat, Kolkata, India. He holds an honorary position of Visiting Scientist at Global Biomedical Technologies Inc., CA, USA and Research Scientist of Laboratory of Applied Mathematical Modeling in Human Physiology, Territorial Organization Of- Scientific And Engineering Unions, BULGARIA, Associate Researcher of Laboratoire RIADI, University of Manouba, TUNISIA. He is the Editor-in-Chief of International Journal of Ambient Computing and Intelligence, International Journal of Rough Sets and Data Analysis (IGI Global), US, Series Editor of Advances in Geospatial Technologies (AGT) Book Series, (IGI Global), US, Executive Editor of International Journal of Image Mining (IJIM), Inderscience, Regional Editor-Asia of International Journal of Intelligent Engineering Informatics (IJIEI), Inderscience and Associated Editor of International Journal of Service Science, Management, Engineering, and Technology, IGI Global, IEEE ACCESS. He has 15 books and 250 international conferences and journal papers. https://www.researchgate.net/profile/Nilanjan_Dey3.

* * *

Sekhar Ranjan Bhadra Chaudhuri is Professor in Electronics and Telecommunication Engineering Department of IIEST, Shibpur. AHis areas of Research include Microstrip Antennas, Metamaterial Inspired antennas and components, Fractal Antennas and Wearable Antennas.

Gaurav Kumar Bharti was born in Bihar, India, in April 1990. He received the Master of Technology (M.Tech) in Electrical Engineering from National Institute of Technical Teachers' Training and Research, Kolkata, India, in 2016, and is currently working toward the Ph.D. degree in Electronics and Instrumentation Engineering at the National Institute of Technology, Agartala, Tripura, India. He is member of Institute of Electrical and Electronics Engineers (IEEE), Institution of Engineers (India) and regular fellow member of Optical Society of India (OSI). His current research interests include Photonics and Optics.

Nandan Bhattacharyya is Assistant Professor, ECE department, RCC Institute of Information Technology with 8 years of teaching experience and 5 years industry experience. Pursuing PhD from Calcutta University. M.Tech (2006) in Radiophysics & Electronics from Calcutta University. B.Tech (2004) in Electronics & Communication from Burdwan University.

Bishanka Brata Bhowmik has done his PhD in the area of optical fiber communication and photonics in 2015 from Indian Institute of Technology Patna. Currently he is working as an Assistant Professor in the Department of Electronics & Communication Engineering at Tripura University (A Central University).

Krishnendu Chattopadhyay is Assistant Professor in Electronics and Communication Engineering Department, MCKV Institute of Engineering, India. He has completed his PhD research work from IIEST, Shibpur, Howrah. His areas of research include Wideband Microstrip Antennas, Fractal Antennas. He is a Life Member of IETE, India and Member of IEEE, USA.

Dilip Kumar Choubey received his Ph.D in Engineering from Birla Institute of Technology (B.I.T), Mesra, Ranchi, India in 2018. He received his M.Tech degree in Computer Science and Engineering from Oriental College of Technology (O.C.T), Bhopal, India in 2012 and has B.E. degree in Information Technology from Bansal Institute of Science and Technology (B.I.S.T), Bhopal, India in 2010. Currently, he is working as an Assistant Professor (On Contract) in NIT Patna, India since March 2017. He worked as an Asst. Prof. in Lakshmi Narain College of Technology (L.N.C.T), Bhopal, India and Oriental College of Technology (O.C.T), Bhopal, India. He has more than 7 years of teaching and research experience. He is a author of 1 book and has more than 19 research publications in international journals, book chapters and conference proceedings to his credit. He has been the member of the organizing and technical program committees of some conferences and Workshop. He has reviewed several research papers in reputed journal. He has been attended so many workshop's/Seminar's/Conference's. He is the lifetime members of several professional bodies such as The International Association of Engineers (IAENG), The Internet Society Switzerland (ISOC), The Society of Digital Information and Wireless Communications (SDIWC), Universal Association of Computers and Electronics Engineers (UACEE), Associate Member (till 4 May, 2019), Computer Science Teachers Association (CSTA), etc. His research interests include Machine Learning, Soft Computing, Bioinformatics, Data Mining, Pattern Recognition, and Database Management System, etc.

Srikanta Das has completed his B.E. in Electronics and Telecommunication Engineering from Tripura Institute of Technology, Tripura in 2016, and M.Tech in Electronics and communication Engineering from Tripura University, Tripura in 2018. Currently he is pursuing his PhD from Tripura University.

Anirban Karmakar received the B.Tech. degree in Electronics & Communication Engineering from West Bengal University of Technology, West Bengal, India, in 2005 and M.Tech (Microwaves) from The University of Burdwan in 2007. He has completed his Ph.D. in Engineering from Jadavpur University, Kolkata, India, in 2015. He has more than 8 years of teaching experience and is currently an assistant professor in the department of Electronics & Communication Engg. at Tripura University, India. He has published about 20 research articles in refereed journals and international conference proceedings. He has served as a reviewer in different international journals. He was awarded best paper award at INDICON 2011. He is a Senior Member of IEEE and has organized workshops in the capacity of workshop convener on Microwaves & Antennas. His areas of interest include planar and fractal wideband antennas, and Fractal array antenna.

Surajit Mandal received his Ph. D. (Tech.) degree from the University of Calcutta, India. He has been in the teaching profession since 2005. Presently he is working as an Associate Professor of Electronics and Communication Engineering at BP Poddar Institute of Management and Technology. He has published a good number of research articles in journals of international repute. He received the 'Teacher of the Year Award' from Maulana Abul Kalam Azad University of Technology, India in 2017.

Ketavath Kumar Naik received a B.Tech in Electronics and Communication Engineering (ECE) from Jawaharlal Nehru Technological University JNTU College of Engineering, Kukatpally, Hyderabad, India; M.Tech in Digital Electronics and Communication Systems from JNTU College of Engineering, Anantapur, India; and Ph.D. in ECE from the College of Engineering, Andhra University, Visakhapatnam, India. He is working as a Professor and Group Head of RF and Microwave Engineering in the Department of ECE, Koneru Lakshamaiah Educational Foundation (deemed to be University), Guntur, India. He has two sponsored research projects from the Department of Science and Technology (DST) worth of Rs. 78,56,000. He also received the "DST Young Scientist Award" in 2014 from the Government of India, and the "Best Paper Award" at the International Conference InCMARS 2008. He is the referee for Sponsored Research Proposals of DST, SERB, Government of India. He is Associate Editor of International Journal of Electronics, Communications, and Measurement Engineering (IJECME), IGI Global, USA. He is a reviewer for various international and national journals and conferences like IEEE, Elsevier, and Springer etc. He has published over 63 research papers in reputed international and national journals and conferences. He is a Fellow IETE and a Senior Member IEEE, etc. His research interests include ring arrays, phased-array antennas, mic microstrip antennas, conformal antennas, SRR, EMI/EMC, RF, and biomedical applications.

Jayanta Kumar Rakshit received his B.Tech and M.Tech degrees from University of Calcutta, India in 2002 and 2004 respectively. He received his PhD degree in engineering from National Institute of Technology Agartala, India, in 2015. Dr. Rakshit is presently working as Assistant Professor in the department of Electronics and Instrumentation Engineering at National Institute of Technology Agartala, an Institute of National Importance in India. His current research interests include Photonics, optical sensors, optical logic-based signal processing, optical computing, and optical communication. He has published/accepted 25 papers in International/national Journals/symposia proceedings. He is member of Institute of Electrical and Electronics Engineers (IEEE), Institution of Engineers (India) and Life member of Optical Society of India (OSI).

Satyabrata Singha has pursued his B.Tech from Regional Institute of Science and Technology in 2015 and M.Tech in Electronics & Communication Engineering from Tripura University in 2018. Currently he is doing his PhD from Tripura University.

Index

A

B

C

D

E

F

I

K

M

O

P

R

S

T

U

W

Purchase Print, E-Book, or Print + E-Book

IGI Global books can now be purchased from three unique pricing formats: Print Only, E-Book Only, or Print + E-Book. Shipping fees apply.

www.igi-global.com

Recommended Reference Books

ISBN: 978-1-5225-3079-4
© 2018; 660 pp.
List Price: $245

ISBN: 978-1-5225-3064-6
© 2018; 230 pp.
List Price: $155

ISBN: 978-1-5225-3176-0
© 2018; 395 pp.
List Price: $235

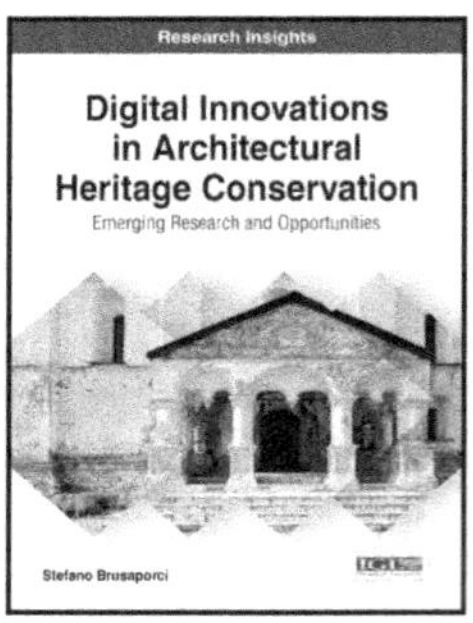

ISBN: 978-1-5225-2434-2
© 2017; 152 pp.
List Price: $115

ISBN: 978-1-5225-0827-4
© 2017; 323 pp.
List Price: $210

ISBN: 978-1-5225-1677-4
© 2017; 417 pp.
List Price: $350

Looking for free content, product updates, news, and special offers?
Join IGI Global's mailing list today and start enjoying exclusive perks sent only to IGI Global members.
Add your name to the list at **www.igi-global.com/newsletters.**

Publisher of Peer-Reviewed, Timely, and Innovative Academic Research

IGI Global
DISSEMINATOR OF KNOWLEDGE

www.igi-global.com Sign up at www.igi-global.com/newsletters facebook.com/igiglobal twitter.com/igiglobal

Ensure Quality Research is Introduced to the Academic Community

Become an IGI Global Reviewer for Authored Book Projects

The overall success of an authored book project is dependent on quality and timely reviews.

In this competitive age of scholarly publishing, constructive and timely feedback significantly expedites the turnaround time of manuscripts from submission to acceptance, allowing the publication and discovery of forward-thinking research at a much more expeditious rate. Several IGI Global authored book projects are currently seeking highly qualified experts in the field to fill vacancies on their respective editorial review boards:

Applications may be sent to:
development@igi-global.com

Applicants must have a doctorate (or an equivalent degree) as well as publishing and reviewing experience. Reviewers are asked to write reviews in a timely, collegial, and constructive manner. All reviewers will begin their role on an ad-hoc basis for a period of one year, and upon successful completion of this term can be considered for full editorial review board status, with the potential for a subsequent promotion to Associate Editor.

If you have a colleague that may be interested in this opportunity, we encourage you to share this information with them.

IGI Global
DISSEMINATOR OF KNOWLEDGE
www.igi-global.com

Celebrating 30 Years of Scholarly Knowledge Creation & Dissemination

CELEBRATING 30 years

InfoSci®-Books

A Collection of 4,000+ Reference Books Containing Over 87,000 Full-Text Chapters Focusing on Emerging Research

This database is a collection of over 4,000+ IGI Global single and multi-volume reference books, handbooks of research, and encyclopedias, encompassing groundbreaking research from prominent experts worldwide. These books are highly cited and currently recognized in prestigious indices such as: Web of Science™ and Scopus®.

Librarian Features:

- No Set-Up or Maintenance Fees
- Guarantee of No More Than A 5% Annual Price Increase
- COUNTER 4 Usage Reports
- Complimentary Archival Access
- Free MARC Records

Researcher Features:

- Unlimited Simultaneous Users
- No Embargo of Content
- Full Book Download
- Full-Text Search Engine
- No DRM

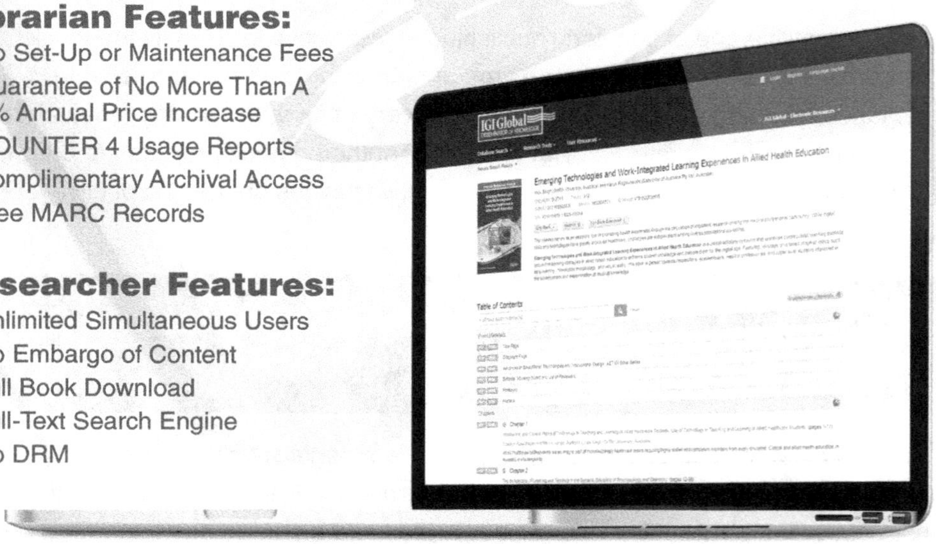

To Find Out More or To Purchase This Database:

www.igi-global.com/infosci-books

eresources@igi-global.com • Toll Free: 1-866-342-6657 ext. 100 • Phone: 717-533-8845 x100

IGI Global
DISSEMINATOR OF KNOWLEDGE
www.igi-global.com

IGI Global Proudly Partners with

eContent Pro International

Expert Copy Editing

eContent Pro International copy editors, with over 70 years of combined experience, will provide complete and comprehensive care for your document by resolving all issues with spelling, punctuation, grammar, terminology, jargon, semantics, syntax, consistency, flow, and more. In addition, they will format your document to the style you specify (APA, Chicago, etc.). All edits will be performed using Microsoft Word's Track Changes feature, which allows for fast and simple review and management of edits.

Additional Services

eContent Pro International also offers fast and affordable proofreading to enhance the readability of your document, professional translation in over 100 languages, and market localization services to help businesses and organizations localize their content and grow into new markets around the globe.

IGI Global Authors Save 25% on eContent Pro International's Services!

Scan the QR Code to Receive Your 25% Discount

The 25% discount is applied directly to your eContent Pro International shopping cart when placing an order through IGI Global's referral link. Use the QR code to access this referral link. eContent Pro International has the right to end or modify any promotion at any time.

Email: customerservice@econtentpro.com

econtentpro.com

ARE YOU READY TO **PUBLISH YOUR RESEARCH?**

IGI Global
DISSEMINATOR OF KNOWLEDGE

IGI Global offers book authorship and editorship opportunities across 11 subject areas, including business, healthcare, computer science, engineering, and more!

Benefits of Publishing with IGI Global:

- Free one-to-one editorial and promotional support.
- Expedited publishing timelines that can take your book from start to finish in less than one (1) year.
- Choose from a variety of formats including: Edited and Authored References, Handbooks of Research, Encyclopedias, and Research Insights.
- Utilize IGI Global's eEditorial Discovery® submission system in support of conducting the submission and blind-review process.
- IGI Global maintains a strict adherence to ethical practices due in part to our full membership to the Committee on Publication Ethics (COPE).
- Indexing potential in prestigious indices such as Scopus®, Web of Science™, PsycINFO®, and ERIC – Education Resources Information Center.
- Ability to connect your ORCID iD to your IGI Global publications.
- Earn royalties on your publication as well as receive complimentary copies and exclusive discounts.

Get Started Today by Contacting the Acquisitions Department at:
acquisition@igi-global.com

IGI Global
DISSEMINATOR OF KNOWLEDGE
www.igi-global.com

Available to Order Now

Order through www.igi-global.com with **Free Standard Shipping**.*

The Premier Reference for Information Science & Information Technology

100% Original Content
Contains 705 new, peer-reviewed articles with color figures covering over 80 categories in 11 subject areas

Diverse Contributions
More than 1,100 experts from 74 unique countries contributed their specialized knowledge

Easy Navigation
Includes two tables of content and a comprehensive index in each volume for the user's convenience

Highly-Cited
Embraces a complete list of references and additional reading sections to allow for further research

Included in:
InfoSci-Books

Encyclopedia of Information Science and Technology Fourth Edition

A Comprehensive 10-Volume Set

Mehdi Khosrow-Pour, D.B.A. (Information Resources Management Association, USA)
ISBN: 978-1-5225-2255-3; © 2018; Pg: 8,104; Release Date: July 2017

For a limited time, receive the complimentary e-books for the First, Second, and Third editions with the purchase of the *Encyclopedia of Information Science and Technology, Fourth Edition* e-book.*

The **Encyclopedia of Information Science and Technology, Fourth Edition** is a 10-volume set which includes 705 original and previously unpublished research articles covering a full range of perspectives, applications, and techniques contributed by thousands of experts and researchers from around the globe. This authoritative encyclopedia is an all-encompassing, well-established reference source that is ideally designed to disseminate the most forward-thinking and diverse research findings. With critical perspectives on the impact of information science management and new technologies in modern settings, including but not limited to computer science, education, healthcare, government, engineering, business, and natural and physical sciences, it is a pivotal and relevant source of knowledge that will benefit every professional within the field of information science and technology and is an invaluable addition to every academic and corporate library.

Scan for Online Bookstore

Pricing Information

Hardcover: **$5,695** E-Book: **$5,695** Hardcover + E-Book: **$6,895**

Both E-Book Prices Include:

- *Encyclopedia of Information Science and Technology, First Edition E-Book*
- *Encyclopedia of Information Science and Technology, Second Edition E-Book*
- *Encyclopedia of Information Science and Technology, Third Edition E-Book*

*Purchase the Encyclopedia of Information Science and Technology, Fourth Edition e-book and receive the first, second, and third e-book editions for free. Offer is only valid with purchase of the fourth edition's e-book through the IGI Global Online Bookstore.

Recommend this Title to Your Institution's Library: www.igi-global.com/books

www.ingramcontent.com/pod-product-compliance
Lightning Source LLC
LaVergne TN
LVHW080339011026
849161LV00009B/2128